Dietary Anticarcinogens and Antimutagens
Chemical and Biological Aspects

Dietary Anticarcinogens and Antimutagens

Chemical and Biological Aspects

Edited by

I.T. Johnson and G.R. Fenwick
Institute of Food Research, Norwich, UK

The Proceedings of the Food and Cancer Prevention III meeting held at the University of East Anglia, Norwich on 5–8 September 1999.

Special Publication No. 255

ISBN 0-85404-815-4

A catalogue record for this book is available from the British Library

Published by The Royal Society of Chemistry,
Thomas Graham House, Science Park, Milton Road, Cambridge CB4 0WF, UK

For further information see our web site at www.rsc.org

Typeset by Paston PrePress Ltd, Beccles, Suffolk
Printed by Athenaeum Press Ltd, Gateshead, Tyne and Wear, UK

Preface

This volume contains papers derived from plenary lectures, oral presentations and posters presented at the third conference on *Food and Cancer Prevention*, held in Norwich in September 1999. In the period which has elapsed since the second conference [Ede, The Netherlands, 1995], epidemiological evidence that diet is a major determinant of cancer risk in both industrialised and developing countries has continued to accumulate.

Two major reviews on diet and cancer published by the *World Cancer Research Fund* [1997] and the UK's *Department of Health* [1998] concluded that the protective effects of diets rich in fruits and vegetables were of particular importance. Indeed, the panel responsible for the report from the *World Cancer Research Fund* concluded that inappropriate diets cause around one third of all deaths from cancer, and one purpose of their work was to identify food and drinks which could protect against cancer.

The *Department of Health* report adopted a slightly more cautious approach. Though it, too, recommended increased consumption of fruits and vegetables in the UK, its committee also concluded that evidence for the mechanisms of action of fruits and vegetables was often inadequate, and that the causal mechanisms linking diet and cancer could often not be identified with any confidence. This is a frustrating situation for regulatory bodies, primary producers and consumers alike. The fundamental research necessary to resolve such difficulties is the subject of this series of international conferences on food and cancer prevention.

The present volume is organised into sections which broadly reflect the structure of the conference; it opens with papers on various aspects of the epidemiology of diet and cancer. Whilst emphasis is placed on the evidence for protective effects of plant foods, with particular emphasis on the phenolic constituents of fruits, vegetables and tea, there are papers on the still controversial issue of meat intake as a cause of bowel cancer. The *Department of Health* report identified measurement error, especially in relation to the intakes of foods and nutrients, as a major factor inhibiting the interpretation of epidemiological evidence. This is partly caused by uncertainties over the biological availability of putative protective factors in plant foods and it is particularly appropriate that a number of papers [Section 2] address the absorption of phytochemicals, their metabolism and delivery to target tissues.

The next four sections describe mechanisms of DNA damage and repair, and the body's various protective mechanisms against the initiation and promotion of neoplasia with which dietary constituents are believed to interact. The final sections deal with experimental approaches to the study of diet and cancer, with especial emphasis [Section 8] on studies with human subjects. There is an acute need for more reliable intermediate biomarkers of cancer risk in human subjects, both to facilitate research and for clinical applications. This is a topic around which we can expect to see much more activity reported at future conferences.

Once again, we would like to express our thanks to the *Food Chemistry Group of the Royal Society of Chemistry* and the *Food Chemistry Division of the Federation of European Chemical Societies* for their encouragement and to the former, the *Biotechnology and Biological Sciences Research Council [BBSRC]* and *The Tetley Group* for financial support. We are grateful to the members of the Scientific and Local Organising Committees for their assistance and support, and to our many colleagues at the *Institute of Food Research* who gave their time and effort to make the conference such a success. Ultimately, of course, our thanks are directed to the delegates of *Food and Cancer Prevention III*, whose research has made the publication of this volume possible.

Ian Johnson
Roger Fenwick

Contents

Section 1

Anticarcinogens and Mutagens in the Human Diet: Epidemiology

1.1

Colorectal Neoplasia and Meat: Epidemiology and Mechanisms

John D. Potter

FRED HUTCHINSON CANCER RESEARCH CENTER,
1100 FAIRVIEW AVE. N., MP 702, PO BOX 19024, SEATTLE,
WASHINGTON 98109-1024, USA

1 Introduction

The earliest dietary risk factor proposed for colorectal cancer was fat. This hypothesis, stated initially on the basis of a series of ecologic studies, was followed up extensively using animal studies that focused not only on the evidence for an effect of fat but also on the idea that the mechanisms might reasonably involve aspects of bile acid metabolism. Specifically, it was postulated that higher dietary fat increases colonic/hepatic bile acid recirculation and thus, via bacterial action on primary bile acids, increases exposure of the colonic mucosa to secondary bile acids with a demonstrated capacity to act as promotional agents.

The early case-control studies showed that there was indeed, an excess risk associated with fat or protein but also that meat, which is strongly correlated with, and a major source of, dietary fat, was also associated with elevated risk. Heterocyclic amines (HA) in the diet are derived exclusively from cooked meat – although tobacco smoke is also a source of these compounds. In 1983, Sugimura and Sato[1] demonstrated that these compounds were potent carcinogens in rats. Subsequently, some tendency to organotropism toward both pancreas and colon has increased suspicions that there are indeed plausible carcinogens in humans.

Meanwhile, though a considerable amount of human epidemiologic data on meat consumption has accumulated, these data have been misinterpreted in some quarters.[2] One of the most egregious claims has been that if meat is associated with colorectal neoplasia, that is only true in the United States (USA) and not in Europe. A second claim is that, although some of the case-control studies show an elevated risk, the cohort studies do not. It is true that there are inconsistencies in the data. This is the case with many exposures associated with difficult-to-measure human behaviors such as diet, physical activity, alcohol intake, and sexual behavior. What is not true is that these findings vary substantially between the USA and Europe. It is also true that the cohort studies show somewhat lower relative risks than the comparable odds ratios seen in case-control studies. However, it is not true that the findings are inconsistent across study designs nor is it true that we should only pay attention to the cohort studies and ignore the case-control studies. There are important differences between cohort and case-control studies, amongst which – after the very real issues of recall and selection bias which favor the cohort data over the case-control design – the age range of the participants is amongst the most important.[3] Cohort studies are focused on a narrow (usually younger) age range and thus may include those with a higher genetically determined risk; case-control studies usually include a much wider age range and thus may provide a better estimate of what is happening across the population and particularly among the less genetically susceptible older population.

2 The Epidemiologic Data

The data have been recently obscured in several ingenious ways, so the data for the USA and Europe from the 1980s onwards are presented here in a straightforward and comprehensive fashion. Table 1 shows that the case-control studies in Europe are largely consistent with an excess risk associated

Table 1 *Red meat and colorectal cancer – case-control studies Europe*

First Author	*Year*	*Comparison*	*OR*	*p value*
Manousos (Greece)	1983	beef: >2/wk vs. <1/wk	1.8	<0.05
Macquart-Moulin (France)	1986	meat: highest vs. lowest quartile	0.9	nr*
LaVecchia (Italy)	1988	beef/veal: highest vs. lowest tertile	2.2	<0.05
Tuyns (Belgium)	1988	beef (colon): >77gm/day vs. <32 gm/day	2.1	<0.05
		beef (rectum): >77 gm/day vs. <32 gm/day	0.7	nr*
Benito (Spain)	1990	meat (colon): >32/month vs. <16/month	2.9	<0.05
		meat (rectum): >32/month vs. <16/month	2.4	nr*
Gerhardsson de Verdier (Sweden)	1991	meat (colon): >7/wk vs. <1.6/wk	1.9	<0.05
		meat (rectum): >7 /wk vs. <1.6/wk	3.2	<0.05
Bidoli (Italy)	1992	meat (colon): highest vs. lowest tertile	1.6	nr*
		meat (rectum): highest vs. lowest tertile	2.0	<0.05
Zaridze (Russia)	1992	meat: highest vs. lowest quartile	1.0	0.5–2.0
Centonze (Italy)	1994	meat ⩾132 gm/day vs. <87 gm/day	0.7	0.4–1.5
Kampmann (Netherlands)	1995	meat (men): >102 gm/day vs. <60 gm/day	0.9	0.4–1.8
		meat (women): >83 gm/day vs. <38 gm/day	2.4	1.0–5.7

*nr: not reported

Table 2 *Red meat and colorectal cancer – case-control studies USA*

First Author	*Year*	*Comparison*	*OR*	*95% CI*
Lyon	1988	fried meat (men): >11.5/wk vs. ⩽6/wk	1.2	0.8–1.9
		fried meat (women): ⩾8.25/wk vs. ⩽3.75/wk	1.3	0.8–2.1
Peters	1989	beef: ⩾5/wk vs. ⩽1/wk	1.0	0.6–1.6
Wohlleb	1990	BBQ'd meat: ⩾1/wk vs. <1/wk	3.3	1.2–9.2
Schiffman	1990	meat: well done vs. less cooked	3.5	1.3–9.6
Peters	1992	meat: per unit	1.16	1.09–1.26
Le Marchand	1997	meat (men): highest vs. lowest quartile	1.8	1.2–2.7
		meat (women): highest vs. lowest quartile	1.0	0.6–1.5

with meat consumption.[4–13] There are 10 studies with a total of 15 estimates of risk, as a result of stratification in some studies on sex or tumor site. Of these, 10 estimates are greater that 1.0 and 8 of these are statistically significantly greater than 1.0. The remaining 5 estimates lie between 0.7 and 1.0 but are not statistically significantly different from 1.0.

Table 2 shows the USA case-control studies.[14–19] There are 6 studies with a total of 8 estimates of risk as a result of stratification on sex in 2 studies. All estimates are greater than or equal to 1.0 and 4 of these have 95% confidence limits that exclude 1.0. In one of these studies,[17] the comparison was between well-done and less cooked meat.

There are 2 European cohort studies (Table 3) and these are null.[20–21] Table 4 shows the 6 USA cohort studies.[22–27] There are 7 estimates of risk. All but one estimate is greater than 1.0 and, in 2 studies the 95% confidence limits exclude 1.0.

Table 5 provides the data on 5 studies, both case-control and cohort, all undertaken in the USA, of adenomatous polyps.[28–32] All 7 estimates of risk are greater than 1.0 and, for 4 of these, the 95% confidence limits exclude 1.0. Again, one study focused on heavily cooked red meat[31] and found a higher risk for this than for less well cooked meat. Table 6 shows that processed meat was associated with an elevated risk in two of the three European case-control studies and that there were no estimates of risk less than 1.0.[8–12]

The five estimates of risk derived from the 4 USA case-control studies of processed meat are all greater than 1.0 and 4 of the 95% confidence limits

Table 3 *Red meat and colorectal cancer – cohort studies Europe*

First Author	*Year*	*Comparison*	*RR*	*95% CI*
Goldbohm (Netherlands)	1994	>158 gm/day vs. <54 gm/day	0.8	0.5–1.4
Knekt (Finland)	1994	highest vs. lowest tertile	1.0	0.6–1.9

Table 4 *Red meat and colorectal cancer – cohort studies USA*

First Author	Year	Comparison	RR	95% CI
Phillips	1985	⩾4/wk vs. <1/wk	0.9	0.6–1.5
Willett	1990	⩾1/day vs. <1/month	2.5	1.2–5.0
Thun	1992	men: highest vs. lowest quintile	1.2	nr*
		women: highest vs. lowest quintile	1.1	nr*
Giovannucci	1994	⩾5/wk vs. 0	3.6	1.6–8.1
Bostick	1994	>3/wk vs. <1/wk	1.2	0.8–2.0
Kato	1997	highest vs. lowest quartile	1.2	0.7–2.2

*nr: not reported

Table 5 *Red meat and colorectal adenomatous polyps*

First Author	Year	Comparison	RR/OR	95% CI
Giovannucci	1992	>110 gm/day vs. <24 gm/day	1.2	0.7–2.1
Sandler	1993	men: ⩾2.3/wk vs. <0.5/wk	1.6	0.7–3.5
		women: ⩾2.6/wk vs. <0.6/wk	2.1	0.8–5.2
Haile	1997	highest vs. lowest quartile	1.6	1.0–2.6
Sinha	1999	per 10 gm of red meat per day	1.11	1.03–1.19
		per 10 gm of well done/very well done meat per day	1.29	1.08–1.54
Potter	1999	⩾7/wk vs. ⩽1/wk	1.9	1.1–3.2

Table 6 *Processed meat and colorectal cancer – case-control studies Europe*

First Author	Year	Comparison	OR	95% CI/p value
Benito (Spain)	1990	>22/month vs. 0	1.4	nr*
Bidoli (Italy)	1992	colon: highest vs. lowest tertile	1.8	nr*
		rectum: highest vs. lowest tertile	1.9	$p < 0.05$
Centonze (Italy)	1994	⩾3 gm/day vs. ⩽2 gm/day	1.0	0.6–1.7

*nr: not reported

exclude 1.0 (Table 7[16,18,19,33]). Table 8 shows that both of the European cohort studies demonstrated statistically significantly elevated risks of colorectal cancer in association with higher intakes of processed meat.[20,21] Table 9 displays the 3 USA cohort studies of processed meat.[23,25,26] All 3 show elevated risks in association with higher consumption and, for one of these, the 95% confidence limits exclude 1.0.

Tables 1 to 9 report on a total of 53 estimates of risk describing the

Table 7 *Processed meat and colorectal cancer – case-control studies USA*

First Author	*Year*	*Comparison*	*OR*	*95% CI*
Young	1988	>20/month vs. <1/month	1.9	1.3–2.6
Wohlleb	1990	⩾1/wk vs. <1/wk	2.9	1.2–7.1
Peters	1992	per unit	1.06	1.01–1.12
Le Marchand	1997	men: highest vs. lowest quartile	2.3	1.5–3.4
		women: highest vs. lowest quartile	1.2	0.8–2.0

Table 8 *Processed meat and colorectal cancer – cohort studies Europe*

First Author	*Year*	*Comparison*	*RR*	*95% CI*
Goldbohm (Netherlands)	1994	>20 gm/day vs. 0	1.7	1.0–2.9
Knekt (Finland)	1999	highest vs. lowest quartile	1.8	1.0–3.5

Table 9 *Processed meat and colorectal cancer – cohort studies USA*

First Author	*Year*	*Comparison*	*RR*	*95% CI*
Willett	1990	⩾2/wk vs. <1/month	1.9	1.2–3.0
Giovannucci	1994	⩾5/wk vs. 0	1.2	0.4–3.0
Bostick	1994	>3/wk vs. 0	1.5	0.7–3.2

consumption of meat and the risk of colorectal neoplasia. Of these 53 estimates, only 6 are less than 1.0 and none is statistically significantly less than 1.0. In contrast, 26 of these estimates are statistically significantly greater than 1.0 or have 95% confidence limits that exclude 1.0. This pattern of estimates of risk is exactly what could be expected, given that there is always misclassification of dietary exposures and lack of power in some studies. If there were truly no association, one would expect to see as many estimates of risks below as above 1.0. The pattern is not plausibly explained by publication bias. The obvious conclusion is that meat is associated with an elevated risk of colorectal neoplasia in the USA and Europe and that this is seen with consumption of both cooked red meat and processed meat.

3 Mechanisms

A variety of mechanisms have been proposed to account for the way in which meat might increase risk of colorectal neoplasia. These have been discussed extensively elsewhere and the reader is referred to some relevant primary and review articles. The mechanisms will be summarized very briefly here.

Meat is a major source of fat and saturated fat. A large number of animal (particularly rat) studies have shown that fat is a strong promoter of rodent colorectal cancer. The most coherent mechanism proposed suggested a role for secondary bile acids. Although this hypothesis has not been disproved, the fact that the epidemiologic evidence for fat is rather weaker than that for meat, coupled with the manifest difficulties of demonstrating this to be a key pathway in humans, has meant that the fat/bile acid hypothesis is currently in limbo. Nonetheless, there are interesting variants on the hypothesis.[34] The reader is referred to Weisburger[35] and the references therein for a review of the fat/bile acid hypothesis.

Since Sugimura and Sato[1] demonstrated that HA were animal carcinogens, evidence for their plausible role in human colorectal carcinogenesis has grown but, at this time, they are not established as the explanation for the elevated risk associated with meat. Nonetheless in rats, dietary HA, particularly PhIP, are standard carcinogens and induce characteristic mutations in APC[36] and frequent mutations in β-catenin.[37] As noted above for humans, there are several studies showing that heavily cooked meat is associated with an elevated risk of both cancer and polyps and there is a variety of data on the amount of HA produced by various cooking practices of different kinds of meat.[31,38]

One piece of the story that is missing is clear evidence that those with both a high intake of heavily cooked meat and the genetically determined metabolic profile that makes likely a higher production of HA-induced mutations (e.g. rapid CYP1A2, rapid NAT1, rapid NAT2 and perhaps GST M1- or P1-null), are at higher risk. Indeed, the data at present suggest that although there is an excess risk associated with meat consumption and even with consumption of heavily cooked meat, NAT phenotype, for instance, does not appear consistently to modify that risk particularly for adenomas,[32,39] although some of the early and very small studies did suggest an elevated risk in association with a rapid NAT2 phenotype.[40–42]

N-nitroso compounds, including nitrosamines, have been invoked as carcinogens in oesophagus and stomach cancers, particularly. Some nitrosamines are established human carcinogens. Exposure to *N*-nitroso compounds has been shown to occur both as a result of ingestion of nitrite-cured processed meats and as the result of ingestion of red meat itself. It is not yet established that *N*-nitroso compounds are colorectal carcinogens but a strong case has been made,[43] and this is consistent with the data implicating both fresh and processed meat noted above.

4 Summary

The epidemiologic data are clear, and largely consistent across study types, different endpoints in the development of colorectal cancer and between the USA and Europe. A variety of plausible mechanisms have been advanced to account for these findings. It is reasonable to conclude that meat consumption is associated with an elevated risk of colorectal neoplasia and there are a number of likely agents. The role of genetic polymorphisms and cooking practices in modifying this risk will probably become clear in the next few years. Despite misleading propaganda to the contrary, it seems prudent at this stage, either to limit, or to eliminate meat consumption in order to lower the risk of colorectal neoplasia.

5 References

1 Sugimura, T., and Sato, S., Mutagens-carcinogens in foods. Cancer Res, *43:* 2415s–2421s, 1983.

2 Hill, M., Meat and colo-rectal cancer: what does the evidence show? Eur J Cancer Prev *6*: 415–417, 1997.

3 Slattery, M. L., Potter, J. D., and Sorenson, A. W., Age and risk factors for colon cancer (United States and Australia): Are there implications for understanding differences in case-control and cohort studies? Cancer Causes Control, *5:* 557–563, 1994.

4 Manousos, O., Day, N. E., Trichopoulos, D., *et al*, Diet and colorectal cancer: A case-control study in Greece. Int J Cancer, *32:* 1–5, 1983.

5 Macquart-Moulin, G., Riboli, E., Cornje, J., *et al.*, Case-control study on colorectal cancer and diet in Marseilles. Int J Cancer, *38:* 183–191, 1986.

6 LaVecchia, C., Negri, E., Decarli, A., *et al.*, A case-control study of diet and colorectal cancer in northern Italy. Int J Cancer, *41:* 492–498, 1988.

7 Tuyns, A. J., Kaaks, R., and Haelterman, M., Colorectal cancer and the consumption of foods: a case-control study in Belgium. Nutr Cancer, *11:* 189–204, 1988.

8 Benito, E., Obrador, A., Stiggelbout, A., *et al.*, Nutritional factors in colorectal cancer risk: A case-control study in Majorca: I Dietary factors. Int J Cancer, *45:* 69–76, 1990.

9 Gerhardsson de Verdier, M., Hagman, U., Peters, R. K., *et al.*, Meat, cooking methods and colorectal cancer: a case-referent study in Stockholm. Int J Cancer, *49:* 520–525, 1991.

10 Bidoli, E., Franceschi, S., Talamini, R., *et al.*, Food consumption and cancer of the colon and rectum in north-eastern Italy. Int J Cancer, *50:* 223–229, 1992.

11 Zaridze, D., Filipchenko, V., Kustov, V., *et al.*, Diet and colorectal cancer: results of two case-control studies in Russia. Eu J Cancer, *29A:* 112–115, 1992.

12 Centonze, S., Boeing, H., Leoci, C., Guerra, V., and Misciagna, G., Dietary habits and colorectal cancer in a low risk area: Results from a population-based case-control study in Southern Italy. Br J Cancer, *69:* 937–942, 1994.

13 Kampman, E., Verhoeven, D., Sloots, L., and van't Veer, P., Vegetable and animal products as determinants of colon cancer risk in Dutch men and women. Cancer Causes Control, *6:* 225–234, 1995.

14 Lyon, J. L., and Mahoney, A. W., Fried foods and the risk of colon cancer. Am J Epidemiol, *128:* 1001–1006, 1988.

15 Peters, R. K., Garabrant, D. H., Yu, M. C., and Mack, T. M., A case-control study of occupational and dietary factors in colorectal cancer in young men by subsite. Cancer Res, *49:* 5459–5468, 1989.

16 Wohlleb, J. C., Hunter, C. F., Blass, B., Kadlubar, F. F., Chu, D. Z. J., and Lang, N. P., Aromatic amine acetyltransferase as a marker for colorectal cancer: environmental and demographic associations. Int J Cancer, *46:* 22–30, 1990.

17 Schiffman, M. H., and Felton, J. Re, Fried foods and risk of colon cancer. Am J Epidemiol, *131:* 376–378, 1990.

18 Peters, R. K., Pike, M. C., Garabrant, D., and Mack, T. M., Diet and colon cancer in Los Angeles County, California. Cancer Causes Control, *3:* 457–473, 1992.

19 Le Marchand, L., Wilkens, L., Hankin, J., Kolonel, L., and Lyu, L.-C., A case-control study of diet and colorectal cancer in a multiethnic population in Hawaii (United States): lipids and foods of animal origin, *8:* 637–648, 1997.

20 Goldbohm, R. A., van den Brandt, P. A., van't Veer, P., Brants, H. A. M., Dorant, E., Sturmans, F., and Hermus, R. J. J., A prospective cohort study on the relation between meat consumption and the risk of colon cancer. Cancer Res, *54:* 718–723, 1994.

21 Knekt, P., Steineck, G., Jaervinen, R., Hakulinen, T., and Aromaa, A., Intake of fried meat and risk of cancer: a follow-up study in Finland. Int J Cancer, *59:* 756–760, 1994.

22 Phillips, R. L., and Snowdon, D. A., Dietary relationships with fatal colorectal cancer among Seventh-Day Adventists. J Natl Cancer Inst, *74:* 307–317, 1985.

23 Willett, W. C., Stampfer, M. J., Colditz, G. A., and etal. Relation of meat fat and fiber intake to the risk of colon cancer in a prospective study among women. N Engl J Med, *323:* 1664–1672, 1990.

24 Thun, M. J., Calle, E. E., Namboodiri, M. M., *et al.*, Risk factors for fatal colon cancer in a large prospective study. J Natl Cancer Inst, *84:* 1491–1500, 1992.

25 Giovannucci, E., Rimm, E. B., Stampfer, M. J., Colditz, G. A., Ascherio, A., and Willett, W. C., Intake of fat, meat, and fiber in relation to risk of colon cancer in men. Cancer Res, *54:* 2930–2997, 1994.

26 Bostick, R. M., Potter, J. D., Kushi, L. H., Sellers, T. A., Steinmetz, K. A., McKenzie, D. R., Gapstur, S. M., and Folsom, A. R., Sugar, meat, and fat intake, and non-dietary risk factors for colon cancer incidence in Iowa women (United States). Cancer Causes Control, *5:* 38–52, 1994.

27 Kato, I., Akhmedkhanov, A., Koenig, K., *et al.*, Prospective study of diet and female colorectal cancer – the New York University Women's Health Study. Nutr Cancer, *28:* 276–281, 1997.

28 Giovannucci, E., Stampfer, M. J., Colditz, G., Rimm, E. B., and Willett, W. C., Relationship of diet to risk of colorectal adenoma in men. J Natl Cancer Inst, *84:* 91–98, 1992.

29 Sandler, R. S., Lyles, C. M., Peipins, L. A., McAuliffe, C. A., Woosley, J. T., and Kupper, L. L., Diet and risk of colorectal adenomas: macronutrients, cholesterol, and fiber. J Natl Cancer Inst, *85:* 884–891, 1993.

30 Haile, R. W., Witte, J. S., Longnecker, M. P., Probst-Hensch, N. M., Chen, M.-J., Harper, J., Frankl, H. D., and Lee, E. R., A sigmoidoscopy-based case-control study of polyps: macronutrients, fiber and meat consumption. Int J Cancer, *73:* 497–502, 1997.

31 Sinha, R., Chow, W.-H., Kulldorff, M., Denobile, J., Butler, J., Garcia-Closas, M., Weil, R., Hoover, R. J., and Rothman, N., Well-done, grilled meat increases the risk of colorectal adenomas. Cancer Res, *59:* 4320–4324, 1999.

32 Potter, J. D., Bigler, J., Fosdick, L., Bostick, R., Kampmann, E., Chen, C., Louis, T., and Grambsch, P., Adenomatous polyps, red meat, and smoking: the role of *N*-acetyltransferases. Proc Am Assoc Cancer Res, *40:* 212, 1999.

33 Young, T. B., and Wolf, D. A., Case-control study of proximal and distal colon cancer and diet in Wisconsin. Int J Cancer, *42:* 167–175, 1988.

34 Morotomi, M., Guillem, J., LoGerfo, P., *et al.*, Production of diacylglycerol, an

activator of protein kinase C, by human intestinal microflora. Cancer Res, *50:* 3595–3599, 1990.

35 Weisburger, J., Causes, relevant mechanisms, and prevention of large bowel cancer. Semin Oncol, *18:* 316–336, 1991.

36 Kakiuchi, H., Watanabe, M., Ushijima, T., Toyota, M., Imai, K., Weisburger, J., Sugimura, T., and Nagao, M., Specific 5′-GGGA-3′ – 5′-GGA-3′ mutation of the Apc gene in rat colon tumors induced by 2-amino-1-methyl-6-phenylimidazo [4,5-b]pyridine. Proc Natl Acad Sci, *92:* 910–914, 1995.

37 Dashwood, R., Suzui, M., Nakagama, H., Sugimura, T., and Nagao, M., High frequency of β-catenin (*Ctnnb1*) mutations in the colon tumors induced by two heterocyclic amines in the F344 rat. Cancer Res, *58:* 1127–1129, 1998.

38 Sinha, R., Rothman, N., Brown, E., Salmon, C., Knize, M., Swanson, C., Rossi, S., Mark, S., Levander, O., and Felton, J., High concentrations of the carcinogen 2-amino-1-methyl-6-phenylimidazo-(4,5-b)pyridine (PhIP) occur in chicken but are dependent on the cooking method. Cancer Res, *55:* 4516–4519, 1995.

39 Probst-Hensch, N. M., Haile, R. W., Li, D. S., Sakamoto, G. T., Louie, A. D., Lin, B. K., Frankl, H. D., Lee, E. R., and Lin, H. J., Lack of association between the polyadenylation polymorphism in the *N*-acetyltransferase-1 gene (*NAT1*) and colorectal adenomas. Carcinogenesis, *17:* 2125–2129, 1996.

40 Lang, N. P., Chu, D. Z. J., Hunter, C. F., Kendall, D. C., Flammang, T. J., and Kadlubar, F. F., Role of aromatic amine acetyltransferase in human colorectal cancer. Arch Surg, *121:* 1259–1261, 1986.

41 Ilett, K. F., Beverly, M. D., Detchon, P., Castleden, W. M., and Kwa, R., Acetylation phenotype in colorectal carcinoma. Cancer Res, *47:* 1466–1469, 1987.

42 Roberts-Thomson, I. C., Ryan, P., Khoo, K. K., Hart, W. J., McMichael, A. J., and Butler, R. N., Diet, acetylator phenotype, and risk of colorectal neoplasia. Lancet, *347:* 1372–1374, 1996.

43 Bingham, S. A., Pignatelli, B., Pollock, J. R. A., Ellul, A., Malaveille, C., Gross, G., Runswick, S., Cummings, J., and O'Neill, I. K., Does increased endogenous formation of N-nitroso compounds in the human colon explain the association between red meat and colon cancer? Carcinogenesis, *17:* 515–523, 1996.

1.2

Green Tea as a Cancer Preventive

Hirota Fujiki, Masami Suganuma, Sachiko Okabe, Eisaburo Sueoka, Naoko Sueoka, Satoru Matsuyama, Kazue Imai and Kei Nakachi

SAITAMA CANCER CENTER RESEARCH INSTITUTE, INA KITAADACHI-GUN, SAITAMA 362-0806, JAPAN

1 Abstract

Green tea is now an acknowledged cancer preventive in Japan and will possibly soon be recognized as such in other countries. Initially, we found that (−)-epigallocatechin gallate (EGCG), the main constituent of green tea inhibited tumor promotion on mouse skin in a two-stage carcinogenesis experiment. Numerous additional studies revealed the anticarcinogenic effects of EGCG and green tea on various organs in rodent experiments. This paper reviews the unique role of green tea in cancer chemoprevention, its anticarcinogenic effects and other preventive activities, bioavailability of tea polyphenols and epidemiological studies with green tea. Of particular interest are studies which showed that daily consumption of green tea delayed clinical onset of various cancers and led to more hopeful prognoses for breast cancer patients in Stage I and II following treatment. Based on these results, I propose two stages of cancer prevention with green tea: prevention before cancer onset, and following cancer treatment. Since green tea is a common beverage, the knowledge that it inhibits cancer will be a great comfort to, especially, aging folk concerned with cancer prevention and any high risk population.

2 Introduction

Green tea is a beverage commonly ingested in Japan every day. The tea plant, Camellia sinensis, was brought from China to Japan by a Japanese Zen priest in the twelfth century for use as a medicine. From that time, we Japanese have simply believed in the benefit of drinking green tea for many years, while not necessarily being aware of its therapeutic effects.

When we started to study cancer chemoprevention in 1983, we had our first scientific contact with (−)-epigallocatechin gallate, the main constituent of green tea, kindly provided by Dr. Takuo Okuda. In collaboration with him, previously Professor on the Faculty of Pharmacology at Okayama University, we studied the anticarcinogenic effects of polyphenols derived from medicinal plants and drugs. In 1987, we first reported, in the new British jounal *Phytotherapy Research*, that repeated topical applications of EGCG to mouse skin treated with 7,12-dimethylbenz[*a*]anthracene (DMBA) as an initiator inhibited tumor promotion in a two-stage carcinogenesis experiment.[1] These results were the first suggestion that EGCG and tea polyphenols might be cancer preventives. Since then, numerous scientists in Japan and the USA have reported the anticarcinogenic effects of EGCG and tea polyphenols on various organs in rodent experiments.[2]

Results from a prospective cohort study with over 8,000 individuals in Saitama Prefecture demonstrated clearly that daily green tea consumption of 10 Japanese-size cups delayed cancer onset as well as death of cancer patients.[3] This paper reviews the study of EGCG and green tea itself as possible cancer preventive prototypes in a practical sense.

3 Cancer Chemoprevention

Cancer chemoprevention is a new strategy of cancer prevention based on the modern understanding of cancer development in humans; it was defined as 'the prevention of the occurrence of cancer by administration of one or more compounds' by Michael Sporn in 1976.[4,5] It is now generally known that cancer development in humans is a multi-stage process associated with numerous genetic changes occurring over an extended period of time, usually from 20 to 30 years.[6] Thus, we have to realize that cancer prevention is essentially different from the prevention of infectious diseases, since cancer is a disease of the aging process, with multistages, while infectious disease is the reaction to a pathogen. My own definition of cancer chemoprevention is, administrating cancer preventives to delay the carcinogenic processes in humans, no matter when the carcinogenesis starts, and thereby block the appearance of clinical symptoms.

If we accept my definition, it can now be shown that daily green tea consumption of over 10 cups per day results in delay of cancer onset for healthy persons and leads to more hopeful prognoses for cancer patients following cancer treatment. Since green tea is a common beverage, at least in Japan, we can bring the concept of cancer prevention into our daily life.

4 Anticarcinogenic Effects and Other Preventive Activities

Initially, repeated topical applications of EGCG resulted in inhibition of tumor promotion in a two- stage carcinogenesis experiment on mouse skin.[1] We proposed as a possible mechanism the sealing effect of EGCG; since topical application of EGCG to mouse skin inhibited interaction of tumor promoters, hormones, and various growth factors with their receptors.[7] Subsequently, green tea and EGCG in drinking water were found to inhibit carcinogenesis in a wide range of target organs in animal experiments, including esophagus, stomach, duodenum, colon, liver, lung, pancreas, skin, breast, bladder and prostate.[2,7,8,9]

Furthermore, EGCG and tea polyphenols have antinitiation, antimutagenic, and antimicrobial activities; they inhibit various enzyme activities along with gene expression of inflammatory cytokines related to carcinogenesis; and they enhance activities of some biochemical reactions supporting anticarcinogenesis.[8,10] Among the demonstrated effects, inhibition of urokinase,[11] telomerase,[12] p38 mitogen-activated protein kinase activation[13] and angiogenesis[14] have attracted attention recently. We also presented evidence that treatment with EGCG induced apoptosis and increased the percentages of cells with G2/M arrest in PC-9 cells.[15]

As can be seen above, tea polyphenols have multifunctions and are therefore quite distinct from enzyme inhibitors, which have specific functions. As for the essential activity of tea polyphenols for cancer prevention, we think their reduction of levels of TNF-α and similar cytokines is a common and key criterion.[16]

Recently we demonstrated the synergistic effects of EGCG with same standard cancer preventive agents:[17] cotreatment with EGCG and sulindac, or EGCG and tamoxifen, synergistically or additively enhanced induction of apoptosis by EGCG in human lung cancer cell line, PC-9 cells. The results indicate the possibility of administering smaller doses of these chemical agents, with fewer adverse effects.

5 Bioavailability of Tea Polyphenols

Our study with ^{3}H-EGCG (48.1 GBq mmol^{-1}), which was administered into mouse stomach, revealed wide distribution of radioactivity in mouse organs.[18] Radioactivity was found in all reported target organs mentioned in Part 4 above, as well as in other organs (brain, kidney, uterus and ovary or testes) in mice (Table 1). We also found, by microautoradiography, that 3H-EGCG had incorporated into the cytosol as well as the nuclei.

6 Epidemiological Studies

The 10-year follow-up results of a prospective cohort study involving 8,552 individuals in Saitama Prefecture identified 164 female and 220 male cancer patients. Our colleagues who did the study found that cancer onset for patients who had consumed over 10 Japanese-size cups of green tea per day, was 8.7 years later among females, and 3.0 years later among males, than for patients who had consumed under three cups per day (Table 2).[19]

The cancer preventive effects of green tea were also found among breast cancer patients following cancer treatment. Our colleagues found that Stage I

Table 1 *Incorporation of ^{3}H-EGCG into target organs*

Organs	*% of total administered radioactivity (24 h after)*	*Reduction of tumor incidence*
Stomach	3.93	62.0 → 31.0
Duodenum	0.35	63.0 → 20.0
Small intestine	5.69	n.d.
Colon	4.52	77.3 → 38.1 67.0 → 33.0*
Liver	0.89	83.3 → 52.2
Brain	0.32	n.d.
Kidney	0.28	n.d.
Lung	0.16	96.3 → 65.5
Pancreas	0.07	54.0 → 28.0
Skin	1.9 × 104/100 mg	65.0 → 28.0*

*: Green tea extract
n.d.: Not determined

Table 2 *Average age at cancer onset and green tea consumption from prospective cohort study*

	Green tea consumption (cups/day)		
	≤3	4–9	≥10
Males (220)	65.3 ± 1.5 (54)	67.6 ± 1.0 (102)	68.3 ± 1.2 (64)
Females (164)	65.7 ± 1.7 (49)	66.8 ± 1.2 (94)	74.4 ± 2.5 (21)

and II patients consuming over 5 cups per day showed a lower recurrence rate, 16.6%, and a longer disease-free period, 3.6 years, than those consuming less than 4 cups per day.[20] The results suggested that green tea is more effective in the early stage of second tumor development, even after the removal of the primary cancer, and that it will result in more hopeful prognoses for breast cancer patients.

7 Two Stages of Cancer Prevention with Green Tea

We have demonstrated the cancer preventive effects of green tea in two stages, cancer prevention before cancer onset and that following cancer treatment. In the former case, the cancer preventive dose of green tea is at least 10 Japanese-size cups per day. Since most Japanese use a medium-size (180 ml) cup for drinking green tea, the effective dose is, then, 360 to 540 mg of EGCG, and 0.8 g to 1.3 g green tea extract, per day.[3] For those who, for whatever reason, are unable to drink 10 cups per day, green tea tablets will be a useful supplement to their beverage intake. For this purpose, we have now prepared green tea tablets in collaboration with Saitama Prefecture Tea Experiment Station, with each tablet being 500 mg. In this way, we hope that cancer prevention can be more practically realized in daily life, the great advantage of green tea in Japan.

8 Summary

The relationship between standard cancer therapy and modern cancer prevention should be briefly discussed. We are now able to consider two stages of cancer prevention for clinically healthy persons: that before cancer onset, and that following cancer treatment. For the former stage, green tea and herbal remedies will be reasonable cancer preventives. For the latter, conventional cancer preventive agents supplemented by green tea and herbal remedies will be suitable agents. It then follows that we should intensively investigate alternative medicine, including green tea and herbal remedies, for the development of cancer preventive agents.

Acknowledgements

This work was supported by the following grants-in-aid: for Cancer Research, Special Cancer Research, and Scientific Research on Priority Areas for Cancer Research from the Ministry of Education, Science, Sports and Culture, Japan; grants for a Second Term Comprehensive 10-year Strategy for Cancer Control and for Comprehensive Research on Aging and Health from Ministry of Health and Welfare, Japan; for Selectively Applied and Developed Research from Saitama Prefecture, Japan; and by grants from the Smoking Research Fund, and the Plant Science Research Foundation of the Faculty of Agriculture, Kyoto University.

9 References

1 S. Yoshizawa, T. Horiuchi, H. Fujiki, T. Yoshida, T. Okuda and T. Sugimura, *Phytother.*, 1987, **1**, 44.
2 J. Weisburger (Ed.) 'Proceedings of The Society For Experimental Biology And Medicine', Blackwell Science, Inc. New York, 1999, Vol. 220.
3 K. Nakachi, K. Imai and K. Suga 'Food Factors for Cancer Prevention' Springer Verlag, Tokyo, Japan, 1997, p.105.
4 M. B. Sporn, *Lancet*, 1993, **342**, 1211.
5 W. K. Hong and M. B. Sporn, *Science*, 1997, **278**, 1073.
6 B. Bogelstein, E. R. Fearon, S. R. Hamilton, S. E. Kern, A. C. Preisinger, M. Leppart, Y. Nakamura, R. White, A. M. M. Smith and J. L. Bos, *N. Engl. J. Med.*, 1988, **319**, 525.
7 H. Fujiki, A. Komori and M. Suganuma, 'Comprehensive Toxicology', Pergamon, Cambridge, 1997, Vol. 12, Chapter 19, p.453.
8 H. Fujiki, M. Suganuma, S. Okabe, N. Sueoka, A. Komori, E. Sueoka, T. Kozu, Y. Tada, K. Suga, K. Imai and K. Nakachi, *Mutat. Res.*, 1998, **402**, 307.
9 C. S. Yang and Z.-Y. Wang, *J. Natl. Cancer Inst.*, 1993, **85**, 1038.

10 NCI, DCPC and Chemoprevention Branch and Agent Development Committee, *J. Cell Biochem*., 1996, **26S**, 236.
11 J. Jankun, S. H. Selmon, R. Swiercz and E. Skrzypczak-Jankun, *Nature*, 1997, **387**, 561.
12 I. Naasani, H. Seimiya and T. Tsuruo, *Biochem. Biophys. Res. Commun*., 1998, **249**, 391.
13 W. Chen, Z. Dong, S. Valcis, B. N. Timmermaun and G. T. Bowden, *Mol. Carcinog*., 1999, **24**, 79.
14 Y Cao and R. Cao, *Nature*, 1999, **398**, 381.
15 S. Okabe, M. Suganuma, M. Hayashi, E. Sueoka, A. Komori and H. Fujiki, *Jpn. J. Cancer Res*., 1997, **88**, 639.
16 M. Suganuma, S. Okabe, E. Sueoka, N. Iida, A. Komori, S. -J. Kim and H. Fujiki, *Cancer Res*., 1996, **56**, 3711.
17 M. Suganuma, S. Okabe, Y. Kai, N. Sueoka, E. Sueoka and H. Fujiki, *Cancer Res*., 1999, **59**, 44.
18 M. Suganuma, S. Okabe, M. Oniyama, Y. Tada, H. Ito and H. Fujiki, *Carcinogenesis (Lond.)*, 1998, **19**, 1771.
19 K. Imai, K. Suga and K. Nakachi, *Prev. Med*., 1997, **26**, 769.
20 K. Nakachi, K. Suemasu, K. Suga, T. Takeo, K. Imai and Y. Higashi, *Jpn. J. Cancer Res*., 1998, **89**, 254.

1.3

Dietary Fibre and Resistant Starch – Do They Protect against Cancer?

Lynnette R. Ferguson[1] and Philip J. Harris[2]

[1] CANCER RESEARCH LABORATORY, THE UNIVERSITY OF AUCKLAND, PRIVATE BAG 92019, AUCKLAND, NEW ZEALAND
[2] SCHOOL OF BIOLOGICAL SCIENCES, THE UNIVERSITY OF AUCKLAND, PRIVATE BAG 92019, AUCKLAND, NEW ZEALAND

1 Introduction

Although it is generally believed that people should eat about 25 g of dietary fibre per day, this is rarely achieved, partly because high fibre foods are often unpalatable. As a consequence, there has been considerable interest in using resistant starch (RS) as a substitute for dietary fibre. RS is defined as the sum of starch and the degradation products of starch that, on average, reaches the large intestine of healthy adult humans.[1,2] Three types of RS are recognised: starch granules physically entrapped in plant cells (RS1); native starch granules that are highly resistant to digestion by α-amylases (RS2); and retrograded starch (RS3) formed when starch is cooled after being heated above its gelatinisation temperature. The amount of RS3 in particular is readily increased in the diet, as for example in various 'high-fibre white breads'. Because RS is fermented by microorganisms in the colon in a similar way to much of the non-starch polysaccharides (NSP) of DF, it is sometimes included in definitions of DF.[3] Furthermore, RS is often assumed to have similar physiological effects on the colon to DF.[4] The assumption that RS prevents cancers in humans, especially colon cancer, provides one of the justifications for increasing the amount of RS in the human diet without extensive additional testing.

DFs have a number of physiological effects on the gastrointestinal tract, some of which may be involved in the prevention of cancer.[5,6] To determine if RS has similar effects on the gastrointestinal tract to DF, we fed three different RS2 preparations and three different types of DF (or sources of DF) to Wistar rats. The RS2 preparations were as follows: native potato starch; native, high-amylose maize starch (Hi-maize™ starch); and Hi-maize™ starch that had been treated with α-amylase. The DFs (or DF sources) were apple pectin, wheat straw, and wheat bran. We fed an AIN-76™ diet in which the normal maize starch was substituted with one of the three sources of RS, or in which 10/35 g of normal maize starch was substituted with DF. Full details will be published elsewhere, but we have used some of these new data plus data in the literature to make the arguments presented below.

2 RS, Soluble DF and Cancer

One mechanism by which DF may prevent cancer is by reducing the exposure of the colon to carcinogens. This may occur by the DF decreasing the transit time, increasing the faecal bulk, and/or adsorbing carcinogens.[5,6] In our rat model, transit time was measured as the time taken for 50% of small glass beads fed in the diet to be excreted. For rats on a control diet, it took just under 30 h for 50% of the beads to be excreted. The RS preparations reduced the transit time by up to one third (potato starch), and the wheat bran reduced the time by about half. The apple pectin had an intermediate effect. However, we found the most marked effect of the RS preparations was on the fresh weight of the faeces. The preparations caused an increase in weight ranging from 3-fold for the native Hi-maize™ starch to 10-fold for the potato starch. The apple pectin caused a similar increase in faecal weight to the Hi-maize™ starch, but the wheat bran caused only a 2-fold increase.

Despite these apparently beneficial effects of RS on gastrointestinal function, we found rats fed RS preparations retained significant amounts of a radioactive carcinogen (^{14}C IQ) in the plasma for much longer than control rats or DF-fed rats (unpublished data, this laboratory). This result suggests that RS does not reduce the probability of tumour induction. More direct support for this hypothesis comes from the studies of Maziere *et al.*[7] who found no effect of RS on the initiation of tumours by 1,2-dimethylhydrazine. Furthermore, Young *et al.*[8] and Sakamoto *et al.*[9] found that when RS (as native potato starch) was present during tumour initiation and promotion, there was enhanced rather than reduced tumour formation.

It has been suggested that later stages of carcinogenesis may be reduced by butyrate in the colon. Butyrate, together with other short-chain fatty acids (SCFAs), is produced in the colon by the fermentation of polysaccharides including RS and the NSPs of DF. *In vitro* fermentation of starch results in the production of a higher concentration of butyrate and a higher yield of butyrate per gram of substrate than does fermentation of the NSPs of DF.[10] Various *in vitro* studies have indicated that butyrate enhances differentiation of cells, affects the expression of various cancer-related genes, and may stimulate apoptosis of damaged cells as a mechanism for preventing their progression to form cancer.[11–13] However, despite the effective production of butyrate by potato starch, RS in the diets of rats exposed to 1,2-dimethylhydrazine resulted in enhanced tumour formation.[9]

Animal studies in which the RS was present only during the tumour promotion phase have suggested that RS may be beneficial.[14,15] However, such experiments are probably poor models for the human situation, where there is continuous exposure to dietary mutagens and carcinogens.[6] In our experiments and those of others, the effects of RS resemble those of soluble DFs such as pectin in increasing faecal weights and enhancing SCFA production in the large intestine. Nevertheless, pectin and other soluble DFs are usually not protective, except in studies where they are present only during tumour promotion. The only evidence that can unarguably be extrapolated to humans comes from human intervention studies, where it is noteworthy that the first soluble DF tested, psyllium, increased rather than decreased the formation of adenomas.[16]

3 Insoluble DF and Cancer

Wheat bran is a good source of insoluble DF. Wheat bran reduces absorption and metabolism of a carcinogen (IQ) in a Wistar rat model, and there is good evidence that it protects against cancer in different animal models.[17] Furthermore, wheat bran is the only DF source that in human intervention studies protects against the formation or increase in size of adenomas (an early marker of cancer).[17] However, it does appear that the effect of wheat bran is variable and not reproducible in all studies (McKeown-Eyssen *et al.*), possibly because of variability in the size of the bran particles. For example, in animal studies[18] and in human studies (MacLennan *et al.*, 1995), it appears that large particles are necessary to show beneficial effects. Wheat brans also differ in their DF content (*c.f.* Refs 18, 19) and this may be important for cancer protection.

However, although wheat bran is a rich source of DF, this makes up less than 50% of its dry weight. Thus, it is unknown whether the DF or components in the cell contents protect against cancer. Some of these components in the cell contents are known to have anti-cancer activity and include phytic acid and various phenolic components, including hydroxycinnamic acids, lignans, and flavonoids, although the bulk of the hydroxycinnamic acids in wheat bran is present in the DF.[17] The components of the cell contents may also vary significantly among different wheat bran sources.

4 Conclusions

We suggest that, although some insoluble DF sources may have cancer protective properties, this is probably not true of soluble DFs and RS. This must raise a concern about the increased consumption of RS and soluble DF in the human diet.

5 References

1 N.-G. Asp, *Eur. J. Clin. Nutr.*, 1992, **46** Suppl. 2, S1.
2 H. N. Englyst, S. M. Kingman and J. E. Cummings, *Eur. J. Clin. Nutr.*, 1992, **46** Suppl. 2, S33.
3 P. A. Baghurst, K. I. Baghurst and S. J. Record, *Food Australia*, 1996, **48**, S1.
4 J. H. Cummings, 'Human Colonic Bacteria. Role in Nutrition, Physiology, and Pathology', CRC Press, Boca Raton, FL, USA, 1995, p. 101.

5 P. J. Harris and L. R. Ferguson, *Mutation Res.*, 1993, **290**, 97.
6 P. J. Harris and L. R. Ferguson, *Mutation Res.*, 1999, **443**, 48.
7 S. Maziere, P. Cassand, J. F. Narbone and K. Meflah, *Nutr. Cancer*, 1998, **31**, 168.
8 G. P. Young, A. McIntyre, V. Albert, M. Folino and J. G. Muir, *Gastroenterol.*, 1996, **110**, 508.
9 J. Sakamoto, S. Nakaji, K. Sugawara, S. Iwane and A. Munakata, *Gastroenterol.*, 1996, **110**, 116.
10 H. N. Englyst, S. Hay and G. T. Macfarlane, *FEMS Micro Ecol.*, 1987, **95**, 163.
11 A. Hague, D. J. Elder, D. J. Hicks and C. Paraskeva, *Int. J. Cancer*, 1995, **60**, 400.
12 W. Scheppach, H. P. Bartram and F. Richter, *Eur. J. Cancer*, 1995, **31A**, 1077.
13 A. Csordas, *Eur. J. Cancer Prev.*, 1996, **5**, 221.
14 I. Thorup, O. Meyer and E. Kristiansen, *Anticancer Res.*, 1995, **15**, 2101.
15 P. Cassand, S. Maziere, M. Champ, K. Meflah and F. Bornet, *Nutr. Cancer*, 1997, **27**, 53.
16 J. Faivre, C. Bonithon-Kopp, O. Kronberg, U. Rath and A. Giacosa, *Gasterero-enterol.* **116**, G0238.
17 L. R. Ferguson and P. J. Harris, *Eur. J. Cancer Prev.*, 1999, **8**, 17.
18 L. R. Ferguson and P. J. Harris, 'Particle Size of Wheat Bran in Relation to Colonic Function in Rats', *Food Sci. Tech.*, 1997, **30**, 735.
19 P. Kestell, L. Zhao, S.-T. Zhu, P. J. Harris and L. R. Ferguson, *Carcinogenesis*, 1999, in press.

1.4

Antioxidant Catechins: Intake and Major Sources in the Dutch Population

I.C.W. Arts,[1,2] P.C.H. Hollman,[2] H.M. van de Putte,[2]
H.B. Bueno de Mesquita,[1] E.J.M. Feskens[1] and D. Kromhout[3]

[1] NATIONAL INSTITUTE OF PUBLIC HEALTH AND ENVIRONMENT (RIVM), DEPARTMENT OF CHRONIC DISEASES EPIDEMIOLOGY, PO BOX 1, NL-3720 BA BILTHOVEN, THE NETHERLANDS
[2] DLO STATE INSTITUTE FOR QUALITY CONTROL OF AGRICULTURAL PRODUCTS (RIKILT-DLO), PO BOX 230, NL-6700 AE WAGENINGEN, THE NETHERLANDS
[3] NATIONAL INSTITUTE OF PUBLIC HEALTH AND ENVIRONMENT (RIVM), DIRECTOR DIVISION OF PUBLIC HEALTH RESEARCH, PO BOX 1, NL-3720 BA BILTHOVEN, THE NETHERLANDS

1 Introduction

Catechins, also referred to as flavanols, are compounds with antioxidant activity which belong to the family of flavonoids (Figure 1). *In vivo* and *in vitro* studies suggest that tea and its main constituents, the catechins, reduce the risk of certain cancers.[1–3] In epidemiological studies, the protective effect of tea consumption is not consistently found.[3,4] An explanation might be that tea is not the only food that contains catechins. However, thus far, reliable data on the catechin content of most foods were lacking. We have determined the catechin content of over 100 plant foods and beverages that are commonly consumed in the Netherlands, and estimated the major sources of catechins in the Dutch population.

2 Catechin Content of Foods

Six major catechins, (+)-catechin, (−)-epicatechin, (−)-epicatechin gallate, (−)-epigallocatechin, and (−)-epigallocatechin gallate were determined by HPLC methods.[5] In brief, the method consisted of freeze-drying of the foods,

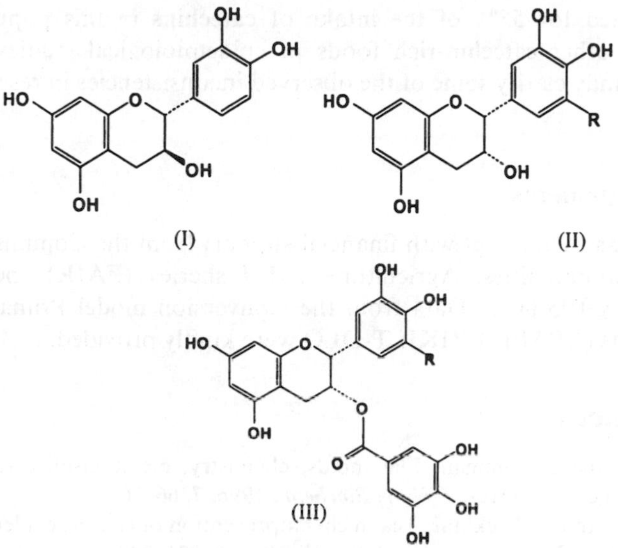

Figure 1 *Chemical structures of catechin: (I) (+)-catechin; (II) R=H, (−)-epicatechin, R=OH, (−)-epigallocatechin; (III) R=H, (−)-epicatechin gallate, R=OH, (−)-epigallocatechin gallate*

followed by extraction with 90% aqeous methanol at room temperature for 1 hour. Catechins were separated by HPLC, and quantified using fluorescence and UV detection. Peak identity was confirmed with diode-array detection.

Total catechin levels in foods ranged from undetectable to 90 mg/100 g food. The total catechin content was highest in chocolate, tea, and certain fruits and legumes.

3 National Food Consumption Survey

The catechin intake in the Netherlands was estimated using data from the Third National Food Consumption Survey.[6,7] This survey was carried out between April 1997 and March 1998 among a representative sample of the Dutch population. We included data from 5908 men and women aged 1–75 years. Information on the diet of the subjects was collected using a 2-day dietary record method.

The average daily catechin intake among adults (over 19 years of age) was approximately 60 mg. Intake ranged from 0 to almost 1000 mg per day. The most important sources of catechins in this population were tea, chocolate, and fruits. On average, tea contributed 55% of the total catechin intake, and chocolate 20%.[8]

4 Conclusions

Black tea and also chocolate and fruits are high in catechins, and at the same time the most important sources of catechins in the Dutch population. Since tea only accounted for 55% of the intake of catechins in this population, the inclusion of other catechin-rich foods in epidemiological studies in similar populations may clarify some of the observed inconsistencies in research on tea and cancer.

Acknowledgements

This work was carried out with financial support from the Commission of the European Communities, Agriculture and Fisheries (FAIR) specific RTD Programme, CT95 0653. Data from the Conversion model Primary Agricultural Products (CPAP) at RIKILT-DLO were kindly provided.

5 References

1 N.C. Cook and S. Samman, Flavonoids: chemistry, metabolism, cardioprotective effects, and dietary sources, *J. Nutr. Biochem.*, 1996, **7**, 66–76.
2 S.K. Katiyar and H. Mukhtar, Tea in chemoprevention of cancer: epidemiologic and experimental studies (review), *Int. J. Oncol.*, 1996, **8**, 221–238.
3 L. Kohlmeier, K.G.C. Weterings, S. Steck and F.J. Kok, Tea and cancer prevention: an evaluation of the epidemiologic literature, *Nutr. Cancer*, 1997, **27**, 1–13.
4 W.J. Blot, W.H. Chow and J.K. McLaughlin, Tea and cancer: a review of the epidemiological evidence, *Eur. J. Cancer Prev.*, 1996, **5**, 425–438.
5 I.C.W. Arts and P.C.H. Hollman, Optimization of a quantitative method for the determination of catechins in fruits and legumes, *J. Agric. Food Chem.*, 1998, **46**, 5156–5162.
6 K.F.A.M. Hulshof and W.A. van Staveren, The Dutch National Food Consumption Survey: design, methods and first results, *Food Policy*, 1991, 257–260.
7 Zo eet Nederland 1998; resultaten van de Voedselconsumptiepeiling 1997–1998. Voedingscentrum, Den Haag, 1998 (in Dutch).
8 I.C.W. Arts, P.C.H. Hollman and D. Kromhout, Chocolate as a source of tea flavonoids, *Lancet*, 1999, **354**, 488.

1.5

Intake of Selected Vitamins and Bioflavonoids from Vitamin and Mineral Supplements by Subjects in the EPIC-NORFOLK Study, UK

Ailsa A. Welch, Angela A. Mulligan, Robert N. Luben and Sheila A. Bingham

EPIC STUDY, INSTITUTE OF PUBLIC HEALTH, UNIVERSITY OF CAMBRIDGE, STRANGEWAYS RESEARCH LABORATORY, CAMBRIDGE CB1 8RN, UK

It is known that substantial contributions to daily nutrient intake can be obtained from vitamin and mineral supplements. We have developed a new database (VIMS; vitamin and mineral supplements) using data supplied by manufacturers which, to the end of 1995, contained 900 individual types and brands. These data have been converted for concentration so that the values are compatible with nutrients as given in the UK food tables.

We have used the VIMS database to allow us to quantify the relative intake of vitamins and bioflavonoids from both dietary and supplemental sources. Preliminary results from the first 1860 participants (coding 98% complete) to enrol in the EPIC-NORFOLK Study (European Prospective Investigations into Cancer, in Norfolk)[1] indicate that 40.6% of participants had consumed a supplement during the past year, 37.5% of them regularly (regular consumption was identified by response to the EPIC Health and Lifestyle Questionnaire).

The Table shows that for vitamins A (retinol), B_6, C, D, E and folic acid, average daily consumption of nutrients from supplements was close to or greater than that from food.

Within the VIMS database many supplements contained amounts greater than average daily intake and the percentage of these within the database were β-carotene (81%), retinol (9%), vitamin C (45%) and vitamin E (55%).

The correlation between plasma vitamin C and intake estimated using the EPIC FFQ was 0.24 in the sample of 204 referred to above and this increased to

Table

Nutrient	Intake from food Mean*	Intake from VIMS		Subjects consuming VIMS	
		Mean	% food intake	N	%
Total carotene (μg)	2917	1956	67	16	1
Vitamin A retinol (μg)	866	806	93	596	32
Vitamin B_6 (mg)	2.1	9.5	453	222	12
Vitamin C (mg)	114	200	176	291	16
Vitamin D (μg)	3.7	6.1	165	600	32
Vitamin E (mg)	6	24.4	407	565	30
Folic acid (μg)	310	258	83	147	8
Bioflavonoids (mg)	26**	15.1	58	25	1

* Values derived from the EPIC food-frequency questionnaire (FFQ) for 204 males and females aged 45–74 years from a stratified random sample of the first 2000 participants to enter the EPIC-NORFOLK Study.[2]

** Values for average intake in males aged 65–74, in the Netherlands.[3]

0.48 when the contribution from supplements was included with the dietary estimates.

These data show that unless nutritional supplements are accounted for this could lead to gross misclassification of exposure in epidemiological studies.

References

1 N.E. Day, S. Oakes, R. Luben, K.T. Khaw, S.A.B. Bingham, A.A. Welch and N. Wareham, *Br. J. Cancer*, submitted.

2 S.A.B. Bingham, C. Gill, A.A. Welch, A. Cassidy, S.A. Runswick, S. Oakes, R. Lubin, D.I. Thurnham, T.J.A. Key, L. Roe, K.T. Khaw and N.E. Day, *Int. J. Epidemiol.*, 1977, **26**, S137–S151.

3 M.G.L. Hertog, E.J.M. Feskens, P.C.H. Hollman, M.B. Katan and D. Kromhout, *Lancet*, 1973, **342**, 1007–1011.

1.6

Starchy Foods and Risk of Cancer of the Upper Aerodigestive Tract: A Case-control Study in Uruguay

Eduardo De Stefani,[1] Paolo Boffetta,[2] Fernando Oreggia,[1] Alvaro Ronco,[1] Hugo Deneo-Pellegrini[1] and María Mendilaharsu[1]

[1] REGISTRO NACIONAL DE CÁNCER, MONTEVIDEO, URUGUAY
[2] UNIT OF ENVIRONMENTAL CANCER EPIDEMIOLOGY, INTERNATIONAL AGENCY FOR RESEARCH ON CANCER, LYON, FRANCE

In 1991, Franceschi *et al.* reported an increased risk of oral cancer associated with high intake of starchy foods.[1] Starchy foods (rice, polenta, pasta, white bread, croissant, potato, sweet potato, and banana) are staple foods in Uruguay, representing 44.8% of total energy intake. Moreover, they are a good source of proteins (34.1%), and, of course, of carbohydrates (76.2%). On the other hand, cancers of the upper aerodigestive tract (UADC) (including carcinomas of the oral cavity, pharynx, larynx, and esophagus) are frequent malignancies in our country, with an age-standardized incidence rate of 36.3 per 100,000 men, occupying the third place among registries in the Americas.[2] Therefore, we decided to carry out a case-control study on this subject, in order to explore the relationship between starchy foods intake and risk of cancers of the upper aerodigestive tract.

In the time period 1996–1998, all cases of squamous cell cancer of the upper aerodigestive tract (UADC) occurring in males and admitted to the four major hospitals in Montevideo were considered eligible as cases for the present study. In total, 243 cases of theses sites with a diagnosis of squamous cell carcinoma were admitted to the hospitals. Of this total number, 12 patients refused the interview; the remaining 231 patients were successfully interviewed (response rate 95.0%). There were 72 patients with cancer of the oral cavity and pharynx (31.2%), 78 patients with laryngeal cancer (33.8%), and 81 patients with esophageal cancer (35.1%).

In the same period, 472 male patients admitted to the same hospitals as the cases, but for conditions not related to tobacco smoking, alcohol drinking or diet, were considered eligible as controls for this study. Fourteen patients refused the interview and the remaining 458 patients were successfully interviewed (response rate 97.0%). Hospital controls were frequency matched to cases on age (10-year group), residence (Montevideo, other counties), and urban/rural status. The main selected conditions were eye disorders (126 patients, 27.5%), abdominal hernia (92 patients, 20.1%), acute appendicitis (57 patients, 12.5%), and traumatic injuries (45 patients, 9.8%).

Both cases and controls were interviewed with a detailed questionnaire, including sections on demographic characteristics, tobacco history, alcohol consumption, 'maté' ingestion, family history of cancer for first degree relatives, height, weight, lifetime occupational history, and a food frequency questionnaire with 64 food items. All cases and controls were interviewed face-to-face in the hospitals. No proxy interviews were performed. The dietary questionnaire used in this study included 64 food items plus vitamin and mineral supplements and questions about alcoholic beverages, soft drinks, coffee, coffee with milk, tea, tea with milk and 'maté' ingestion. For each food, a commonly used unit or portion size was specified, and participants were asked how often, on average, over the past year or the year prior to onset of symptoms for their cases, they had consumed that amount of each food. The responses were open-ended allowing to treat each food as a continuous variable. Responses were converted to times per year, multiplying by the appropriate time units. The following food groups were created: (1) cereals (rice, polenta, pasta, bread and croissant), (2) tubers (potato, sweet potato) and (3) starchy foods (cereals, tubers and bananas).

The distribution of all study subjects (cases and controls) was categorized in quartiles for each food group or nutrient. Crude and adjusted odds ratios (OR) among quartiles were estimated through unconditional logistic regression.[3] Potential confounders were included in all models. These were age, residence, urban/rural status, education, body mass index, tobacco smoking (in pack-years), total alcohol intake, total energy and total carbohydrate intake.

The distribution of cases and controls by sociodemographic characteristics and selected risk factors showed similar proportions regarding age, residence, and urban/rural status. Cases were less educated and leaner than controls. On the other hand, there was a significanly higher proportion of heavy smokers and heavy alcohol drinkers among cases compared with controls. Finally, there was a higher proportion of maté drinkers among cases compared with controls.

Odds ratios of cancer of the upper aerodigestive tract for food groups are shown in Table 1. Grains were associated with a moderate increase in risk (OR 1.36, 95% CI 0.78–2.39), whereas tubers intake displayed an increased risk of 2.22 (95% CI 1.36–3.65). Starchy foods (grains plus tubers) displayed a significantly increased risk (third tertile of intake OR 2.75, 95% CI 1.45–5.18). Finally, starch intake was associated with a positive association (OR 2.74, 95% CI 1.49–5.03), with a significant dose-effect trend ($p = 0.001$).

When food groups were examined by cancer site (oral cavity and pharynx,

Table 1 *Odds ratios of cancer of the upper aerodigestive tract for food groups*[a]

Food groups	*Tertiles* *I*	*II*	*III*	*p-value for trend*
Cereals	1.0	1.28 (0.78–2.09)	1.36 (0.78–2.39)	0.29
Tubers	1.0	1.50 (0.94–2.40)	2.22 (1.36–3.65)	0.001
Starchy foods	1.0	1.75 (1.03–2.99)	2.75 (1.45–5.18)	0.002
Starch	1.0	1.64 (0.99–2.75)	2.74 (1.49–5.03)	0.001

[a] Adjusted for age, residence, urban/rural status, education, body mass index, tobacco smoking (pack-years), alcohol drinking, total energy, and total carbohydrate intakes.

larynx, esophagus) the results were mostly similar (Table 2). The effect of grain intake was more marked in esophageal cancer (OR for the third tertile of intake 1.86, 95% CI 0.82–4.21), compared with the risk of cereals in oral cancer (OR 0.91, 95% CI 0.39–2.12). On the other hand, tubers showed a strong positive association with oral cancer (OR for the highest tertile of intake 2.75, 95% CI 1.25–6.03). When both food groups were combined as starchy foods, the highest increase in risk was observed for laryngeal cancer (OR 3.48, 95% CI 1.32–9.16). Finally, starch intake displayed high risks for oral and esophageal cancers (OR for esophageal cancer 3.53, 95% CI 1.49–8.37).

Previous studies are non-consistent. According to the World Cancer Research Fund, both cereals and tubers should be studied in detail, given the possibility of discriminating the effects according to the degree of refinement.[4] Whole-grains are infrequently consumed in the Uruguayan population, repre-

Table 2 *Odds ratios for food groups by cancer site*[a]

Variable	*Cancer site* *Oral cavity*	*Larynx*	*Esophagus*
Grains	1.0	1.0	1.0
	0.83 (0.38–1.78)	1.41 (0.66–2.98)	1.83 (0.92–3.62)
	0.91 (0.39–2.12)	1.59 (0.68–3.72)	1.86 (0.82–4.21)
Tubers	1.0	1.0	1.0
	1.28 (0.59–2.81)	1.28 (0.61–2.66)	1.84 (0.96–3.51)
	2.75 (1.25–6.03)	2.02 (0.96–4.25)	1.90 (0.94–3.83)
Starchy foods	1.0	1.0	1.0
	1.20 (0.50–2.85)	2.19 (0.94–5.06)	2.12 (1.01–4.49)
	2.04 (0.77–5.42)	3.48 (1.32–9.16)	3.18 (1.28–7.90)
Starch	1.0	1.0	1.0
	1.58 (0.69–3.61)	1.58 (0.77–3.24)	1.76 (0.86–3.60)
	3.26 (1.22–8.67)	1.92 (0.77–4.81)	3.53 (1.49–8.37)

[a] Adjusted for age, residence, urban/rural status, education, body mass index, tobacco smoking (pack-years), total alcohol consumption, total energy and total carbohydrate intakes.

senting less than 2.0% of all cereals intake. Thus, in our study, the effect of cereals should be interpreted as due to refined grain intake.

Mechanisms of action of starchy foods are presently unclear. According to the World Cancer Research Fund, increases in risk are probably due to a diet low in micronutrients.[4] This is in agreement with the profile of the Uruguayan population, which is characterized by a high intake of tubers and a low intake of vegetables and fruits. On the other hand, tubers are ingested at high temperatures, mainly as constituents of stew. Thus, the possibility of thermal injury of mucosa of the upper aerodigestive tract is biologically plausible. Finally, it is not possible to rule out completely a role of starch in carcinogenesis of cancers of the upper aerodigestive tract. Against this mechanism is the lack of experimental studies on starch and cancer.[4] Starch could act as an abrasive agent, leading to cell damage, inflammatory changes and eventually to epithelial displasia.

References

1 Franceschi, S., Bidoli, E., Barón, A.E., Barra, S., Talamini, R., Serraino, D. and La Vecchia, C., Nutrition and cancer of the oral cavity and pharynx in north-east Italy, *Int. J. Cancer*, **47**, 20–25, 1991.

2 Parkin, D.M., Whelan, S.L., Ferlay, J., Raymond, L. and Young, J., Cancer incidence in five continents. Volume VII, IARC Scientific Publications No. 143, IARC, Lyon, 1997.

3 Breslow, N.E. and Day, N.E., Statistical Methods in Cancer Research. Vol. I – The analysis of case-control studies, IARC Scientific Publications No. 32, IARC, Lyon, 1980.

4 World Cancer Research Fund in association with American Institute for Cancer Research, Food, Nutrition and the Prevention of Cancer: a global perspective, World Cancer Research Fund, Washington DC, USA, 1997.

1.7

Fruit and Vegetable Consumption and Lung Cancer among Smokers in Finland, Italy and The Netherlands

Margje C.J.F. Jansen,[1,2,*] H. Bas Bueno-de-Mesquita, Aulikki M. Nissinen, Flaminio Fidanza, Alessandro Menotti, Frans J. Kok and Daan Kromhout, for the Seven Countries Study Research Group

[1] DEPARTMENT OF CHRONIC DISEASES EPIDEMIOLOGY, NATIONAL INSTITUTE OF PUBLIC HEALTH AND THE ENVIRONMENT, PO BOX 1, 3720 BA BILTHOVEN, THE NETHERLANDS
[2] DIVISION OF HUMAN NUTRITION AND EPIDEMIOLOGY, WAGENINGEN AGRICULTURAL UNIVERSITY, WAGENINGEN, THE NETHERLANDS

Extended Abstract

Unquestionably, cigarette smoking is the dominant risk factor for lung cancer. In addition, epidemiological studies consistently show an inverse association between fruit and vegetable consumption and lung cancer risk.[1,2,3] However, critics point at the possible role of residual confounding by smoking, and suggest that results from cohort studies may be less consistent compared with those from case-control studies.[4] Therefore, we examined the relationship between plant food consumption and lung cancer mortality in a cohort of men, focusing on smokers at baseline.

Around 1970, dietary intake of Finnish, Italian and Dutch middle-aged men participating in the Seven Countries Study was assessed using a cross-check dietary history. Smoking habits and other lifestyle factors were determined by a standardised questionnaire. For 3108 men complete baseline information was available, of which 1578 were baseline smokers. During 25 years of follow-up, 187 lung cancer deaths occurred, of which 149 were smokers at baseline.

* Corresponding author: E-mail margje.jansen@rivm.nl

Relative risks and 95% confidence intervals for consumption in country-specific tertiles were estimated by Cox proportional hazard analyses, and statistical significance was determined by two-sided tests. Risk estimates were calculated for the total study population and per country, both univariately and adjusted for potential confounders, including pack-years of smoking.

Fruit consumption was inversely associated with lung cancer mortality in the total group of baseline smokers. Among Finnish smokers, lung cancer risk tended to be reduced in men with a medium or high fruit intake compared with those with the lowest fruit intake; however, estimates were not statistically significant. In Italian smokers, fruit consumption and lung cancer mortality were not related. Fruit consumption was only statistically significant inversely associated in the Dutch cohort. Stratifying the total cohort of smokers according to intensity of smoking, revealed the most pronounced relationship among the heaviest smokers. Vegetable intake and lung cancer mortality were not statistically significant related, neither in the total study population nor in the countries separately.

In conclusion, this study shows a lower lung cancer risk among heavier smokers consuming more fruits. For vegetable consumption no such relationship was observed.

References

1 K.A. Steinmetz and J.D. Potter, *Cancer Causes Control*, 1991, **2**, 325.
2 G. Block, B. Patterson and A. Subar, *Nutr. Cancer*, 1992, **18**, 1.
3 R.G. Ziegler, S.T. Mayne and C.A. Swanson, *Cancer Causes Control*, 1996, **7**, 157.
4 L.C. Koo, *Int. J. Cancer*, 1997, **10**, 22.

1.8

K-ras Mutations Rectal Crypt Cell Proliferation and Diet: A Study in Patients with Left-sided Colorectal Carcinoma

J.A. Matthew,[1] H. O'Brien,[1] M. Watson,[2] J.M. Gee,[1] M. Rhodes,[2] C. Speakman,[2] H.J. Kennedy,[2] W. Stebbings[2] and I.T. Johnson[1]

[1] INSTITUTE OF FOOD RESEARCH, NORWICH RESEARCH PARK, COLNEY, NORWICH, NR4 7UA, UK
[2] NORFOLK & NORWICH HOSPITAL, BRUNSWICK ROAD, NORWICH, NR1 3SR, UK

1 Introduction

Colorectal carcinoma is one of the most common causes of death from cancer in industrialised countries. It is widely accepted that dietary factors are important, with protective effects shown for increased fruit and vegetable intake and a possible adverse effect of red meat and fat consumption.[1] Previous reports have suggested that k-ras mutations may be more common in patients with relatively high meat intake and fat consumption.[2] In this study we explored some relationships between eating habits, crypt cell proliferation and the pattern of k-ras mutations in patients undergoing surgery for colorectal cancer.

2 Methods

Fifty one patients (26 male, 25 female; mean age 71.8 ± 1.1 years) diagnosed with left-sided colorectal carcinoma gave informed consent and were recruited into the study by a clinician. A rectal biopsy was taken prior to surgery, and patients received a self-administered food frequency questionnaire. A sample of tumour tissue was taken at surgery for k-ras mutation studies. Rectal biopsies were also obtained for assessment of crypt cell proliferation from a group of volunteers (12 male, 12 female; mean age 60.1 ± 0.5 years) who came from the same catchment area, and who attended a screening clinic (ICRF/MRC Flexiscope Sigmoidoscopy Screening Trial). A healthy age- and gender-matched control group of thirty subjects was also recruited for completion of the food intake questionnaire only.

Crypt cell proliferation (CCP). Rates were measured using a direct visual assessment method for intact crypts.[3,4] Crypt morphology was noted and where there were abnormal features the specimen was photographed.

The food intake questionnaire. The questionnaire, based on a previously validated design,[5] listed 184 items, including single foods and complete dishes. The analysis template was based on information from McCance & Widdowson's 'The Composition of Foods' 4th Edition.

K-ras mutations. K-ras mutations were detected using a PCR-based oligonucleotide hybridisation assay. DNA was extracted from stored tumour tissue. Oligonucleotide primers were prepared and used to amplify a region of exon 1 of k-ras 2 spanning codons 12 and 13 by PCR. Analysis was by oligonucleotide hybridisation using a panel of ^{32}P-labelled probes. DNA for the control panels was produced by amplification of wild-type human genomic DNA, using a set of primers designed to produce fragments containing target mutations. To confirm the specificity of the hybridisation, a positive control filter was included for every hybridisation with each probe consisting of a panel of PCR-amplified fragments corresponding to each possible mutation for codons 12 and 13. To test the efficiency of mutation detection, filters consisting of a panel of different proportions of mutant to wild-type DNA (0.5 μg total DNA; 1%–100% mutant DNA) were prepared for several mutations.

3 Results

Crypt morphology and proliferation. Crypts from the cancer patients were frequently larger than those of controls and many had morphological abnormalities such as multiple side-branches. The mean frequency of mitosis per crypt for the cancer patients, measured on apparently normal cylindrical crypts (12.66 ± 0.93), was almost twice that of healthy subjects (7.13 ± 0.72; $p < 0.0001$). However, there was no significant displacement of cell division towards the luminal end in crypts from the patients. There was no significant difference in mean frequency of crypt mitosis in patients with or without k-ras mutations.

K-ras mutations. A total of 43 tumours were available for detection of K-ras mutations. The upper bound for the sensitivity of the method, as estimated using sensitivity blots, was approximately 5% mutant : wild type DNA. Of the 43 tumours successfully screened, 15 (34.9%) were found to have K-ras mutations. There were 13 mutations in codon 12 and 3 in codon 13, one tumour having mutations in both codon 12 and 13. All the observed mutations were G to A or G to T.

Dietary intake. We did not find any significant difference in the red meat intake of patients with (92.4±9.7 g/d) or without (82.3±7.7 g/d) k-ras mutations, nor was there any detectable correlation within the patient group between frequency of mitosis per crypt and red meat intake or fat consumption. The healthy control group completing the dietary questionnaire consumed significantly more fruit and vegetables (643±114 g/d) than the cancer patients (343±31 g/d; $p < 0.011$).

4 Conclusions

In this study, a group of patients with left-sided colorectal cancer had higher rates of crypt cell proliferation than healthy volunteers with no rectal polyps. There was no significant displacement of proliferation within crypts, but we noted a much higher incidence of abnormally shaped crypts in biopsies from these patients than we found in controls.

Within the patient group there were no significant correlations of red meat intake with presence or absence of a k-ras mutation in the tumour tissue and there was no relationship between rates of crypt cell proliferation and either red meat or total fat consumption. However, there were significant differences in the amounts of fruits and vegetables eaten by these patients compared with an age- and sex-matched control group with no known colorectal disease. This observation tends to support previous findings suggesting a protective effect of fruit and vegetable consumption against colorectal cancer. It should be noted, however, that any case-control study of this type is subject to bias. When completing the questionnaire, patients were asked to try to disregard any effects of their current symptoms and to describe their eating habits prior to their illness, but we cannot be sure that our subjects had not unknowingly changed their meat and vegetable consumption as a result of their condition. Further studies to determine the relationship between the pattern of somatic mutations in sporadic colorectal cancers and endogenous or environmental variables seem warranted.

Acknowledgements

This work was funded by the Biotechnology and Biological Sciences Research and the Ministry of Agriculture.

5 References

1 World Cancer Research Fund (1997), Food, Nutrition and The Prevention of Cancer: A Global Perspective, American Institute of Cancer Research, Washington, DC, USA.

2 Freedman, A.N., Michalek, A.M., Muro, K., *et al.*, Meat consumption is associated with k-ras mutations in tumours of the distal rectum. *Proc. Am. Assoc. Cancer Res.*, 1996, **38**, 457A.

3 Matthew, J.A., Pell, J.D, Prior, A., Kennedy, H.J., Fellows, I.W., Gee, J.M., Burton,

J. and Johnson, I.T., Validation of a simple technique for the detection of abnormal mucosal cell replication in humans, *Eur. J. Cancer Prevention*, 1994, **3**, 337–344.

4 Matthew, J.A., Fellows, I.W., Prior, A., Kennedy, H.J., Bobbin, R. and Johnson, I.T., Habitual intake of fruits and vegetables amongst patients at increased risk of colorectal neoplasia, *Cancer Lett.*, 1994, **114**, 255–258.

5 Lloyd, I.I.M., Paisley, C.M. and Mela, D.J., Changing to a low fat diet: Attitudes and beliefs of UK consumers, *Eur. J. Clin. Nutr.*, 1994, **47**, 361–373.

1.9

The Colon as a Heat Producing Organ – Implications for Tumour Growth

Olov H. Holmqvist

SWEDISH MEATS R&D (PREVIOUSLY SWEDISH MEAT RESEARCH INSTITUTE), PO BOX 504, S-244 24 KÄVLINGE, SWEDEN

1 Ecological Factors of the Colon Related to Tumour Growth

1.1 Traditionally Observed Functions of the Colon

Humans and other mammals have a large colon, which functions as a route for excretion of wastes and water. It also houses a microflora that ferments carbohydrates, modifies chemicals, restricts invasion of harmful bacteria and produces vitamin K2. None of these functions, however, is necessary for life. Vitamin K2 can be had with diet; germ-free animals in fact live longer and reptiles have only a diminutive colon.[1,2] This suggests that some other function of importance may be in operation.

1.2 Microbially Generated Heat of the Colon

In farmed animals it has been estimated that 6–8% of the energy content of the feed may be lost as heat produced by microbial fermentation.[3] It may be argued that the figure should be at least as high for adult humans who, as distinct from farmed animals, are not on an optimal diet. Microbially generated heat in the colon is not under direct somatic control. Hence factors which favour microbial metabolism may increase the temperature of the colon.

1.3 Temperature of the Colon and Tumour Development

A detectable tumour may contain 10^9 cells and derive from one transformed cell after 30 divisions, corresponding to five years.[4] If a slightly higher temperature, by *e.g.* 0.25 °C, over the years will decrease division time by 10%, detection will be possible half a year earlier. Such a difference may well surface in epidemiological studies.

2 Risk Factors of Colon Cancer and Relation to Colon Heat

A number of associations frequently occur in epidemiological studies.[5] These are discussed in relation to microbially generated colonic heat.

Height. A taller person has a larger colon harbouring more microbes and a greater body volume to body surface ratio. Hence more heat is produced and less is evaporated. It fits the relation.

Obesity. Body fat thermally isolates and is also indicative of a poor balance between energy intake and expenditure. More energy will be available for the microbes of the colon. It fits the relation.

Low physical activity. Low physical activity may reasonably associate with a poor energy balance and also with more sitting which thermally isolates and allows for less colonic heat to evaporate. It fits the relation.

Smoking. Smoking increases the number of red blood cells, reduces the fluidity of blood and consequently reduces the chilling capacity of the colonic epithelium. It fits the relation.

Constipation. Constipation is synonymous with increased transit time, more microbial mass and more time for colonic heat production. It fits the relation.

Westernization. Migration or a switch from a traditional to a Western life style may mean transit from manual or farm labour to more of white collar jobs meaning more sitting. This may be especially pronounced in transitions from a traditional tribal life as in Africa, where squatting was replaced by sitting. Additionally a switch to poor energy balance and a diet with a lesser content of plant material not digested by the colonic microbes may occur. It fits the relation.

Gender difference. Men and women locate body fat differently with age. Men develop a 'beer-belly' covering the colon whereas women locate their fat on their hips. The male development isolates the colon but not so the female one. It fits the relation.

Family history. Not only genetic traits for obesity or unfavourable basic metabolic rate are transferred from older generations, but also social traditions, *e.g.* of little physical activity. It fits the relation.

Alcohol. A higher consumption of alcoholic beverages may well associate with habits of sitting. Alcohol also stimulates appetite and may indicate a shift towards a poor energy balance and, of course, a 'beer belly' in men. It fits the relation.

Sucrose. Sucrose consumed in large amounts may not be completely degraded in the small intestine but end up in the colon and become a microbial substrate. Habitual consumption may associate with habits of sitting and a poor energy balance. It fits the relation.

Fat. Fat is energy dense and may contribute to a poor energy balance and also end up in the colon and become a microbial substrate. It fits the relation.

Fibre. Fibre, unlike fat or sucrose, is low in energy. Burkitt's classic observation of a gradient of increasing stool size from traditional, via westernized Africans to Occidentals in Africa must have reflected fibre that were not digested by the colonic microbes and hence did not increase their metabolism and heat generation. It fits the relation.

Vitamins. Vitamins indicate fruits, vegetables and fibre. It is therefore possible that the effect is indirect. If this is correct vegetables and fruits, but not pills, will associate with reduced risk.

Meat. Meat is a major contributor of heme iron. By poor energy balance less heme iron is absorbed in the small intestine and more ends up in the colon. Interestingly the dominant flora of the colon is hemin dependent.[6] More heme iron and hemin may also be available to colonic microbes after browning of meat as this denatures the heme iron structure and makes iron less absorbable upstream the colon. It fits the relation.

Dairy products. Dairy products provide sugars like lactulose which tend to shift the colonic microbial flora towards lactobacillii which grow slowly and possibly generate less heat.[7] It fits the relation.

Calcium. Calcium indicates dairy products. It fits the relation.

Fruits. Fruits provide fructose, which is not digested upstream the colon and tends to shift the colonic flora towards slow growing Bifidobacterium.[8] It fits the relation.

In summary, many of the associations between risk factors and colon cancer can be, at least in part, explained by assuming that microbially generated heat in the colon affects tumour growth.

3 References

1 M. Pollard and B.S. Wostman, *Progr. Clin. Biol. Res.*, 1985, **181**, 75.
2 E. Jacobshagen, 'Handbuch der Vergleichende Anatomie der Wirbeltiere', ed. L. Bolk, E. Göppert, E. Kallius and W. Lubosch, Urban Schwarzenberg, Berlin und Wien, 1937, p. 640.
3 R. Braude, Chairman, Commonwealth Agricultural Bureaux. Farnham Royal, Slough SL2 3BN, UK. The Nutrient Requirements of Pigs. Technical Review by an Agricultural Research Council Working Party, Norwich, 1981, Chapter 1, p. 43.
4 I.J. Fidler, Cancer of the Colon, Rectum and Anus, ed. A.M. Cohen and S.J. Winawer, McGraw-Hill, New York, 1995, Chapter 17, p. 176.
5 J.D. Potter, *Cancer Causes Control*, 1996, **7**, 127.
6 L.V. Holdeman, R.W. Kelley and W.E.C. Moore, 'Bergey's Manual of Determinative Bacteriology', ed. N.R. Krieg and J.G. Holt, Williams & Wilkins, Baltimore/London, 1984, Section 6, p. 604.
7 S. Salminen and E. Salminen, *Scand. J. Gastroenterol.*, 1997, **222** (Suppl.), 45.
8 D.J. Jenkins, C.W. Kendall and V. Vuksan, *J. Nutr.*, 1999, **129** (7 suppl.), 1431.

Section 2

Bioavailability of Dietary Anticarcinogens and Mutagens

2.1

Bioavailabilities of Bioactive Components in the Human Diet

Susan Southon, Richard Faulks and Annette Fillery-Travis

INSTITUTE OF FOOD RESEARCH, NORWICH RESEARCH PARK, COLNEY, NORWICH NR4 7UA, UK

1 From Epidemiology to Experiment: A Leap of Faith

Scientific groups throughout the world are convinced of, and are investigating, the link between diet, or more likely specific dietary components, and cancer. In the main, the hypotheses being tested have arisen from epidemiologic studies that indicate significant associations between the consumption of certain diets, food groups, or specific foods and reduced incidence of cancers.[1] Thus, communications of research in this area usually commence with a brief and apparently convincing resume of the epidemiologic data from which a particular hypothesis and programme of experimentation has been developed. However, in most such communications, there is a great leap of faith between the description of the observation and the definition of the experiment.

The fact that the dietary intake data used in epidemiological studies is notoriously difficult to collect with any degree of accuracy, that the validity and reproducibility of dietary survey methods may not have been tested within the studies cited, and that interpretation of epidemiological studies in terms of likely protective dietary components depends upon the state-of-the art as far as chemical analysis of the food is concerned is rarely discussed, even briefly. In addition, the natural bias of some researchers to interpret observational studies in favour of their compound(s) of interest is rarely confessed, and whether or not the putative protective agents in the health promoting diet can be absorbed from the dietary mix and effectively targeted to their proposed site of action, in a chemical form that could elicit the suggested beneficial response, is surprisingly not always considered when interpreting and discussing results from model systems *in vitro*.

Let us take one group of compounds, the carotenoids, as an example to illustrate some of the above points briefly. This provides the authors with an opportunity to confess 'compound' bias, and puts the importance of understanding absorption, metabolism and tissue targeting (bioavailability) of bioactive components in the diet into the context of other important factors to be taken into account when developing hypotheses about diet-cancer relationships. Whilst we will naturally rely on our own expertise and experience as a basis for this discussion, the points raised are salient to most, if not all, dietary derived microcomponents.

First, determination of the availability of the compound from the diet and more specifically the collection of *dietary intake data* in population studies. Food Frequency Questionnaires (FFQ, the most commonly used method for obtaining food intake data in larger population studies) have been documented to over-estimate the intake of fruits and vegetables, and hence are likely to over-estimate the compounds found within these foods,[2] most probably because the carotenoid-containing fruits and vegetables are perceived as healthy. Other components of fruits and vegetables may suffer the same effect. Carotenoid intake versus plasma concentration relationships become more tenuous the greater the variety of food sources consumed (because there is more scope for error in calculating intake), which means that cross-cultural studies using questionnaires may also be flawed because of the difference in types and amounts of foods consumed within those populations. Amounts of carotenoids (and phytochemicals in general) present in the same plant food can vary hugely depending on variety, soil, climate, growing conditions and preparation and so current food tables cannot hope to provide a particularly accurate estimate of amounts consumed. Even direct analysis of diets may not improve matters particularly when between laboratory (between population) data are being compared, since there can be a coefficient of variation of up to 40% in the carotenoid content of the same food measured in different laboratories.[3] It is not surprising that data arising from observational studies of food component-health relationships can be variable and even contradictory. What is surprising is the consistency and strength of the relationship between the consumption of fruit and vegetable-rich diets and reduced risk of various chronic diseases including cancers.

Second, chemical analysis. Question: Why do we see more associations between the dietary intake of β-carotene and reduced risk of cancers than for other carotenoid, or other phytochemical? Answer: Because β-carotene is a highly effective anti-carcinogen, *or* because, unlike some of the other bioactive compounds in our diet, β-carotene is relatively chemically stable in food samples if kept cold and dark, and we can quantify β-carotene more readily than many other plant food constituents for which it may be acting as a marker?[4] The 'marker' hypothesis is now in general use for a range of compounds but before judgement is passed on 'active' versus 'marker' compounds it would seem sensible to ascertain if the compound of interest is actually *absorbed* from the supposed health promoting diet in which it is contained and, if so, whether it actually *targets* the proposed site(s) of action within the body. Without this information, observational data can be highly misleading and 'active' versus 'marker' impossible to determine. For example, there is an inverse association between plasma β-carotene and risk of cataract.[5]

Since β-carotene is now known not to be present in human lenses[6] then it would seem appropriate to assume that, if there is a reduced risk of cataract related to carotenoids, the relationship is with lutein and/or zeaxanthin, both of which are found in this region and are consumed along with β-carotene in yellow/green vegetables. In the 'phytochemical world', there are many postulates of such diet-health relationships, despite the fact that there is only minimal evidence that some of these compounds are absorbed from the diet and, if they are, there is little (or no) knowledge of their chemical form(s) *in vivo*, their ability to target body tissues, or their residence time within the body (rates of disposal). If the bioactivity and cellular roles of these compounds in human tissues are to be proposed, then the above information is essential to allow design and execution of appropriate experiments.

2 Determining Bioavailability is Essential to Progress

Information on the extent of absorption and metabolism of putative bioactive components of the human diet is essential to allow interpretation of observational data, to develop appropriate hypotheses and to direct experimental studies. Providing such information is by no means trivial and has its own areas of uncertainty, as exemplified by the wide range of values quoted in the literature for percentage absorption of various vitamins. Table 1 shows values of 0–95% for β-carotene, 20–80% for vitamin E, and 25–90% for folates. These huge ranges probably reflect differences in the matrix in which the compound was delivered to the human volunteer, the physiological state of the volunteer, the experimental approach used to obtain the data, and the set of assumptions used to interpret the data obtained. The following sections discuss: the main experimental approaches used to measure absorption and predict metabolism; the limitations of these approaches; how they might be misinterpreted to give inappropriate answers; and how they might be improved to reduce the

Table 1 *Bioavailability: estimated efficiency of vitamin absorption*

Vitamin	*% Absorption*	*Mechanism*
Retinol	>80	active, diffusion, micelles
β-Carotene	0–95	diffusion, micelles
Vitamin E	20–80	diffusion, micelles
Vitamin D	50	diffusion, micelles
Thiamin (B1)	>80	active, diffusion
Riboflavin (B2)	>60	active, diffusion
Vitamin B6	60–90	diffusion
Vitamin B12	20–75	active, diffusion
Niacin	>80	active, diffusion
Folate	25–90	active, diffusion
Biotin	20–30	active, diffusion
Pantothenate	40–60	??
Vitamin C	80–100	active, diffusion

(apparent) wide variation between experiments described in the literature. This discussion will naturally veer towards the compounds of interest to the authors, but the points raised have general relevance to the biavailability of all bioactive species found within the human diet.

3 Experimental Approaches

3.1 Simple Faecal Metabolic Balance Approach

Broadly speaking, this technique relies on measurement of the difference between input and excretion of the compound of interest, and the approach is used to attempt to quantify amount absorbed. Analysis of intake and faecal content of the compound, over the balance period, or after an acute dose must be extremely accurate because differences may be relatively small. The diet needs to be carefully controlled both during, and sometimes for periods before and after the dose, faecal markers need to be employed to determine the start and end of the balance period and, to ensure complete collection, it must be assumed that the faeces are the only significant excretory mechanism for the compound and that none of the unabsorbed compound has undergone biotransformation (due to digestive processes or microbial action) or otherwise been lost (again possibly by microbial action). It is also essential to consider enterohepatic circulation because losses to faeces may have been absorbed, re-excreted in the bile and then lost to faeces. Similarly, enteric recycling can occur within the gut involving absorption, excretion and re-absorption of nutrients over several cycles. In both cases, the nutrient enters the body pool, where it will have an impact on the metabolic kinetics, and would, therefore, be considered to be bioavailable.

It is likely that, for bioactive microcomponents of current interest (*e.g.* folate vitamers, vitamins C & E, carotenoids and what are referred to as non-nutrient phytochemicals), these assumptions cannot be made. Such components are likely to be susceptible to microbial and oxidative degradation in the gut and the unabsorbed portion will not be quantitatively recovered from the faeces. Nevertheless, much of our information on the absorption of the micronutrients is based on either acute or chronic faecal mass balance methods that, unsurprisingly, show great variability in results. Are there alternative approaches to getting better mass balance measurements?

3.2 The Gastrointestinal Lavage Technique

This approach has been used in an attempt to overcome some of the problems inherent in the 'simple' mass balance approach. In the gastrointestinal lavage technique, the entire gastrointestinal track is washed out by consuming a large volume (1 gallon/4.5 litres) of 'Colyte' containing polyethylene glycol (PEG) and electrolyte salts. Washout is complete with the production of clear rectal effluent (2.5–3.5 hours). Volunteers then consume the test meal and are permitted only water or 'diet' soft drinks (non-caloric) for the next 24 hours.

All the effluent is collected and pooled with the following day's effluent collected after a further dose of 'Colyte', which washes out the remainder of the test meal. The compound recovered in the stool is subtracted from that fed to obtain an absorption figure.[7]

The advantage of this approach is that it the residence time of the compound in the GI tract (particularly the large bowel where fermentation occurs) is reduced and standardised. The disadvantages are that the method is relatively time consuming, it may give an underestimation if absorption is compromised by the 'Colyte' (*e.g.* excessive transit rates) and, as with the faecal mass balance, the method depends upon there being no degradation, or loss, of unabsorbed compound.

3.3 The Ileostomy Mass Balance

In individuals who have undergone ileostomy, the colon has been surgically removed and the terminal ileum brought to a stoma on the abdominal wall. Ingested food passes through the stomach and ileum in around 6 hours as it would in the intact individual. The digesta (ileal effluent) can be collected at regular intervals (2 hours) from a test meal (given in the morning after an overnight fast) and can be recovered over the next 12 hours. During this time, the volunteers can consume meals and beverages, which obviously should not contain the compound being studied. Using this approach, the unabsorbed compound can be recovered from the ileal effluent in real time, without the delay of the colon and rectum, or the confounding influence of the colonic microflora. The model also has the added advantage that an excretion profile can be obtained, the timing of which gives the time span for the absorption, which can in turn be compared to changes in plasma concentration over the test period.[8] One of the disadvantages with this method is that volunteers are generally restricted to older adults. However, results obtained from ileostomy studies can be compared with those of other techniques (like plasma response) within an individual, and results used to validate plasma-based methods that could then be applied to other populations groups.

Recently, we have used the ileostomy mass balance approach to quantify the percentage absorption of the carotenoids lutein and β-carotene from spinach. The study was performed, partly, in response to results from blood response-based studies that suggested that there was little, if any, absorption of β-carotene from green leafy vegetables. Figure 1 shows the percentage loss of β-carotene – in the ileal effluent – following an oral dose of β-carotene (a test meal of 150 g of cooked spinach) given after an overnight fast. The loss was determined by analysis of the ileal effluent, over 12 hours. In the three volunteers analysed thus far, 20–30% of the β-carotene and lutein from spinach disappeared from the ileal effluent, indicating significant absorption. However, this significant 'absorption' was not associated with a change in plasma carotenoid concentration in any of the volunteers.

A lack of plasma response has been interpreted in terms of responders and non-responders; that is, for some dietary compounds in some volunteers, no

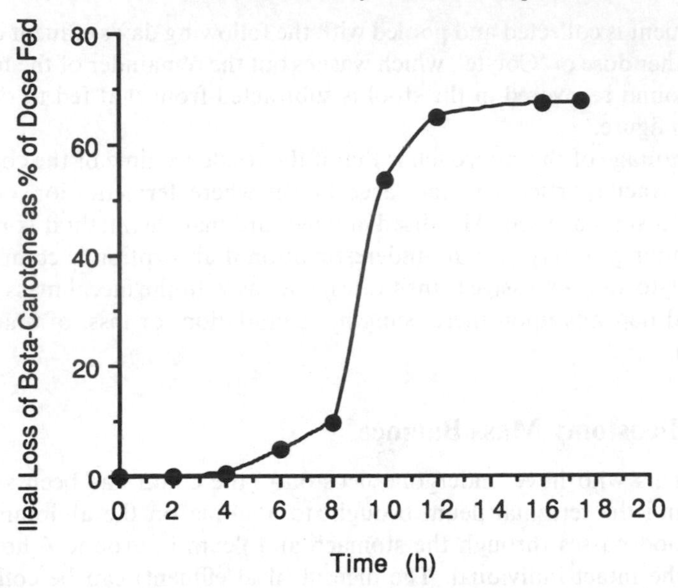

Figure 1 *Percentage of β-carotene lost to ileal effluent in ileostomy subject fed 150 g of cooked whole spinach*

change in plasma concentration is observed even after a substantial acute oral dose and such individuals are termed non-responders.[9] It has been postulated that this non-response is because the compound (in this case the carotenoid) remains in the mucosal enterocytes for some time before transport into the body and/or loss in faeces via mucosal cell turnover. Evidence cited for this hypothesis, with respect to the carotenoids, is a second peak of carotenoid found frequently following a second carotenoid-free meal. If the meal was carotenoid free, where has the carotenoid come from if not from mucosal storage? However, there is no known mucosal storage mechanism. Our ileostomy studies provide no support for the temporary storage hypothesis since there is no tailing of carotenoid in the effluent after 12 hours as would be expected if the compound had been retained in the enterocyte and was lost during normal cell turn over in the gut. Further, radiolabelled β-carotene absorption has been shown to be complete in less than 12 hours.[8,10] This discrepency between results obtained from ileostomy and plasma-response studies will be discussed again after description of plasma response methods.

3.4 Plasma Response – Acute Doses

Measurement of changes in plasma, serum or whole blood concentration following acute or chronic dosing is currently the most well-used approach for estimating absorption and clearance. However, changes in blood concentration, particularly following a single acute dose of dietary (rather than supplement) amounts of a compound, can be difficult to detect and the results from such studies are those most frequently misunderstood and mis-interpreted. The plasma response approach depends upon frequent blood sampling after an acute dose of an isolated compound or after a single meal. Changes in concentration of the compound of interest are measured and plotted against time to produce a response curve. The area under the curve (AUC) is then calculated and used to determine (qualitatively) the extent of absorption.

At its simplest, the AUC is dependent upon the rate at which the newly absorbed compound enters the blood and the rate of disposal to other body compartments, both of which are occurring at the same time. Changes in the characteristics of the AUC within a study may therefore be due to differences in absorption, differences in kinetics of disposal or both. There may also be a problem of re-exportation of the compound between the plasma and other compartments, so that the area under the curve now consists of three components: newly absorbed compound entering the plasma; disposal of the compound to other tissues and possibly urinary excretion; and re-exportation of the compound from a tissue (or tissues) into the plasma. In isolation, and without a mechanistic understanding of the processes of absorption and disposal, the AUC cannot provide quantitative data on extent of absorption. A rise in plasma concentration demonstrates that 'some' of the compound is absorbed. A subsequent decrease in plasma concentration demonstrates that 'some' of the compound is cleared from circulation. Peak plasma concentration multiplied by the estimated plasma volume gives the lower boundary of the amount absorbed and the response curve gives some indication of when maximum absorption is likely to occur. However, if sufficient time points are obtained and metabolic modelling techniques employed, it is possible to get more useful information from the plasma response than a simple area value. The declining part of the curve can be used to calculate clearance kinetics of the compound; this in turn can be used to interpret the rising part of the curve (which is a balance between absorption and clearance) and both the rate and extent of absorption can be calculated.[11]

This approach is particularly useful for hydrophilic compounds present in the aqueous plasma phase. For lipophilic compounds quantitation of amounts absorbed is more complex because the whole plasma response is a multi-component response involving the transfer of the compound (and/or its metabolites) to, and between, carriers in the blood, which have their own 'absorption' and 'clearance' kinetics. For example, the lipoproteins have half-lives ranging from minutes[12] to days.[13] Where there is complex partitioning of a compound between blood fractions then whole blood or plasma provides a composite response and cannot be used to quantify amounts absorbed. A better approach, in this case, is examination of the fraction containing the newly absorbed compound. This is discussed in the next section but before passing on it is important to re-iterate that if there is no observable change in plasma concentration this does not mean there is no absorption. If the rate of absorption of a compound is equivalent to its rate of clearance from the plasma there may still be substantial absorption from an acute dose with no observable shift in plasma concentration.

3.5 Plasma Triglyceride Rich Lipoprotein (TRL) Fraction

For some compounds life can be made much easier by forgetting whole plasma response and concentrating on that fraction that will contain, primarily, the newly absorbed compound, if there is such a fraction. For example, for carotenoids and other lipophilic compounds, response in the triglyceride-rich fraction of plasma can be measured. Newly absorbed carotenoids are initially present in chylomicrons before they are sequestered by body tissues and re-exported in, or transferred to, other lipoprotein fractions. Thus, measurement of carotenoids in this fraction and a knowledge of the rate of clearance of the chylomicrons should permit the calculation of rates of absorption and tissue disposal, and overall absorption based on AUC measurement.[14] This method has the advantage that chylomicrons present in fasting plasma are few and they are almost devoid of carotenoids. The disadvantage is that the plasma has to be ultracentrifuged to separate the lipoprotein classes and this does not normally permit the separation of the chylomicron fraction totally free of other low density lipoproteins, particularly the VLDL (which may be the primary vehicle for hepatic re-exportation of absorbed carotenoids).

The advantage of using a chylomicron response curve rather than a plasma response is apparent from Figure 2, which shows an observable perturbation of the chylomicron pool of lutein and β-carotene following a spinach meal whilst no change in whole plasma concentration is observed. The use of TRL has been primarily applied to carotenoids and is judged as one of the more useful options for the calculation of β-carotene bioavailability. Its use for other lipophilic compounds is being tested.

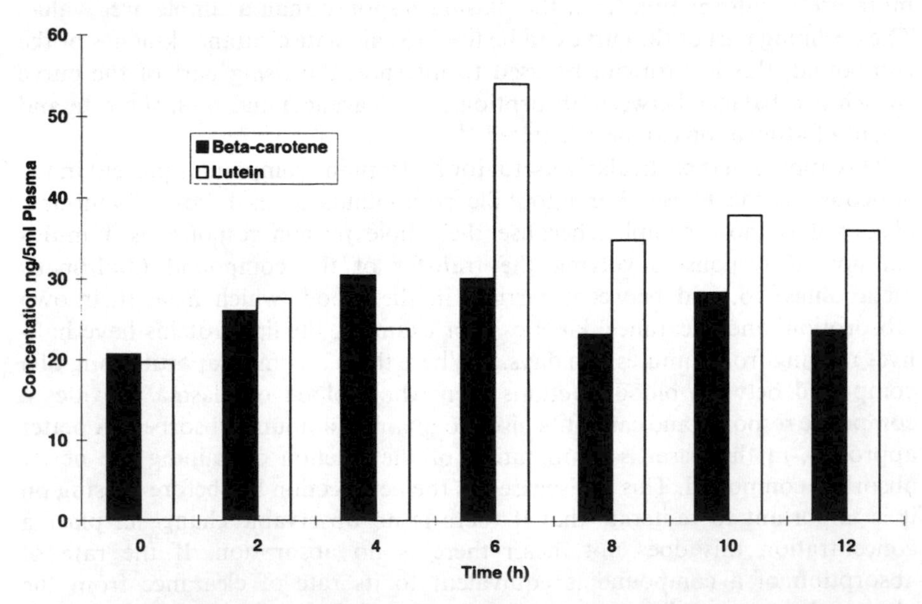

Figure 2 *Chylomicron response of β-carotene and lutein to a single meal of 150 g cooked whole leaf spinach*

3.6 Plasma Response – Chronic Doses

Chronic dosing with foods or supplements needs to be carried out until the plasma concentration reaches a plateau, being careful not to 'flood' the system (*e.g.* being careful not to exceed the renal threshold in the case of vitamin C). Absolute absorption cannot be determined by chronic dosing (only relative absorption) and differences between individuals in disposal rate to the tissues can confound interpretion of changes in plasma concentration. The chronic dosing technique can only be used to obtain comparative absorption of the same compound from different matrices. As with acute plasma response AUC, chronic response cannot be used to compare different compounds whose absorption and clearance kinetics are not known.

3.7 Isotope Methods

One of the major problem with any stable isotope approach is the supply of appropriately labelled isolated compounds and the production of foods where the compound of interest is labelled within the food structure – termed intrinsic labelling. It is also essential that the labelled compound does not behave differently, chemically or physically, from those naturally abundant. For example, ^{13}C labelled carotenoids are used in preference to deuterated compounds because the deuterated compound can be separated chromatographically from normal β-carotene raising questions about possible difference in metabolism *in vivo*.[15] Also, measurement of the isotope by mass spectrometry can require many months of development and the equipment is costly. Having said that, if facilities are available the isotope approach is often the only way to track physiological amounts of a compound within a much larger pool which does not have to be manipulated (*e.g.* by deprivation or preloading). The use of labelled material, either as an isolate or within a food, should help the measurement of absolute absorption and the kinetics of disposal from plasma to tissues and conversion to other metabolites, which in turn will elicit simpler and less costly approaches. There are several good examples, in the literature, of the use of stable isotopes in the study of bioavailability, largely in the field of mineral nutrition, but application to organic microcomponent research is widening and strengthening.[15–18]

3.8 *In vitro* Methods

Apart from the obvious influence of chemical form of the compound(s) on absorption and subsequent metabolism, we hypothesise that the bioavailability of dietary components is markedly influenced by their physicochemical availability from the food structure. The release of individual components from the food source occurs primarily upon ingestion and initial digestion within the stomach, and the factors influencing their release are: localisation within the food matrix; physical break-up of the food; chemical/enzymatic digestion; and, in the case of lipophyllic compounds, the presence of a suitable lipid phase predominately emulsion or free lipid phase. The task of determining and quantifying such factors is best approached, at least initially, by performing measurements *in vitro* of the release under conditions that mimic ingestion and passage through the human stomach and small intestine. This approach can provide results more quickly, under a range of conditions and at a lower cost, than studies with human volunteers. In addition, if performed in conjunction with human studies, methods *in vitro* can be validated (or not) and thus possibly provide more routine methods of predicting the availability of dietary components for absorption. A description of the use of *in vitro* techniques for predicting the bioavailability of inorganic dietary components, including iron, can be found in the literature.[19] Methods for predicting the availability for absorption of organic components, which may undergo substantial biotransformation and/or which are lipophilic in nature, are more complex and are only now under development. In our laboratory methods for carotenoids and related compounds are being devised.

Investigations to date demonstrated that the rate of transfer of carotenoid to the oil phase, during simulated digestion, is significantly enhanced with small particle sizes; the amount of carotenoid transferred during the initial two hours of an digestion *in vitro* is linearly dependent upon the initial carotenoid concentration; the rate and extent of partitioning of the carotenoids from a carrot juice is highly dependent upon the pH of the medium (low pH reversibly enhanced the extent of partitioning); the transfer of carotenoid from a carrot juice to the oil phase is reduced by the presence of soluble material from the juice; there is no evidence of solubilisation of the carotenoid within the continuous phase by the formation of carotene/protein complexes; and significant improvement in transport into the oil phase at both high and low pH is facilitated by the enzymatic breakdown of the tissue.[20] The mechanism of carotenoid transfer appears to be by direct contact between the chromoplasts and the lipid droplets. The surface charge on the oil droplets varies with pH passing through a minimum at pH 2.5. The reduction in electrostatic repulsion between the tissue fragments and the lipid droplets allows aggregation to occur with a corresponding increase in the transfer of carotenoid to the lipid. This is in agreement with the influence of gastric acidity on carotenoid absorption measured *in vivo* with human volunteers (Fillery-Travis, Faulks & Southon, unpublished data). Such an effect is observed for plant tissues containing crystalline carotenoid with relatively high protein concentrations (carrot and tomato). For spinach the effect of pH is still present but to a lesser extent. In brief, the partitioning of the carotenoid was found to be dependent upon: tissue particle size, lipid type, carotene concentration, pH and processing of the plant tissue prior to digestion. Other factors influencing carotenoid release from the food matrix and subsequent transfer to the oil and micellar phases are becoming apparent and will be published shortly. This brief description merely serves to illustrate the relatively rapid progress that can be made using *in vitro* approaches *vs.* human studies and how measurements *in vitro* might aid interpretation of observations in humans. Nevertheless great care has to be taken to ensure that model systems *in vitro*, for predicting the bioavailability of bioactive components in the human diet, reflect the physiological environment as far as is practicable and that their validity is checked by reference to human studies.

4 References

1 Block, G., Patterson, B. and Subar, A. (1992) Fruits, vegetables and cancer: a review of the epidemiological evidence. *Nutr Cancer* **18**, 1–29.

2 Feskanich, D., Rimm, E.B., Giovanucci, E., Colditz, G.A., Stampfer, M.J., Litin, L.B. and Willett, W.C. (1993) Reproducibility and validity of food intake measurements from a semiquantitative food frequency questionnaire. *J Am Dietetic Assoc* **93**, 790–796.

3 Scott, K.J., Finglas, P.M., Seale, R., Hart, D.J. and Froidmont-Gortz, I. (1996) Inter-laboratory studies of HPLC procedures for the analysis of carotenoids in foods. *Fd Chem* **57**, 85–90.

4 Scott, K.J., Thurnham, D.I., Hart, D.I., Bingham, S.A. and Day, K. (1996). The correlation between the intake of lutein, lycopene and β-carotene from vegetables and fruits, and blood plasma concentrations in a group of women aged 50–65 years in the UK. *Br J Nutr* **75**, 409–418.

5 Knekt, P., Heliovaara, M., Rissanen, A. Aromaa, A. and Aaren, R-K. (1992) Serum antioxidant vitamins and risk of cataract. *BMJ* **305**, 1392–4.

6 Yeum, K.J., Taylor, A., Tang, G. and Russell, R.M. (1995) Measurement of carotenoids, retinoids and tocopherols in human lenses. Invest *Ophthalmol Vis Sci* **36**, 2756–61.

7 Shiau, A., Morbarhan, S., Stacewicz-Saponzakis, M., Benya, R. Liao, Y., Ford, C., Bowen, P., Friedman, H. and Frommel, T.O. (1994) Assessment of the intestinal retention of β-carotene in humans. *J Am Coll Nutr* **13**, 369–375.

8 Faulks, R.M., Hart, D.J., Wilson, P.D.G., Scott, K.J. and Southon, S. (1997) Absorption of all trans and 9-cis β-carotene in human ileostomy volunteers. *Clin Sci* **93**, 585–591.

9 Johnson, E.J. and Russell, R.M. (1992) Distribution of orally administered β-carotene among lipoproteins in healthy men. *Am J Clin Nutr* **56**, 128–135.

10 Blomstrand, R. and Werner B. (1967) Studies on the intestinal absorption of radioactive β-carotene and vitamin A in man. *Scand J Clin Lab Invest* **37**, 250–261.

11 Wilson, P.D.G. and Dainty, J.R. (1999) Modelling in nutrition: an introduction. *Proc Nutr Soc* 29 June–2 July 1998, 133–138.

12 Grundy, S.M. and Mok, H.Y.I. (1976) Chylomicron clearance in normal and hyperlipidemic man. *Metabolism* **25**, 1225–1239.

13 Langer, T., Strober, W. and Levy, R.I. (1972) The metabolism of low density lipoprotein in familial type II hyperlipoproteinemia). *J Clin Invest* **51**, 1528–36.

14 Van Vliet, T., Schreurs.W.H.P. and van den Berg, H. (1995) Intestinal β-carotene absorption and cleavage in men: response of β-carotene and retinyl esters in the triglyceride-rich lipoprotein fraction after a single oral dose of β-carotene. *Am J Clin Nutr* **62**, 110–116.

15 Dueker, S.R., Jones, A.D., Smith, G.M. and Clifford, A.J. (1994) Stable isotope methods for the study of β-carotene-d_8 metabolism in humans utilising tandem mass spectrometry and high performance liquid chromatography. *Analytical Chemistry* **66**, 4177–4158.

16 Parker, R.S., Swanson, J.E., Marmor, B., Goodman, K.J., Spielman, A.B., Brenna, J.T., Viereck, S.M. and Canfield, W.K. (1993) Study of β-carotene metabolism in

humans using ^{13}C-β-carotene and high precision isotope ratio mass spectrometry. *Ann N Y Acad Sci* **691**, 86–95.

17 Novotny, J.A., Dueker, S.R., Zech, L.A. and Clifford, A.J. (1995) Compartmental analysis of the dynamics of β-carotene metabolism in an adult volunteer. *J Lipid Res* **36**, 1825–1838.

18 Parker, R.S., Brenna, J.T., Swanson, J.E., Goodman, K.J. and Marmor, B. (1997) Assessing metabolism of β-[^{13}C] carotene using high precision isotope ratio mass spectrometry. *Methods Enzymol* **282**, 130–139.

19 Glahn, R.P., Wien, E.M., Van Campen, D.R., Miller, D.D. (1996) Caco-2 cell iron uptake from meat and casein digests parallels *in vivo* studies: use of a novel *in vitro* method for rapid estimation of iron bioavailability. *J Nutr* **126**, 332–339.

20 Rich, G., Fillery-Travis, A. and Parker, M. (1998) The influence of pH on carotene transfer from carrot juice to olive oil. *Lipids* **33**, 985–992.

2.2

The First Step in the Metabolism of Flavonoid Glycosides

Andrea J. Day, Michael R.A. Morgan, Michael J.C. Rhodes and Gary Williamson

INSTITUTE OF FOOD RESEARCH, NORWICH RESEARCH PARK, COLNEY, NORWICH NR4 7UA, UK

Dietary flavonoids exist in nature almost exclusively as β-glycosides.[1] Normal cooking methods do not modify the flavonoid glycosides, although the aglycone is released by fermentation or processes involving autolysis.[2] Thus in general, the aglycone is of little dietary importance. Until recently it has been assumed that the glycosides cannot be absorbed from the small intestine and, therefore, absorption of these compounds will not occur until they reach the large intestine.[3] In the colon, microflora will not only hydrolyse the β-glucoside link releasing the aglycone, but will also degrade the flavonoids further producing many ring fission products. Recently, Hollman *et al.* provided evidence for rapid absorption of onion flavonol glycosides from the small intestine;[4] however, as samples were hydrolysed before analysis the presence of the glycoside in the plasma was not confirmed. Although others[5,6] have reported quercetin glucosides in plasma, positive identification has not been provided. This is a limitation of the methodology as flavonol glucosides and glucuronides have very similar retention times on HPLC and identical UV absorption spectra. Further evidence, such as mass spectrometry data, is required before it can be concluded that the flavonol glycoside remains intact on absorption.

Flavonoid glucosides are too hydrophilic to diffuse through biological membranes. The sodium-dependent glucose transporter (SGLT1) from rat small intestine has been shown to interact with some flavonol glycosides, such as quercetin 3-glucoside, quercetin 4′-glucoside and quercetin 3,4′-glucoside.[7] However, this does not prove transport as phenolic glucosides can also interact by inhibiting SGLT1 without being transported across the membrane. For example, phloridzin is an inhibitor of SGLT1 but is not transported across the brush-border due to the bulky nature of the aglycone.[8] As the flavonoid glucosides are similar to phloridzin in size (Figure 1), it would seem likely that

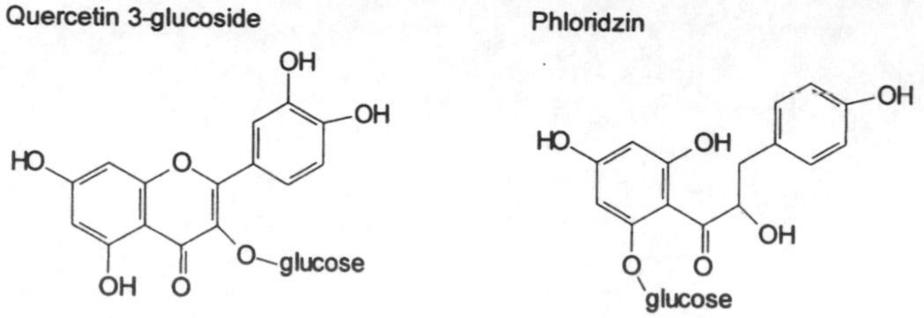

Quercetin 3-glucoside

Phloridzin

Figure 1 *Related structures of phloridzin and quercetin 3-glucoside*

they also would not be transported by SGLT1. However, subtle structural changes in sugar or phenolic can convert one type into another[9] and there is considerable inter-species variation in SGLT1 properties even though sequence identities can be relatively high.[8]

Deglycosylation is likely to be the first step of flavonoid glycoside metabolism, and this is required if further metabolism is to occur. In addition to microflora β-glucosidase activity, four mammalian β-glucosidases have been identified so far: lactase phloridzin hydrolase, glucocerebrosidase, broad-specificity cytosolic β-glucosidase and pyridoxine glucoside hydrolase.[10,11] Using cell-free extracts of human small intestine and liver various flavonoid glucosides were hydrolysed.[12] The enzyme has high affinities for quercetin 4′-glucoside and genistein 7-glucoside with K_m values of 32 and 14 μM, respectively. By distinguishing between the β-glucosidases using various inhibitors/activators, the broad-specificity cytosolic β-glucosidase was found to be responsible for most of the activity towards the flavonoid glycosides.[10]

The hepatic β-glucosidase was inactive on quercetin-3-glucoside, although the small intestine extract exhibited some activity toward quercetin 3-glucoside. This activity may be due to a different β-glucosidase in the small intestine, such as lactase phloridzin hydrolase (LPH).[13] LPH is an enzyme complex, present in the brush-border, capable of hydrolysing phloridzin to the aglycone phloretin. Mammalian LPH hydrolyses various flavonoid glycosides,[14] and hence may be responsible for the hydrolysis of quercetin 3-glucoside observed in the small intestine cell-free extract. As LPH acts on the apical side of the brush-border the released aglycone may passively diffuse into the enterocyte. Thus, active transport of the flavonol glycoside will not be the only mechanism of absorption in the small intestine.

Figure 2 shows a the potential mechanism for absorption of flavonoid glycosides. Hydrolysis of the sugar may occur in the small intestine lumen by action of LPH, or after transport into the enterocyte by the broad-specificity cytosolic β-glucosidase. Flavonoid glycosides reaching the colon will be hydrolysed by the gut microflora, but will also undergo further metabolism and degradation. Once absorbed dietary flavonoids may be methylated, and conjugated to sulfate and/or glucuronic acid prior to excretion in the urine or bile. Hence, these metabolites will be responsible for any biological activity demonstrated *in vivo* and should be the subject of further investigations.

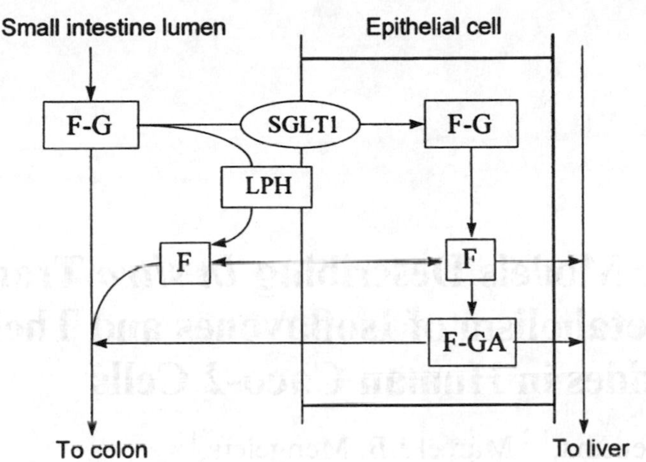

Figure 2 *Potential mechanism for absorption of the flavonoid glycosides* (Adapted from Day and Williamson, 1999.[15] F-G, *flavonoid glycoside*; SGLT1, *sodium dependent glucose transporter*; LPH, *lactase phloridzin hydrolase*)

References

1 J.B. Harborne, T.J. Mabry and H. Mabry, 'The Flavonoids', Chapman and Hall, London, 1975.
2 K.R. Price, J.R. Bacon and M.J.C. Rhodes, *J. Agric. Food Chem.*, 1997, **45**, 938.
3 L.A. Griffiths and A. Barrow, *Biochem. J.*, 1972, **130**, 1161.
4 P.C.H. Hollman, J.H. Van Trijp, M.N.C.P. Buysman, M.S. Gaag, M.J.B. Menglers, J.H.M. de Vries and M.B. Katan, *FEBS Lett.*, 1997, **418**, 152.
5 G. Paganga and C. Rice-Evans, *FEBS Lett.*, 1997, **401**, 78.
6 A.A. Aziz, C.A. Edwards, M.E.J. Lean and A. Crozier, *Free Rad. Res.*, 1998, **29**, 257.
7 J.M. Gee, M.S. DuPont, M.J.C. Rhodes and I.T. Johnson, *Free Rad. Biol. Med.*, 1998, **25**, 19.
8 B.A. Hirayama, M.P. Lostao, M. Panayotova-Heiermann, D.D.F. Loo, E. Turk and E.M. Wright, *Am. J. Physiol.*, 1996, **270**, G919.
9 M.P. Lostao, B.A. Hirayama, D.D.F. Loo and E.M. Wright, *J. Mem. Biol.*, 1994, **142**, 161.
10 L.B. Daniels, P.J. Coyle, Y. Chiao and R.H. Glew, *J. Biol. Chem.*, 1981, **256**, 13004.
11 L.G. McMahon, H. Nakano, M-D. Levy and J.F. Gregory, *J. Biol. Chem.*, 1997, **272**, 32025.
12 A.J. Day, M.S. DuPont, S. Ridley, M. Rhodes, M.J.C. Rhodes, M.R.A. Morgan and G. Williamson, *FEBS Lett.*, 1998, **436**, 71.
13 H.J. Leese and G. Semenza, *J. Biol. Chem.*, 1973, **248**, 8170.
14 A.J. Day, F.J. Cañada, J.C. Díaz, P.A. Kroon, R. Mclauchlan, C.B. Faulds, G.W. Plumb, M.R.A. Morgan and G. Williamson, *FEBS Lett.*, 2000, **468**, 166.
15 A.J. Day and G. Williamson, 'Plant polyphenols 2: Chemistry, Biology, Pharmacology, Ecology', ed. G.G. Gross, R.W. Hemingway and T. Yoshida, Plenum Press, New York, 1999, 415–434.

2.3

Kinetic Models Describing *In vitro* Transport and Metabolism of Isoflavones and Their Glycosides in Human Caco-2 Cells

Aukje Steensma,[1,2] Marcel J.B. Mengelers,[1]
Hub P.J.M. Noteborn[1] and Harry A. Kuiper[1]

[1] STATE INSTITUTE FOR QUALITY CONTROL OF AGRICULTURAL PRODUCTS (RIKILT), PO BOX 230, 6700 AE WAGENINGEN, THE NETHERLANDS
[2] WAGENINGEN AGRICULTURE UNIVERSITY, DEPARTMENT OF FOOD TECHNOLOGY AND NUTRITIONAL SCIENCES, SUB DEPARTMENT OF TOXICOLOGY, TUINLAAN 5, 6703 HE WAGENINGEN, THE NETHERLANDS

1 Introduction

Kinetic models can be used to describe the *in vitro* transport and metabolism of isoflavones by intestinal epithelial cells (Caco-2) grown on semi-permeable filters. These models enable a quantitative comparison of *in vitro* and *in vivo* bioavailability parameters. Frequently, a simple model with limited applicability is used as shown in Figure 1.[1] An *in vitro* permeability parameter (P_{app}) of this model is compared with certain *in vivo* absorption parameters. The P_{app} is determined from relatively early time periods and neither information on transport during later periods nor information on metabolism is used. Therefore, extended models were developed, as shown in Figures 2 and 3, to progress more information relevant for the *in vitro* transport and metabolism of isoflavones and their glycosides in human intestinal epithelial (Caco-2) cells.

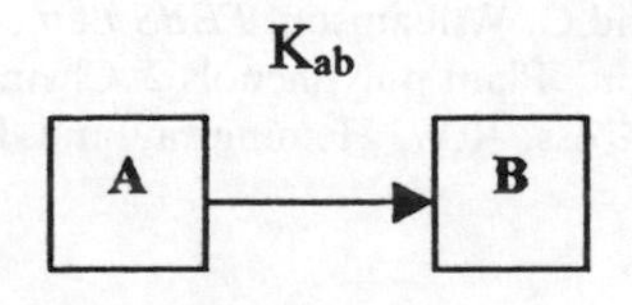

Figure 1 *A = apical side and B = basolateral side*

2 Materials and Method

2.1 Materials

Caco-2 cells originating from human colorectal carcinoma were obtained from the American Type Culture Collection (ATCC 37-HTB, Rockville, USA). Genistein, daidzein, genistin and tissue culture media were obtained from Sigma Chemical Co. (St. Louis, USA). Daidzin was purchased from Plantech (Reading, UK).

2.2 Cell culture

Caco-2 cells were grown in Dulbecco's modified Eagle's medium (DMEM) in tissue culture flasks (75 cm^2) at 37°C in an atmosphere of 5% CO_2 and 95% relative humidity. The DMEM contained glucose (4.5 $g\,l^{-1}$) supplemented with 10% heat inactivated foetal calf serum (FCS), 50 IU ml^{-1} penicillin, 50 $\mu g\,ml^{-1}$ streptomycin and 1% non-essential amino acid (NEAA). When the cells were 90% confluent, the cells were detached from the flasks by treatment with trypsin (0.25% in Hank's balanced salt solution without calcium, magnesium and phenol red) and 0.05% w/v EDTA.

2.3 Metabolism studies

Cells were cultivated for three weeks in 24-well culture plates (Costar, Cambridge, USA) at a density of 1.10^5 viable cells/ml (250 μl per well) using DMEM with 10% FCS, 1% NEAA, 50 IU ml^{-1} penicillin and 50 $\mu g\,ml^{-1}$ streptomycin. The medium was changed twice a week. Cells were incubated at 37°C in an atmosphere of 5% CO_2 and high humidity. Stock solutions of isoflavones were prepared in dimethylsulfoxide (DMSO) and were diluted in Eagle's Minimum Essential Medium (EMEM without phenol red and supplemented with L-glutamine (0.29 $g\,l^{-1}$)) or Hank balanced salt solutions (HBSS) to a final concentration of 0.2% DMSO. Cells were washed once with EMEM or HBSS and incubated for 30 hours with test compounds used seperately at a concentration of 5 μM in EMEM or HBSS. At the end of the incubation period medium was removed and analysed with HPLC. Two experiments were carried out in triplicate. From metabolism studies the constants K_{ac}, K_{ca} and K_m were calculated. The value of the metabolic constant K_m was used in the transport studies. The kinetic model that was used is shown in Figure 2.

2.4 Transport experiments

For transport experiments Caco-2 cells were cultivated on Transwell-clear, tissue culture treated polyester membrane filter inserts (pore size 0.4 μm, diameter 24 mm, apical volume 2 ml, basolateral volume 3 ml, Costar, Badhoevedorp, the Netherlands) at a density of 2.10^5. Between passage number 35–

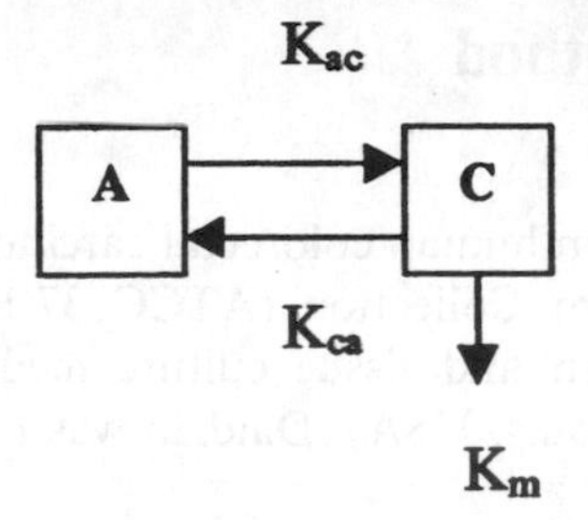

Figure 2 *A = medium, C = Caco-2 cells, K_{ac} = constant for transport from medium, K_{ca} = constant for transport from cells to medium and K_m = metabolic constant*

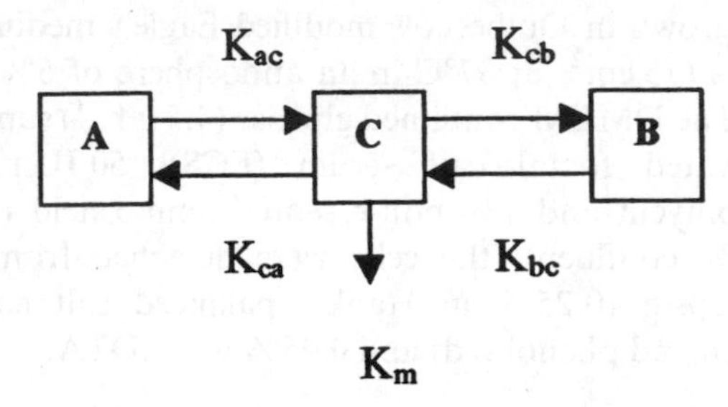

Figure 3 *A = apical side, C = Caco-2 cells, B = basolateral side, K_{ac} = constant for transport from apical side to cells, K_{cb} = constant for transport from cells to basolateral side, K_{ca} = constant for transport from cells to apical side, K_{bc} = constant for transport from basolateral side to cells and K_m = metabolic constant*

60 the cells were allowed to grow and differentiate to confluent monolayers for 20–22 days. The medium was changed twice a week.

All transport experiments were performed at 37°C in an atmosphere of 5% CO_2 and 95% relative humidity and in HBSS or EMEM. The transport experiments were initiated by washing monolayers with buffer before test solutions were added to the apical side of the cells. Stock solutions of genistein, daidzein and their glycosides genistin and daidzin were prepared in dimethylsulfoxide (DMSO) and were diluted in medium at a maximum of 0.2% (v/v) DMSO. Test solutions of genistein, daidzein and their glycosides were used separately at a concentration of 5 μM. Two experiments were carried out in triplo. Samples of 100 μl were taken from the basolateral side for up to 24 hours and analysed with HPLC. The kinetic model that was used is shown in Figure 3.

2.5 HPLC

Aliquots of cell culture and perfusion media were analysed by reversed-phase HPLC (Waters, Etten-Leur, the Netherlands) equipped with a Supelcosil LC-ABZ (250 × 4.6 mm, 5 μm) column (Supelco, Sigma-Aldrich Chemie BV, Zwijndrecht, the Netherlands) and UV photodiode array detector (Waters,

Etten-Leur, the Netherlands). The column was eluted at a flow rate of 1 ml min^{-1} employing a gradient of two eluents: eluent A: 10 mM ammonium-acetate pH 6.5; eluent B: acetonitrile at 30 °C for 60 minutes. The eluent was monitored at 260 nm. Routinely 50 μl of medium was injected without further pre-treatment and metabolites were identified by comparison with retention times and UV spectra of known standards.[2]

2.6 Kinetic analysis

Metabolism studies. Data were fitted according to an equation derived from a 2-compartment model. From these studies the constants K_{ac}, K_{ca} and K_m were calculated. K_{ac} and K_{ca} were used as initial values for the parameter estimation in the transport studies. K_m was used in the transport studies without alteration.

Transport experiment. Data were fitted using continuous dynamic simulation of a 3-compartment model.

3 Results

Genistein and daidzein were metabolised by the Caco-2 cells. From metabolism studies the constants K_{ac}, K_{ca} and K_m were calculated and are shown in Figure 4. K_{ac} and K_{ca} values of genistein and daidzein were higher in HBSS than in EMEM. K_m values of genistein and daidzein were comparable in both

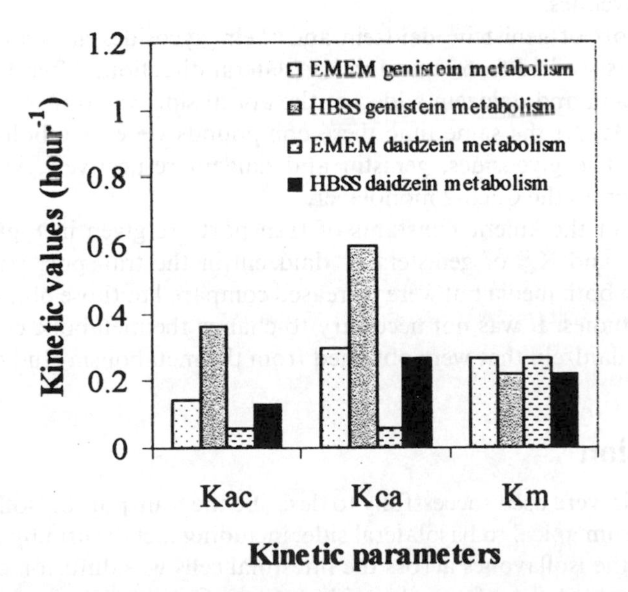

Figure 4 *Values of the kinetic constants obtained from metabolism studies of genistein and daidzein in different media by Caco-2 cells. Initial concentrations of the compounds were 5 μM*

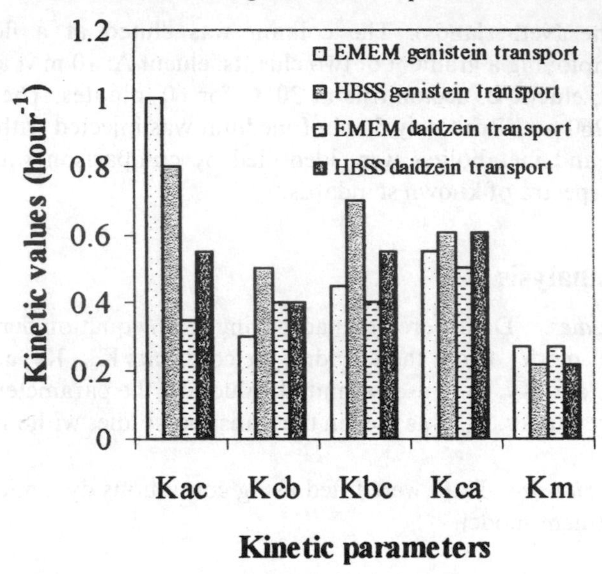

Figure 5 *Values of the kinetic constants obtained from transport studies of genistein and daidzein in different media across Caco-2 cells. Initial concentrations of the compounds were 5 μM*

media. The value of the metabolic constant K_m was used in the transport studies. The glycosides, genistin and daidzin were metabolised to their respective aglycones.

The transport of genistein, daidzein and their glycosides across the Caco-2 monolayer was studied from apical to basolateral direction. After 4 hours 15–20% of genistein and daidzein added at the apical side was transported to the basolateral side. At the same time these compounds were metabolised by the Caco-2 cells. The glycosides, genistin and daidzin respectively, were hardly transported across the Caco-2 monolayer.

The values of the kinetic constants of transport are given in Figure 5. The constants K_{ac} and K_{ca} of genistein or daidzein in the transport studies were comparable in both media but were increased compared to those obtained in the metabolism studies. It was not necessary to change the metabolic constants of genistein and daidzein that were obtained from the metabolism studies.

4 Discussion

Kinetic models were used successfully to describe the transport of isoflavones by Caco-2 cells from apical to basolateral side, including metabolism by these cells. Transport of the isoflavones across the intestinal cells was diffusion controlled. However, the transport of the glycosides across Caco-2 cells was too low to enable kinetic modelling. Moreover, the glycosides were metabolised to their respective aglycones. Additional metabolism studies were carried out in order to incorporate metabolic rates in the kinetic models. The metabolic rates obtained from the metabolism studies could be incorporated in the model used for describing the transport experiments without alterations.

5 References

1 P. Artursson and J. Karlsson, *Biochem. Biophys. Res. Commun.*, 1991, **175**, 880–885.
2 A. Steensma, H.P.J.M. Noteborn, R.C.M. Van der Jagt, T.H.G. Polman, M.J.B. Mengelers and H.A. Kuiper, *Env. Toxicol. Pharmacol.*, 1999, **7**, 209–212.

2.4

Transport and Metabolism of Genistein, Daidzein and Their Glycosides in Caco-2 Cells and in Perfused Gut Segments

Aukje Steensma,[1,2] Hub P.J.M. Noteborn,[1]
Marcel J.B. Mengelers[1] and Harry A. Kuiper[1]

[1]STATE INSTITUTE FOR QUALITY CONTROL OF AGRICULTURE PRODUCTS (RIKILT-DLO), PO BOX 230, 6700 AE WAGENINGEN, THE NETHERLANDS
[2]WAGENINGEN AGRICULTURE UNIVERSITY, DEPARTMENT OF FOOD TECHNOLOGY AND NUTRITIONAL SCIENCES, SUB DEPARTMENT OF TOXICOLOGY, TUINLAAN 5, 6703 HE WAGENINGEN, THE NETHERLANDS

1 Introduction

Genistein and daidzein have received much attention because they may help to prevent hormone-related cancers and cardiovascular disease.[1] Genistein and daidzein are present in minor amounts in soybean and soy derived foods, whereas their sugar conjugated forms occur in relatively high amounts (1–3 mg/g product).[2,3] Data on the bioavailability of these isoflavones are scarce. The aim of the present study was to obtain information on the transport, metabolism and mechanism of action of genistein, daidzein and their glycosides. Therefore human intestinal epithelial cells derived from a human colon adenoma carcinoma (Caco-2) grown on semi-permeable filters and the perfusion of rat segments of the intestinal tract were used as models for intestinal absorption studies.[4,5]

2 Materials and Method

2.1 Materials

Caco-2 cells originating from a human colorectal carcinoma were obtained from the American Type Culture Collection (ATCC 37-HTB, Rockville, USA).

Genistein, daidzein, genistin and tissue culture media were obtained from Sigma Chemical Co. (St. Louis, USA). Daidzin was purchased from Plantech (Reading, UK). [4-^{14}C]Genistein (specific activity 0.2 mCi ml^{-1}) was obtained from Moravek Biochemical (Brea, USA).

2.2 Cell culture

Caco-2 cells were grown in Dulbecco's modified Eagle's medium (DMEM) with high glucose (4.5 g l^{-1}) supplemented with 10% heat inactivated foetal calf serum (FCS), 2% penicillin /streptomycin solution and 1% non-essential amino acid at 37°C in an atmosphere of 5% CO_2 and 95% relative humidity. The cells were expanded in tissue culture flasks (75 cm^2) and until 90% confluence was reached where after the cells were detached from the flasks by treatment with trypsin (0.25% in Hank's balanced salt solution without calcium, magnesium and phenol red) and 0.05% (w/v) EDTA.

2.3 Animals

Male rats (Wistar, 200–250 g) were obtained from the Laboratory Animal Centre (Wageningen Agricultural University, the Netherlands).

2.4 Transport in cell culture

Caco-2 cells (passage number between 35–60) were cultivated on Transwell-clear, tissue culture treated polyester membrane filter inserts (pore size 0.4 μm, diameter 24 mm, Costar, Badhoevedorp, the Netherlands) with an apical volume of 2 ml and a basolateral volume of 3 ml at a density of 2.10^5 cells/filter. Cells were allowed to grow and differentiate to confluent monolayers for 20–22 days by changing the medium twice a week.

All transport studies were performed in Hank's balanced salt solution at 37°C in an atmosphere of 5% CO_2 and 95% relative humidity. The transport experiments were initiated by washing monolayers with buffer before test solutions were added to the apical side of the cells. Stock solutions of genistein, daidzein and their glycosides genistin and daidzin prepared in dimethylsulfoxide (DMSO) were diluted in medium at a maximum of 0.2% (v/v) DMSO. Exposures of genistein, daidzein and their glycosides were performed separately in three different experiments at a substance concentration of 50 μM. Aliquot samples of 0.1 ml were taken from the basolateral side for up to 24 hours. Transport of the test compounds across the Caco-2 monolayer was expressed as a fraction of the initial amount applied to the apical side.

2.5 Transport in gut segments

The preparation of intestinal segments was carried out according to the method of Richter and Strugala.[5] Rats were kept under ether anaesthesia while the segments of jejunum, ileum or colon were fixed to glass cannulae. Subsequently, the blood supply was interrupted by carefully dissecting the segments and immediately mounting the segments into the lower chamber of the perfusator. The perfusate contained the marker [^{14}C]polyethyleneglycol, 50 μM [4-^{14}C]genistein or 50 μM genistin dissolved in Tyrode buffer. Aliquots of resorbate and perfusate were taken at various time points during 2 hours perfusion and analysed by HPLC. After 2 hours of perfusion glucose levels were measured both in perfusate aand resorbate fluid using the GOD-period method (Boehringenger Mannheim GmbH, Germany).

2.6 Isoflavone detection

Aliquotes of cell culture and perfusion media were analysed by reversed-phase HPLC (Waters, Etten-Leur, the Netherlands) equipped with a Supelcosil LC-ABZ (250 × 4.6 mm, 5 μm) column (Supelco, Sigma-Aldrich Chemie BV, Zwijndrecht, the Netherlands) and a UV photodiode array detector (Waters, Etten-Leur, the Netherlands). The column was eluted at a flow rate of 1 ml min^{-1} employing a gradient of two eluens (eluent A: 10 mM ammonium acetate pH 6.5; eluent B: acetonitril) at 30°C for 60 minutes. The eluent was monitored at 260 nm. Routinely 50 μl of media was injected without further pre-treatment and metabolites were identified by comparison with retention times of known standards.[6]

3 Results

The transport of genistein, daidzein and their glycosides across the Caco-2 monolayer was studied in apical to basolateral direction. There was a significant difference in the transport rate and metabolism of genistein and daidzein compared to their glycosides in Caco-2 cells. After 6 hours 30–40% of genistein and daidzein added at the apical side was transported to the basolateral side (Figure 1 and see also reference 6). Thereafter the rate of transport remained constant for the next 24 hours. The glycosides, genistin and daidzin respectively, were hardly transported across the Caco-2 monolayer. Moreover, genistein and daidzein were only slightly metabolised in the Caco-2 cells into sulphates and glucuronides, whereas genistin and daidzin were mainly metabolised to genistein and daidzein for approximately 60% of the added parent compound.

In all perfused gut segments the transport of genistein was higher when compared with its glycoside (Figure 2). Furthermore, it appeared that the transport of genistein was the highest in ileum segments, whereas there was no difference in transport of genistin in the various other segments tested. In addition, the glycoside was metabolised in all gut segments having no microfloral contents, with no difference in rates between the segments prepared (Figure 3). Jejunal segments showed a high metabolic activity in case of genistein yielding metabolites that are structurally unknown as yet.

4 Discussion

In Caco-2 cells and the perfusion model the transport of genistein and daidzein appeared to be much higher than that of their related glycosides. In both model

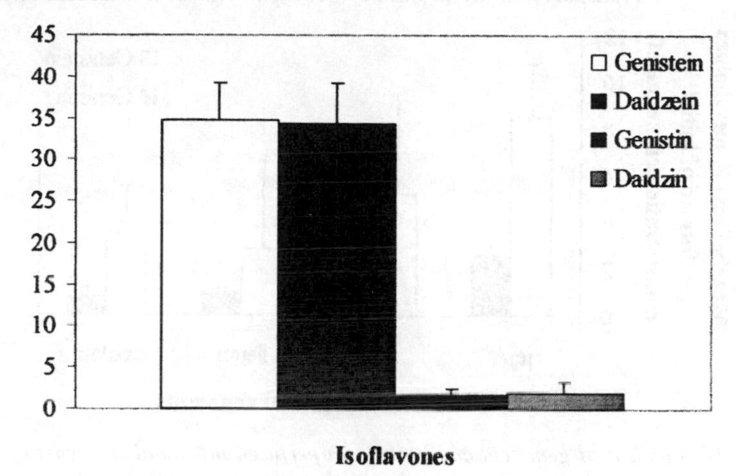

Figure 1 *Transport of genistein, daidzein and their glycosides genistin and daidzin across Caco-2 cells at 6 hours after initiating incubation. The initial substance concentration of genistein, daidzein, genistin and daidzin was 50 μM. Data are expressed as mean ±SD of three triplicate measurements (n = 9)*

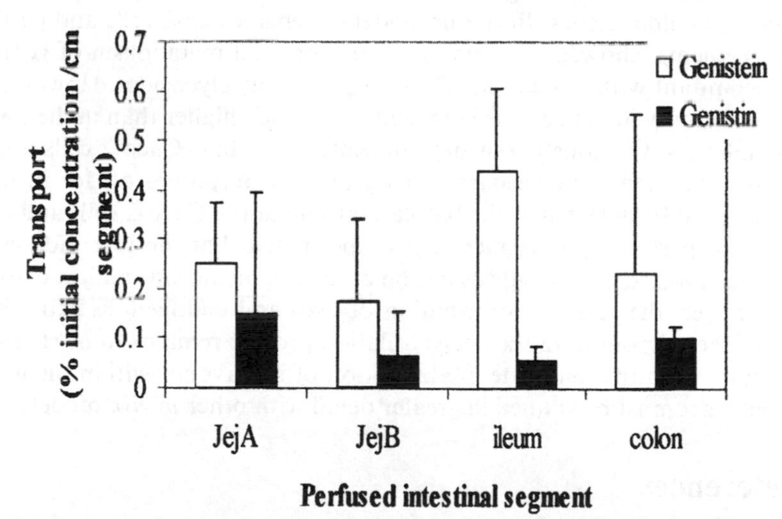

Figure 2 *Transport of genistein and genistin across perfused intestinal segments. The initial substance concentration was 50 μM. Data are expressed as mean ±SD for triplicate measurements (n = 3)*

systems the glycosides were first metabolised to their respective aglycones. It could be demonstrated that genistein metabolised mainly to sulphates and glucuronides in the Caco-2 cells and to glucuronides in the perfused rat gut segments.

Moreover, in the perfusion model it was possible to study in more detail the behaviour of isoflavones in the different segments of the rat. Focus was on the

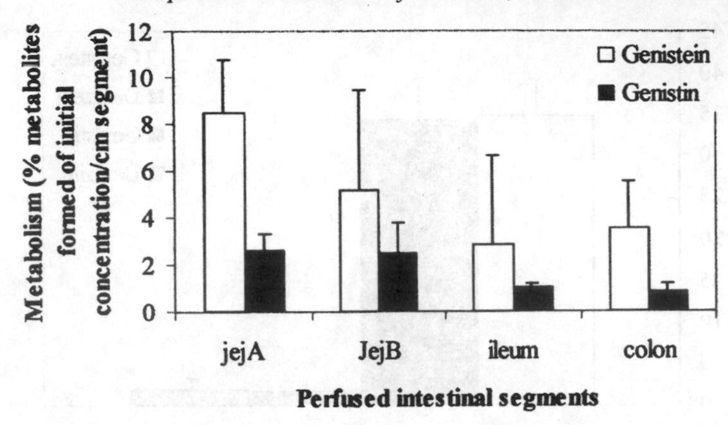

Figure 3 *Metabolism of genistein and genistin in perfused intestinal segments of the rat. The initial substance concentration of the compounds was 50 µM (i.e. perfusate compartment). Data are expressed as mean ±SD for triplicate measurements (n = 3)*

transport and metabolism of genistein and their glycosides in jejunum, as well as in ileum and colon tissues. Both our model systems, Caco-2 cells and perfused rat gut segments, showed a substantial transport and metabolism of genistein and concomitant with a much smaller transport of its glycosides. However, the metabolic activity in the perfused segments was much higher than in the Caco-2 cell monolayers. Obviously, our data indicated also that Caco-2 cells and rat segments contained a glucosidase activity as only aglycones of, for instance, genistein could be detected at the basolateral side of the Caco-2 cells and in the resorbate of perfused gut segments. It is speculated that the glycosidases can only be transported after deglycosidation at the cell membrane and/or cytosolic side. However, the relative contribution of exo- and endogenous activities as well as of the microflora in the deglycosilation process remains to be studied in more detail. Also the main site of absorption of isoflavones within the gastrointestinal tract must be studied in greater detail with other *in vivo* models.

5 References

1 S. Barnes, J. Sfakianos, L. Coward and M. Kirk, in: Dietary Phytochemicals in Cancer Prevention Treatment, Butrum R. ed., Plenum Press New York, 1996, pp. 87–100.

2 H. Adlercreutz, T. Fostis, S. Watanbe, J. Lampe, K. Wähälä, T. Mäkelä and T. Hase, *Cancer Detect. Prevent.*, 1994, **18**, 259–271.

3 M.J. Messina, V. Persky, K.D.R. Setchell and S. Barnes, *Nutr. Cancer*, 1994, **21**, 113–131.

4 P. Artursson and J. Karlsson, *Biochem. Biophys. Res. Commun.*, 1991, **175**, 880–885.

5 E. Richter and G.J. Strugala, *J. Pharmacol. Methods*, 1985, **14**, 297–304.

6 A. Steensma, H.P.J.M. Noteborn, R.C.M. Van der Jagt, T.H.G. Polman, M.J.B. Mengelers and H.A. Kuiper, *Environ. Toxicol. Pharmacol.*, 1999, in press.

2.5

Stability of Flavonol Glycosides During Digestion and Evidence for Interaction with the Sodium Dependent Glucose/Galactose Transport Pathway

J.M. Gee, M.S. DuPont and I.T. Johnson

INSTITUTE OF FOOD RESEARCH NORWICH LABORATORY, NORWICH RESEARCH PARK, COLNEY, NORWICH NR4 7UA, UK

1 Introduction

Flavonol mono- and di-glycosides, natural antioxidants in the human diet, also have antiproliferative activity with potentially protective effects against cancer. There have been reports of flavonol glycosides in human plasma following consumption of foods such as onions and apples,[1] but their stability in the gastrointestinal lumen and the site and mechanisms involved in their transfer to the circulation are not yet known. In the present study we explored the survival of these compounds *in vitro*, under conditions simulating those in the stomach and small intestine, in order to determine if any are available for transport across the brush border. We have also used everted segments of rat jejunum to obtain further evidence of interactions between quercetin glucosides and the small intestinal brush border transporter SGLT1.[2]

2 Methods

2.1 Simulated Digestion

Compounds, at nutritionally relevant concentrations, were solubilised in Krebs solution, adjusted to pH 2.1, and incubated at 37°C with pepsin (5000 U/9 ml) and occasional mixing to simulate gastric conditions. After 2 hours the pH was raised to 7.6 and trypsin (2000 U), lipase (1200 U), protease (600 U) and α-amylase (1200 U) added to simulate conditions in the small intestine. Incubation was continued for a further 5 hours. Aliquots (0.5 ml) were removed at intervals throughout the procedure and adjusted to pH 7. These were filtered and analysed by reverse phase HPLC using a C18 ODS3 column with a water:tetrahydrofuran:trifluoracetic acid (98:2:0.1) and acetonitrile gradient. Detection and quantification were performed using diode array detection and Packard Chemstation software.

2.2 Stimulated Efflux

Everted sacs (5 cm) of rat proximal jejunum were canulated on to 1 ml disposable syringes containing physiological saline (0.5 ml) and filled by depression of the plunger. The tissue was pre-loaded with radiolabelled galactose by suspending the sacs in oxygenated (95% CO_2:5% O_2) buffered Krebs Ringer (pH 7.2–7.4) containing high specific activity U-^{14}C galactose (10.5 GBq mmol^{-1}; 43.5 KBq) for 10 minutes. Sacs were subsequently rinsed gently and then incubated with Krebs buffer containing the flavonoid (0.1–1.0 mM) for 20 minutes. A control sac with no flavonoid addition was included for each animal and the procedure repeated five times. During efflux the mucosal medium was sampled (100 μl) at 2.5 minute intervals. Aliquots were added to scintillation cocktail for radioactive counting, to assess the rate and degree of displacement of radiolabelled galactose from the mucosal epithelium.

2.3 Competitive Inhibition of Transport

Everted sacs, prepared as described above, were suspended in gassed (95% CO_2:5% O_2) Krebs Ringer containing 1 mM U-^{14}C-labelled galactose either alone, or in combination with (1) 1 mM mannitol, (2) 1 mM mannitol + 10 μM phloridzin or (3) 1 mM quercetin-3-O-glucoside at 37°C for 15 minutes. After incubation, the serosal solutions were collected, mucosal solutions sampled and the tissue dried and acid digested prior to liquid scintillation counting, to assess the uptake and transfer of the radiolabel.

3 Results

As shown in Figure 1, quercetin-3-O-glucoside, quercetin-4′-O-glucoside and quercetin-3,4′-O-di-glucoside remained essentially intact during simulated digestion. The following compounds were also assessed under these conditions and remained unaffected by the digestive process: quercetin-3-O-galactoside, quercetin-3-O-rhamnoside, quercetin-3-O-arabinoside, quercetin-3-O-rhamnosyl-glucoside (rutin), quercetin-3-O-galactosyl-rhamnosyl-glucoside, quercetin-3-O-glucosyl-rhamnosyl-glucoside and quercetin-3-O-sophoroside. Thus it would appear that the major dietary flavonol glycosides are stable under conditions simulating those in the lumen of the upper gastrointestinal tract.

We have previously reported that, at equi-osmolar concentrations, quercetin 3-O-glucoside stimulates galactose efflux more than the 4′-monoglucoside or the 3,4′-diglucoside *in vitro*.[1] Results reported here demonstrate that the kinetics of stimulated efflux for the 3-O monoglucoside also appear saturable with similar J_{max} but much lower K_m (Figure 2).

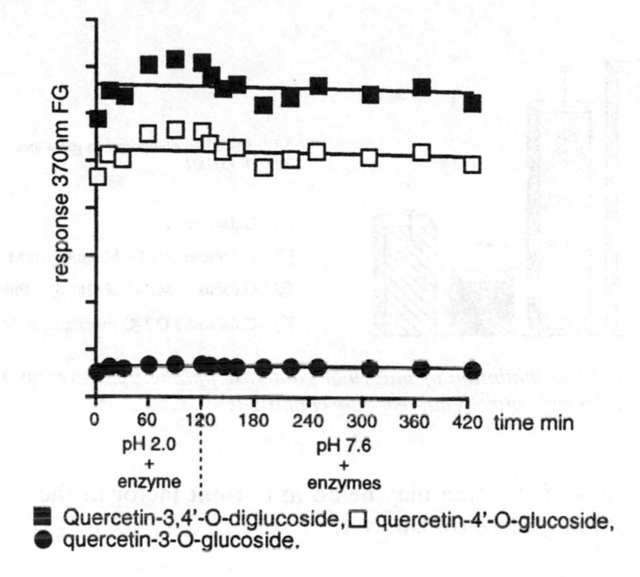

Figure 1 *Time-course of simulated digestion of flavonol glucosides found in brown onion*

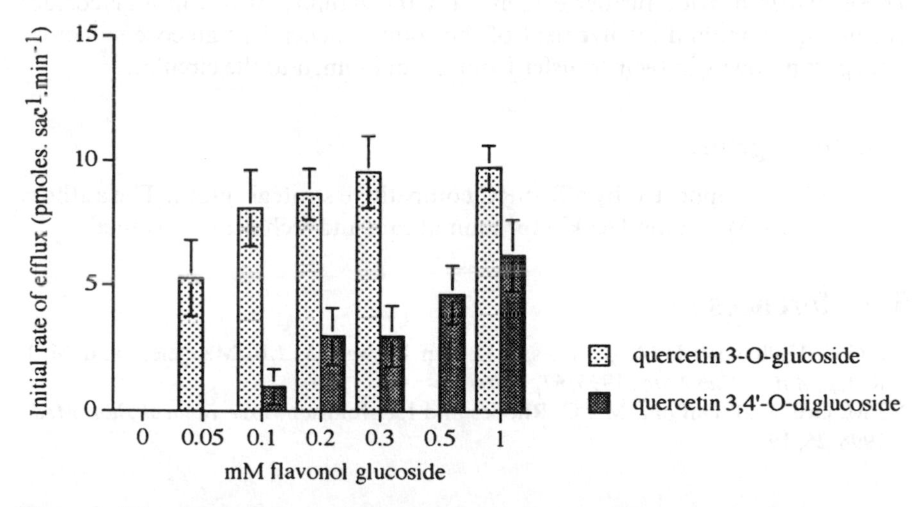

Figure 2 *Stimulated efflux of galactose from intestinal everted sacs* in vitro

In a separate experiment the simultaneous presence of quercetin 3-O-glucoside (1 mM) during intestinal galactose transport demonstrated that this compound competitively and significantly inhibits the active transport of galactose in the small intestine *in vitro* by approximately 55% ($P < 0.01$) (Figure 3).

Although such competitive inhibition studies have not yet been completed using other flavonol glycosides, the results from stimulated efflux studies suggest

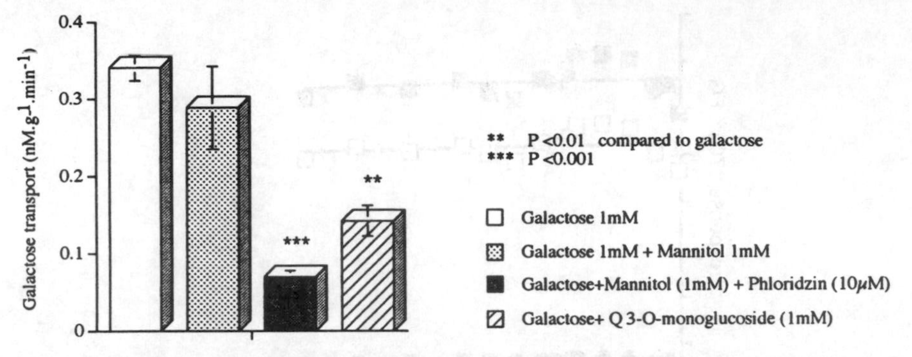

Figure 3 *Competitive inhibition of intestinal galactose uptake by quercetin 3-O-glucoside and phloridzin under equi-osmolar conditions*

that glycosidic configuration may be an important factor in these interactions with the intestinal glucose transporter.

4 Conclusions

These studies provide further evidence for the stability of flavonol glycosides during digestion, and involvement of the sodium dependent glucose/galactose transport pathway in their transfer from the gut lumen to the circulation.

Acknowledgements

This work was supported by a BBSRC competitive strategic grant. The authors wish to thank Mr Simon Deakin for animal care and technical assistance.

5 References

1 P.C.H. Hollman, J.H.M. de Vries, S.D. van Leewen, M.J.B. Mengelers and M.B. Katan, *Am. J. Clin Nutr.*, 1995, **62**, 1276.

2 J.M. Gee, M.S. DuPont, M.J.C. Rhodes and I.T. Johnson, *Free Radical Biol. Med.*, 1998, **25**, 19.

2.6

Metabolism of Chlorogenic Acid, Quercetin-3-rutinoside and Black Tea Polyphenols in Healthy Volunteers

Margreet R. Olthof,[1] Els Siebelink,[1] Peter C.H. Hollman[2] and Martijn B. Katan[1]

[1] WAGENINGEN UNIVERSITY AND RESEARCH CENTRE, DIVISION OF HUMAN NUTRITION AND EPIDEMIOLOGY, PO BOX 8129, 6700 EV WAGENINGEN, THE NETHERLANDS
[2] DLO STATE INSTITUTE FOR QUALITY CONTROL OF AGRICULTURAL PRODUCTS, PO BOX 230, 6700 AE WAGENINGEN, THE NETHERLANDS

1 Introduction

Polyphenols in foods are antioxidants *in vitro*, and might therefore contribute to the prevention of cardiovascular disease and possibly some forms of cancer.[1,2] Flavonoids, a subclass of polyphenols, are extensively metabolised.[3] These metabolites might also have antioxidant properties, and therefore play a role in the potential health effects of polyphenols in humans. However, data on metabolites of polyphenols in humans are scarce. Therefore the aim of this study was to identify in urine of humans the metabolites of the following polyphenols: chlorogenic acid, quercetin-3-rutinoside and black tea polyphenols (Figure 1). Chlorogenic acid consists of caffeic acid esterified with quinic acid and occurs mainly in coffee: coffee drinkers ingest 0.5–1 g of it per day.[4] Quercetin-3-rutinoside is a major flavonol in tea and tea consumption is the main source of quercetin intake.[5] In addition to quercetin-3-rutinoside tea also contains a lot of other polyphenols, such as catechins.[6]

2 Subjects and Study Design

2.1 Subjects

Twenty healthy men and women with a mean age of 24 ± 8 y ($\pm$SD) and a mean body mass index of 22.2 ± 2.5 kg m^{-2} participated in this study.

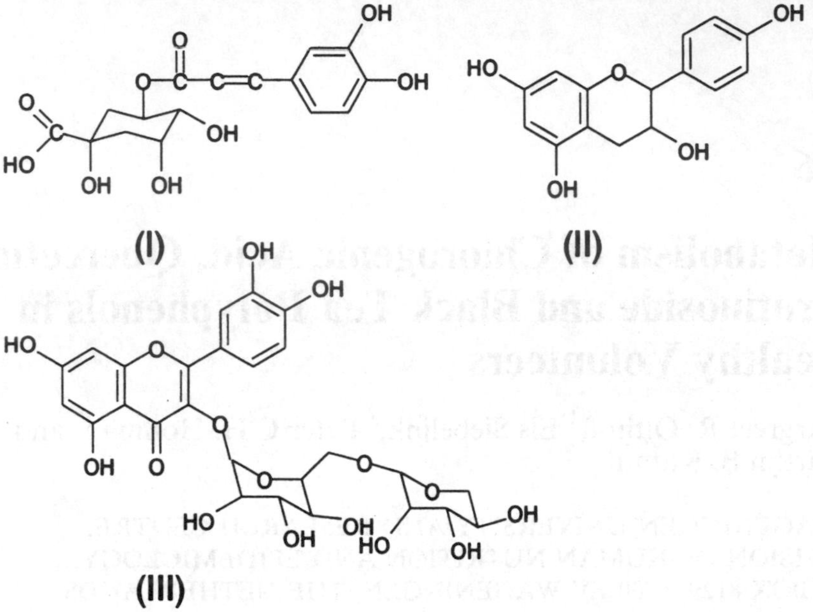

Figure 1 *Structure of chlorogenic acid (I), catechins (II), and quercetin-3-rutinoside (III)*

2.2 Study Design

Throughout the 4 week study subjects consumed a controlled diet which was low in polyphenols. In addition to the diet subjects ingested daily one of the following supplements: either chlorogenic acid, quercetin-3-rutinoside, black tea extract, or citric acid (placebo). Subjects ingested each supplement for seven consecutive days in random order. On day 7 of each supplement period volunteers collected urine during 24 hours. We measured approximately 60 phenolic acids in urine.

3 Results and Conclusion

We found that chlorogenic acid, quercetin-3-rutinoside and black tea polyphenols are extensively metabolised in humans into a limited number of phenolic acid metabolites. Therefore we conclude that circulating metabolites might play an important role in the potential health effects of polyphenols from foods. Information on the metabolism of polyphenols from foods is essential for a proper evaluation of their health effects.

4 References

1 P.C. Hollman, E.J. Feskens and M.B. Katan, *Proc. Soc. Exp. Biol. Med.*, 1999, **220**, 198.

2 B. Halliwell, *Lancet*, 1994, **344**, 721.
3 P.C.H. Hollman and M.B. Katan. 'Flavonoids in Health and Disease', Marcel Dekker Inc, New York, 1998, p. 483.
4 M.N. Clifford, *J. Sci. Food Agric.*, 1999, **79**, 362.
5 M.G. Hertog, P.C. Hollman, M.B. Katan and D. Kromhout, *Nutr. Cancer*, 1993, **20**, 21.
6 D.A. Balentine, S.A. Wiseman and L.C. Bouwens, *Crit. Rev. Food Sci. Nutr.*, 1997, **37**, 693.

2.7

Bioavailability of Blackcurrant Anthocyanins in Humans

M. Netzel,[1] G. Strass,[2] M. Janssen,[2] I. Bitsch[2] and R. Bitsch[1]

[1] INSTITUTE OF NUTRITIONAL SCIENCES, FRIEDRICH-SCHILLER-UNIVERSITY, JENA, GERMANY
[2] INSTITUTE OF NUTRITIONAL SCIENCE, JUSTUS-LIEBIG-UNIVERSITY, GIESSEN, GERMANY

1 Introduction

Anthocyanins are a group of very efficient bioactive components which are widely distributed in plant food.[1,2] Several fruits (blackcurrant, blackberry, red grape, cherry, plum) and some vegetables (eggplant, onion, red radish) are rich sources of these natural pigments and extracts of some of them are used as food colorants as well as in pharmaceutical preparations and functional food.[3] Furthermore, anthocyanins are considered to exert several protective effects in the human body, via their ability to inhibit radical reactions. But up to now there have been only few data available demonstrating their capability to reach the systemic circulation of humans in intact or metabolised form. The present study was designed to determine the potential bioavailability in humans of delphinidine-3-glucoside (del-3-gluc), delphinidine-3-rutinoside (del-3-rut), cyanidine-3-glucoside (cya-3-gluc) and cyanidine-3-rutinoside (cya-3-rut), being the most important anthocyanins of blackcurrants (Figure 1).

2 Method

Urinary samples of three healthy male volunteers (aged 25–31; non-smokers) were collected before (baseline) and over a period of 5 hours with intervals of 30 minutes after ingestion of 200 ml of blackcurrant juice (containing 30 mg del-3-gluc, 78 mg del-3-rut, 4 mg cya-3-gluc and 41 mg cya-3-rut). The fresh urine samples were mixed with formic acid (1:0.28), evaporated under vacuum with a rotary evaporator (50 °C) and analysed under isocratic conditions with high-performance liquid chromatography (HPLC).[4]

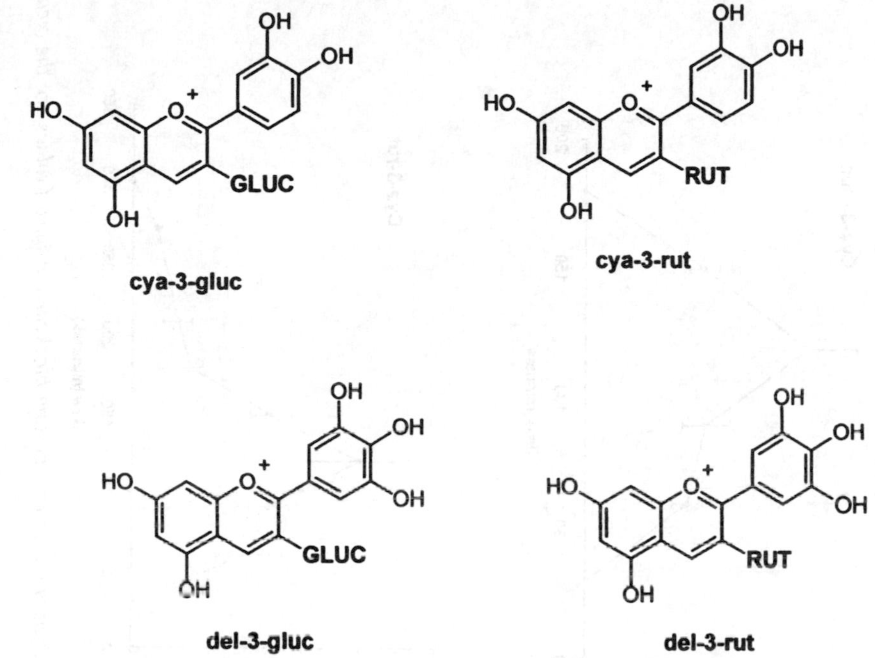

Figure 1 *Chemical structures of blackcurrant anthocyanins*

3 Results

With HPLC analysis it was possible to quantify the 4 anthocyanins of blackcurrants, excreted unchanged in the urine. These results demonstrate that a fraction of orally administrated anthocyanins is systemically available in humans (Table 1). Plots of the urinary anthocyanin concentrations against time obtained after ingesting blackcurrant juice are shown for each individual in Figure 2.

Table 1 *Excretion parameters of anthocyanins in human urine after ingesting blackcurrant juice*

Anthocyanins	t_{max} *(min)*	c_{max}* *(μg/urine fraction)*	*Cumulative amount excreted unchanged (μg)**
del-3-gluc	120	4.85 ± 1.46	9.84 ± 2.73
del-3-rut	120	5.75 ± 1.39	20.48 ± 3.55
cya-3-gluc	120	0.33 ± 0.07	0.83 ± 0.141
cya-3-rut	120	6.0 ± 1.32	20.63 ± 2.98

* Values are the mean ±SD of n = 3 volunteers.

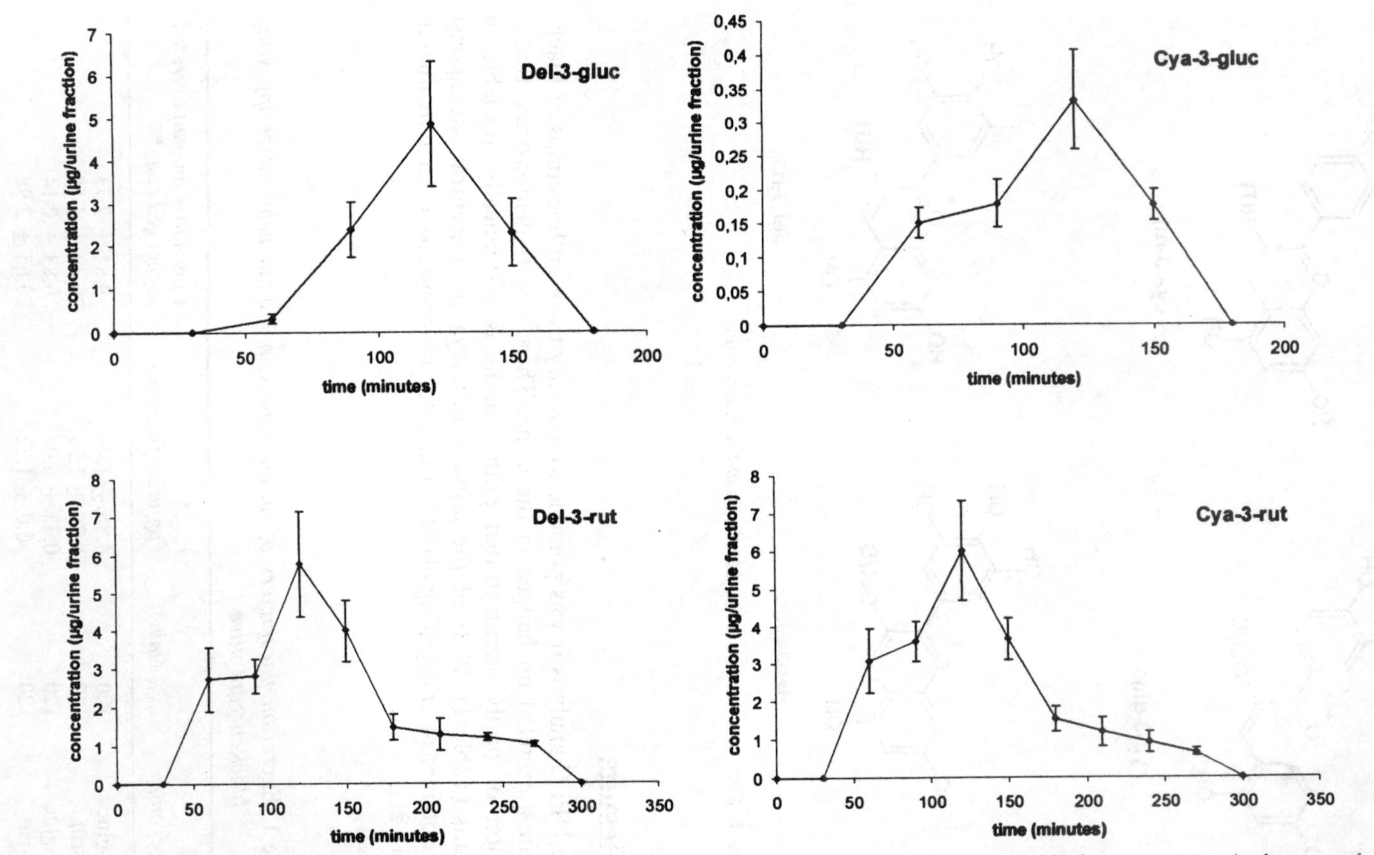

Figure 2 *Time-course plots of the concentrations of anthocyanins in human urine obtained after ingesting blackcurrant juice (values are the mean ±SD of $n = 3$ volunteers)*

4 Discussion

Our results demonstrate for the first time that humans can absorb (and excrete) remarkable amounts of blackcurrant anthocyanins as unmodified glycosides, being possible of relevance for biological effects. The cumulative amounts however excreted unchanged in the urine, were only 0.02–0.05% of the oral doses. Possible losses of these unstable compounds due to incomplete absorption, decomposition in the lumen, elimination in the faeces or substantial first-pass elimination may contribute to their low urinary excretion rate. In conclusion, more research is required for further elucidation of the mechanisms of anthocyanin absorption, metabolism, disposition and biochemical action in the body after ingestion of anthocyanin rich fruits and juices.

Acknowledgement

The study has been carried out with financial support of the AIF/FEI (project-No. 11160 B).

5 References

1 H. Tamura and A. Yamagami, *J. Agric. Food Chem.*, 1994, **42**, 1612.
2 T. Tsuda, M. Watanabe, K. Ohsima, S. Norinobu, S. Choi, S. Kawakishi and T. Osawa, *J. Agric. Food Chem.*, 1994, **42**, 2407.
3 M. Tits, L. Angenot, J. Damas, Y. Dierckxens and P. Poukens, *Planta Med.*, 1997, **57**, 134.
4 M. Netzel, M. Janssen, I. Bitsch and R. Bitsch, Proceedings of the 3rd Karlsruhe Nutrition Symposium: European Research towards Safer and Better Food, 1998, **2**, 183.

2.8

Gallic Acid in Black Tea and Its Bioavailability for Man

M. Netzel, S. Shahrzad, A. Winter and I. Bitsch

INSTITUTE OF NUTRITIONAL SCIENCE, JUSTUS-LIEBIG-UNIVERSITY, GIESSEN, GERMANY

1 Introduction

In free or bound forms gallic acid is found in significant amounts in tea leaves, from which it is extracted in hot water infusions. Gallic acid (Figure 1) is a strong antioxidant and possesses antiallergic, anti-inflammatory, antimutagenic and anticarcinogenic activities.[1–4] In spite of these protective effects, there are only few data available about the extent of its absorption and metabolism in humans. Therefore our study was performed with adult humans to estimate the absorption and metabolism of gallic acid after oral administration of a tea infusion.

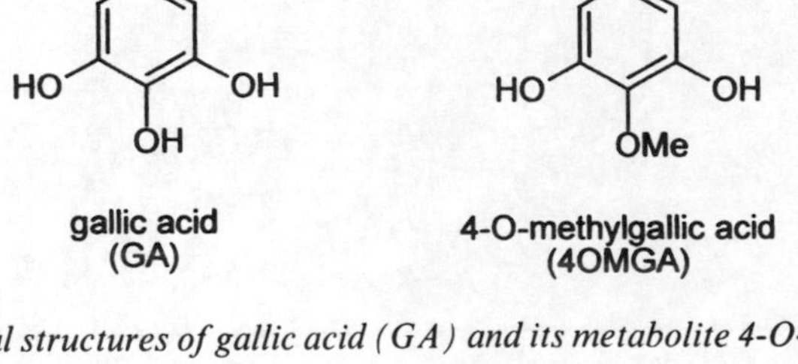

Figure 1 *Chemical structures of gallic acid (GA) and its metabolite 4-O-methyl-gallic acid (4OMGA)*

2 Method

After following a GA free diet for 24 hours, plasma (respectively urine) of three healthy male volunteers (aged 24–29; non-smokers) were collected before (baseline) and 0.75 (3), 1.5 (5), 2.25 (7), 4 (9), 8 (11) and 10 (24) hours after an oral ingestion of black tea (containing 50 mg GA). The samples (plasma and urine) were mixed with an equal volume of 1 M hydrochloric acid and hydrolysed in a boiling water bath under argon for 30 minutes. After hydrolysis the samples were cooled to room temperature and extracted with ethyl acetate, centrifuged (to separate the organic fractions) and evaporated to dryness under vacuum with a rotary evaporator. Each extract was redissolved in HPLC mobile phase (4.4 mM phosphoric acid in water) and analysed with an isocratic reversed phase HPLC method.[5]

3 Results

After black tea consumption, GA and its metabolite 4OMGA, could be quantified in plasma and urine (Table 1). Plots of the plasma and urinary GA and 4OMGA concentrations against time are shown in Figure 2. Before ingestion of black tea (baseline), plasma and urine did not contain any detectable concentrations of GA respectively 4OMGA.

4 Discussion

In this paper we have first established that GA from black tea is absorbed and excreted in humans in remarkable amounts. We can therefore reasonably assume, that the *in vivo* antioxidant effect of black tea as it was demonstrated by Serafini *et al.*[6] might be caused by GA and its metabolite 4OMGA.

Table 1 *Biokinetic parameters of GA and 4OMGA in humans obtained after ingesting black tea*

		t_{max} (h)	c_{max}* ($mg\ l^{-1}$ respect. mg/urine fraction)	Cumulative amount* (mg over 24 h)
Plasma	GA	1.5	0.324 ± 0.090	–
	4OMGA	2.25	0.512 ± 0.101	–
Urine	GA	3	1.46 ± 0.21	4.05 ± 1.05
	4OMGA	7	4.12 ± 1.33	12.86 ± 3.84

* Values are the mean ± SD of n = 3 volunteers.

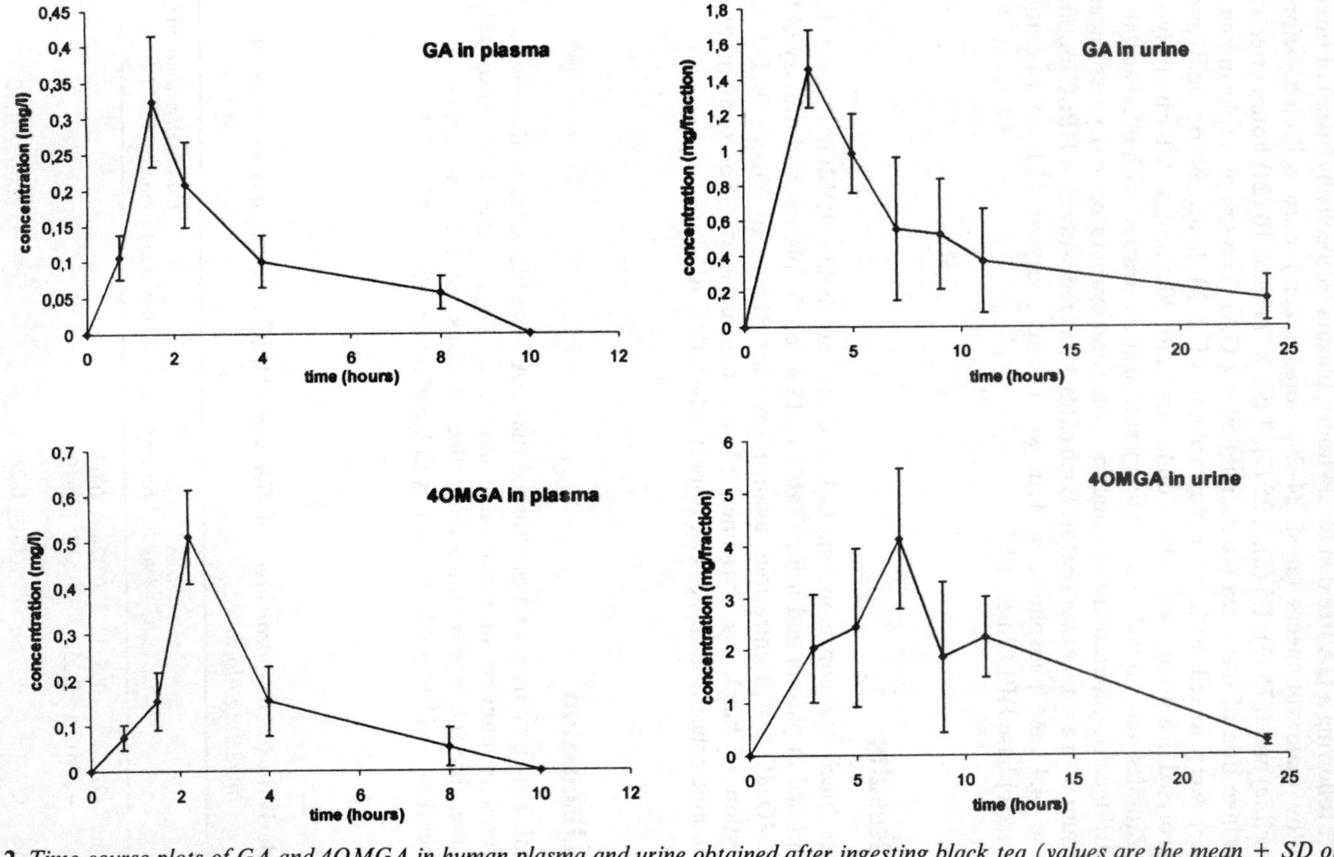

Figure 2 *Time-course plots of GA and 4OMGA in human plasma and urine obtained after ingesting black tea (values are the mean* $\pm$ *SD of n = 3 volunteers)*

5 References

1 B. Fuhrmann, A. Lavy and M. Aviram, *Am. J. Clin. Nutr.*, 1995, **61**, 549.
2 T. P. Whitehead, D. Robinson, S. Allaway, J. Syms and A. Hale, *Clin. Chem.*, 1995, **41**, 32.
3 K. Kondo, A. Matsumoto, H. Kurata, H. Tanahashi, H. Koda, T. Amachi and H. Itakura, *Lancet*, 1994, **334**, 1152.
4 P. Li, H. Z. Wang, X. Q. Wang and Y. N. Wu, *Biomed. Environ. Sci.*, 1994, **7**, 68.
5 S. Shahrzad and I. Bitsch, *J. Chromatogr. B*, 1998, **705**, 87.
6 M. Serafini, A. Ghiselli and A. Ferro-Luzzi, *Eur. J. Clin. Nutr.*, 1996, **50**, 28.

2.9

Absorption and Modification of Rutin in the Human Stomach

H. Pforte,[1] T. Näser,[2] G. Jacobasch[1] and H.J. Buhr[2]

[1] GERMAN INSTITUTE OF HUMAN NUTRITION, POTSDAM-REHBRÜCKE, GERMANY
[2] CLINICAL AND MOLECULAR GASTROENTEROLOGY AND SURGERY, FREIE UNIVERSITÄT BERLIN, GERMANY

1 Introduction

Flavonoids have a variety of protective effects on human health.[1] Biological activities result from both systemically generated actions and influences connected with the intestinal microflora.[2] Antiatherosclerotic and anticarcinogenic effects are realized mainly systemically.[3] It is necessary to know the therapeutic doses in order to optimize a continous gastrointestinal absorption of the flavonoids. Since quercetin is the most abundant flavonoid in the human diet and characterized by a high protective metabolic effectiveness, we focussed our investigations on the absorption of this flavonol compound.

Recent findings suggest that quercetin is relatively unstable under milieu conditions of about pH 7.0 appearing in the human small intestine. It is therefore difficult to achieve a defined continuous dose of flavonoid absorption. The natural dietary compounds of quercetin in plants are stable glycosides. The carbohydrate chain can be cleaved microbially in the gastrointestinal tract, for instance by *Enterococcus casseliflavus*, liberating the aglycon quercetin.[4] As a result, the question arises whether and how effectively quercetin rutinoside can be absorbed in the different parts of the stomach?

2 Methods

The absorption process of rutin by the stomach mucosa was studied in Ussing chambers. Stomach wall samples of healthy areas of the antrum and corpus were taken from patients during surgical removal of malignomes and immediatly used for the experimental studies. In cold isotonic solution the samples were stripped and clamped within 10 minutes post-operatively as separating wall between two compartments of the Ussing chamber. The chambers were filled with Ringer solution and additionally containing albumin (5 g%) in the serosal side. The chambers had a temperature of 37 °C and were gassed continously with a mixture of 95% O_2 and 5% CO_2. The vitality of the tissue samples was evaluated on the basis of voltage, short circuit and epithelial resistance. Absorption and transport experiments were started by adding quercetin rutinoside (0.8 and 1.6 μM) into the mucosal part of the chamber. Samples from both solutions of the serosal and the mucosal sides were freeze-dried after 0, 15, 30, 60 and 90 minutes respectively. After that, they were stored together with the tissue samples at −80 °C until use.

Flavonoid analysis: The frozen mucosa was lyophilized and pulverized in a mortar and defatted with heptane by sonicating for 10 minutes and Soxhlet-extraction with methanol for 4 h under reduced pressure at about 45 °C. From the mucosa and from the two solutions flavonoids were extracted and dissolved for analysis by HPLC in DMF/H_2O (v/v 2:1). The stationary phase was a Nucleosil RP-18 endcapped column (Macheray-Nagel, Düren Germany) (250 mm × 4.6 mm; 5 μM) at 30 °C with a 4 × 4 mm pre-column. Flavonoid glycosides and aglycones were eluted with a gradient consisting of water, 80% methanol, buffered with phosphate pH 3.4. Detection was carried out by an ESA coularray detector containing 12 pairs of electrodes and a detection range between 0 and 825 mV. To identify the flavonoids retention times and voltamogrammes of standards were used. The flavonoids were quantified by using an external calibration curve.

3 Results and Discussion

Only a small number of studies have been carried out to determine where and how effectively flavonoids are absorbed in the gastrointestinal tract. Results of our Ussing chamber experiments prove that human stomach mucosa absorbs quercetin rutinoside. Both absorption rate of rutin and enzymatic release of the aglycon quercetin differ in dependence of the sample localization in the stomach. Data given in Figure 1 show that generally the antrum takes in more rutin than the corpus, but we identified higher and lower absorption rates in both areas. These results do not agree with the conclusion of Kuhnau that only aglycons could pass through the mucosa.[5] Hydrolysis of rutin does not occur as a result of the intestinal microorganisms but is caused by a stomach specific β-glucosidase. Our own unpublished data of germfree rats also confirm that the low activity of the microbial β-glucosidase in stomach is nearly without any physiological relevance. The conclusion that absorption of flavonoid does not require their aglycone formation furthermore agrees well with results of the Hollman group, who demonstrated in a human study with ileostomy subjects an absorption of quercetin glycosides from onions of 52% and of pure rutin of 17%.[6] The pattern of flavonoid compounds in both the serosal fluid and the stomach walls suggest a higher enzymatic activity of β-glucosidase in the corpus than in the antrum (Table 1). The corpus is also characterized by a higher activity of enzymatic O-methylation of quercetin. Formation of isorhamnetin is

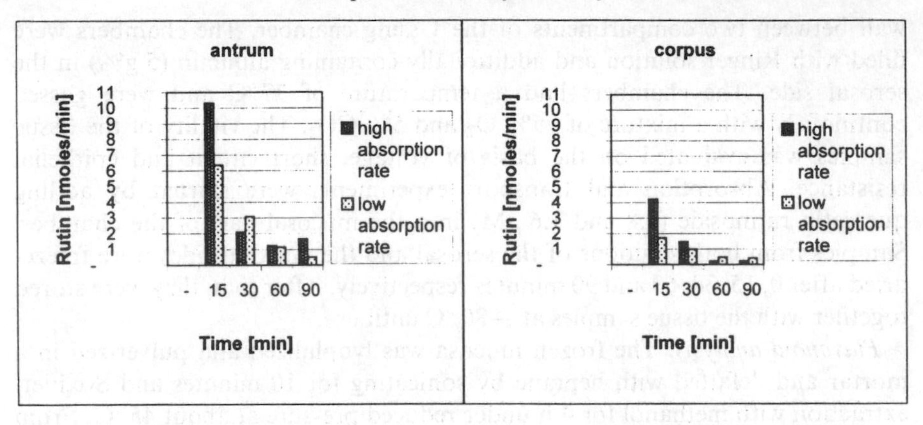

Figure 1 *Time dependence of the rutin transport in the human corpus and antrum mucosa of the stomach*

Table 1 *Flavonoid concentrations in the human corpus mucosa of the stomach 90 minutes after addition of rutin on the luminal side*

	Rutin [nM]		*Quercetin [nM]*		*Isorhamnetin [nM]*	
Transport rate	$\bar{x}$	$\pm s$	$\bar{x}$	$\pm s$	$\bar{x}$	$\pm s$
High	2.79	1.20	0.31	0.02	0.02	0.01
Low	1.75	0.50	0.17	0.13	0.02	0.01

believed to be catalyzed by distinct classes of O-methyltransferases (OMTs). Until now no OMTs of the human stomach mucosa have been isolated and characterized. Therefore, we can not decide whether the different rates of the isorhamnetin formation reflect a distinct local expression of OMT-isoenzymes or only a higher O-methylation capacity in the corpus. From these results, the following conclusion is drawn, that the absorption rates of rutin in the stomach correspond with those of the small intestine. The absence of flavonoid degrading bacteria in the stomach favours this part of the gastrointestinum for a monitoring of systemic therapeutic effects of quercetin.

4 References

1 J.V. Formica and W. Regelson, Review of the biology of quercetin and related bioflavonoids, *Food Chem. Toxicol.*, 1995, **33**, 1061–1080.

2 H. Schneider, A. Schwiertz, M.D. Collins and M. Blaut, Anaerobic transformation of quercetin-3-glucoside by bacteria from the human intestinal tract, *Arch. Microbiol.*, 1999, **171**, 81–91.

3 D.D. Schramm and J.B. German, Potential effects of flavonoids on the etiology of vascular disease, *J. Nutr. Biochem.*, 1998, **9**, 560–566.

4 R. Simmering, B. Kleessen and M. Blaut, Quantification of the flavonoid-degrading bacterium *Eubacterium ramulus* in human fecal samples with a species-specific oligonucleotide hybridization probe, *Appl. Environ. Microbiol.*, 1999, **54**, 1079–1084.
5 J. Kuhnau, The flavonoids: a class of semi-essential food components: their role in human nutrition, *World Rev. Nutr. Diet.*, 1976, **24**, 117–120.
6 P.C.M. Hollman, Bioavailability of flavonoids, *Eur. J. Clin. Nutr.*, 1997, **51**, 66–69.

2.10

Production of Allyl Isothiocyanate from Sinigrin in the Distal Gut of Gnotobiotic Rats Harbouring a Human Colonic Bacterial Strain of *Bacteroides Thetaiotaomicron*

L. Elfoul,[1] S. Rabot,[1] N. Khelifa,[2] S. Garrido[1] and J. Durao[1]

[1] INSTITUT NATIONAL DE LA RECHERCHE AGRONOMIQUE, UNITÉ D'ECOLOGIE ET DE PHYSIOLOGIE DU SYSTÈME DIGESTIF, F-78352 JOUY-EN-JOSAS, FRANCE
[2] FACULTÉ DES SCIENCES PHARMACEUTIQUES ET BIOLOGIQUES, LABORATOIRE DE MICROBIOLOGIE, F-75006 PARIS, FRANCE

1 Introduction

Sinigrin (SIN) is a simple aliphatic glucosinolate found in *Brassica* vegetables such as Brussels sprouts or cabbage. Its enzymatic hydrolysis under the action of plant myrosinase (EC 3.2.3.1) leads to the formation of allyl isothiocyanate (AITC), allyl cyanide, 1-cyano-2,3-epithiopropane and other minor products. Among these derivatives, AITC may contribute to the protective effect of *Brassica* vegetables against colorectal cancer, chiefly by a suppressing mechanism involving a selective toxicity towards colon cancer cells.[1] If the vegetables are cooked, the myrosinase is inactivated and glucosinolates reach the distal gut where they are broken down by the resident microflora into still ill-known compounds.[2,3] Therefore the present experiment was designed to investigate the ability of the digestive microflora to produce AITC from an oral dose of SIN. Gnotobiotic rats harbouring a *Bacteroides thetaiotaomicron* strain, isolated from human faeces and shown to be a SIN-degrader *in vitro*,[4] were used as a study model.

2 Materials and Methods

Thirty F344 male germ-free rats, housed in sterile isolators, were orally inoculated with the *B. thetaiotaomicron* strain. To simulate a realistic dietary environment, rats were offered a feed containing glucosinolates (3.9 μmol g^{-1}). These were provided by a dehulled myrosinase-free and sinigrin-free rapeseed meal (*cv* Darmor).

After 2 weeks adaptation to the flora and the diet, 28 rats were randomly allocated to 7 groups of 4 animals, and dosed with 50 μmol SIN dissolved in 0.5 ml distilled water (6 treated groups) or with 0.5 ml distilled water only (1 control group). Gavage was performed by stomach tubing under light ether anaesthesia. After dosing, rats were killed by CO_2 inhalation, at 0 h for the control group and at 3, 6, 12, 18, 24 and 36 h for the treated groups. The digestive tract was removed and the contents of the stomach, small intestine, cecum and colon were collected separately and stored at −20 °C until SIN and AITC analyses. Bacterial counts were performed in the digestive contents of the 2 extra rats to assess the establishment of the bacterial strain.

SIN analysis was performed by HPLC.[5] AITC was analysed by GC, following headspace solid-phase microextraction (SPME).[6] Briefly, this method involved trapping AITC present in the sample onto a 75 μm carboxen-poly(dimethylsiloxane)-coated silica-fibre (Supelco) and desorbing it in the hot injection port of the chromatograph prior to GC separation. Identification of AITC was based on its retention time and confirmed by GC-MS analysis.

3 Results

Average bacterial counts, expressed as $\log_{10}$ of the number of bacteria per g of content, were 2.58, 7.00, 9.60 and 9.48 in the stomach, small intestine, cecum and colon, respectively.

Neither SIN nor AITC were detected in the gastrointestinal tract of control rats. In treated rats, the total recovery of residual intact SIN averaged 28 μmol, 3 h after the gavage; the greatest amount, accounting for 46% of the ingested dose, was measured in the cecum and colon, with the stomach and small intestine containing only low levels (10% of the ingested dose). After 6 h, SIN steadily decreased in all compartments ; from 18 h onwards, it was virtually absent in the stomach and small intestine and did not exceed 2 μmol in the cecum and colon (Figure 1).

AITC appeared in the gastrointestinal tract of treated rats as soon as 3 h after dosing. In the stomach, trace amounts were detected temporarily, between 3 and 6 h after the gavage, whereas, in the small intestine, AITC was present in low amounts until the end of the experiment. The greatest amounts were measured in the distal gut. In the cecum, the formation of AITC peaked between 6 and 12 h after SIN administration and substantial amounts (25% of the peak amount) could still be detected until 24 h. The kinetics of AITC formation in the colon followed a similar pattern (Figure 2).

4 Conclusion

The results of this experiment show that the *B. thetaiotaomicron* strain, established at a high level in the distal gut, extensively degraded SIN available

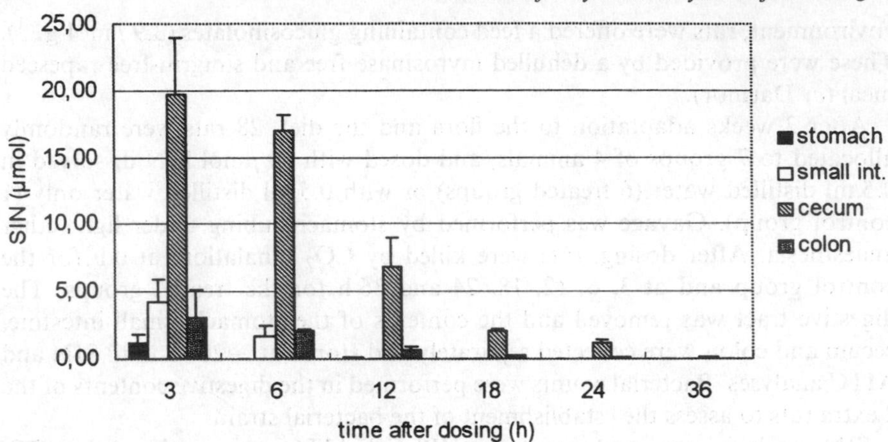

Figure 1 *Kinetics of SIN disappearance in the gastrointestinal tract of gnotobiotic rats harbouring a* B. thetaiotaomicron *strain of human origin and orally dosed with 50 µmol SIN*

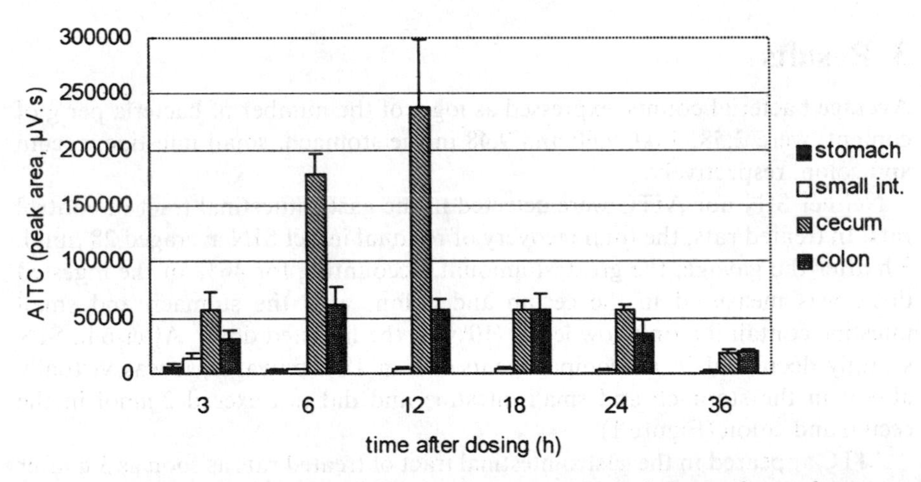

Figure 2 *Kinetics of AITC formation in the gastrointestinal tract of gnotobiotic rats harbouring a* B. thetaiotaomicron *strain of human origin and orally dosed with 50 µmol SIN*

in it and converted this compound, at least partly, into AITC. These results support the hypothesis that the colonic microflora may play a key role in the protective effects of SIN against the development of colorectal cancer. Further experiments using gnotobiotic rats harbouring a whole human faecal flora and consuming SIN in its natural plant matrix, *e.g.* Brussels sprouts, would be beneficial to extend the significance of these findings.

Acknowledgements

The authors thank Dr Jacques Evrard, CETIOM, Pessac, France, for the generous gift of the rapeseed meal, and Dr Jean-Luc Luisier and Mr Ramin Azodanlou, Ecole d'Ingénieurs du Valais, Sion, Switzerland, for their most helpful advice on the use of SPME. This research was supported by the European Community under the programme FAIR CT97 3029 entitled 'Effects of food-borne glucosinolates on human health'.

5 References

1 S.R.R. Musk and I.T. Johnson, *Carcinogenesis*, 1993, **14**, 2079.
2 S. Michaelsen, J. Otte, L.-O. Simonsen and H. Sorensen, *Acta Agric. Scand.*, 1994, **44**, 25.
3 S. Rabot, L. Nugon-Baudon, P. Raibaud and O. Szylit, *Br. J. Nutr.*, 1993, **70**, 323.
4 S. Rabot, C. Guérin, L. Nugon-Baudon and O. Szylit, Proceedings of the 9th International Rapeseed Congress, GCIRC, Paris, 1995, Vol. 1, p. 212.
5 AFNOR, 1995, NF EN ISO 9167–1.
6 A. Steffen and J. Pawliszyn, *J. Agric. Food Chem.*, 1996, **44**, 2187.

2.11

Factors Influencing the Release of Cancer-protective Isothiocyanates in the Digestive Tract of Rats Following Consumption of Glucosinolate-rich Brassica Vegetables

G. Rouzard,[1] A.J. Duncan,[1] S. Rabot,[2] B. Ratcliffe,[3] J. Durao,[2] S. Garrido[2] and S. Young[1]

[1] THE MACAULAY LAND USE RESEARCH INSTITUTE, CRAIGIEBUCKLER, ABERDEEN AB15 8QH, UK
[2] UNITÉ D'ECOLOGIE ET DE PHYSIOLOGIE DU SYSTÈME DIGESTIF, INSTITUT NATIONAL DE LA RECHERCHE AGRONOMIQUE, 78352 JOUY-EN-JOSAS, FRANCE
[3] THE ROBERT GORDON UNIVERSITY, QUEEN'S ROAD, ABERDEEN AB15 4PH, UK

1 Introduction

Glucosinolates (GLS) are thioglucosides present in cruciferous plants including brassica vegetables such as Brussels sprouts. Hydrolysis of GLS by plant or microbial myrosinase following ingestion leads to the formation of a range of compounds among which are the cancer-protective isothiocyanates (ITCs). Little is known about the fate of GLS *in vivo*. The effects of plant myrosinase and gut microbial myrosinase have been described separately.[1,2] In human diets, however, both plant and bacterial myrosinase are present and the extent of their respective involvement in GLS breakdown is unclear. The current experiment was designed to investigate the respective influence of plant myrosinase, microbial myrosinase and their interactions on the release of the anticarcinogenic compound benzyl ITC from an oral dose of benzyl GLS.

2 Experimental Design

To study the role of microbial myrosinase, two sets of eight Fischer 344 rats, differing in their microbial status, were used. One set of rats harboured a whole human faecal microflora (Flora + treatment) while the second set was germ-free

(Flora – treatment). To investigate the hydrolysis catalysed by plant myrosinase, two diets were sequentially offered to rats. One diet contained plant myrosinase (Myro + treatment) while the other was plant myrosinase-free (Myro – treatment). The order of offering the diet was reversed in half the animals in each set of rats. The Myro + diet contained 15% Brussels sprouts (*Brassica oleracea* var. *cyrus)* to provide a source of plant myrosinase. The Myro – diet had the same composition as Myro + diet but Brussels sprouts were previously treated for inactivation of plant myrosinase by soaking them in hot 70% ethanol.

Rats were kept in individual metabolism cages inside sterile isolators for the entire experiment. The experiment lasted 6 weeks and was divided into two equal feeding periods. Each period began with a 9 day adaptation phase. On days 10 and 17, animals were dosed with either 25 μmol of benzyl GLS (GLS dose) or 25 μmol of benzyl ITC (ITC dose). At each dosing all rats also received 25 μmol of butyl ITC as an internal standard. Urine and faeces samples were collected at 0 h, 6 h, 24 h, 48 h, 72 h, 144 h and 168 h after dosing and stored at $-20\,^{\circ}$C until analysis. The order of administration of GLS and ITC was randomised within each feeding period and within each set of rats.

Benzyl ITC released and absorbed in the intestinal tract was quantified using urinary end products of metabolism as biomarkers in a similar fashion to previous experiments.[2] Butyl ITC was used to allow correction for different post-absorptive excretion rates between animals. The animal-specific rates of isothiocyanate excretion were estimated in ITC dosed animals. The corrected relative recovery of benzyl ITC in rats dosed with GLS *vs.* ITC was used to estimate the percentage of conversion of benzyl GLS to its metabolite benzyl ITC *in vivo*. Benzyl GLS excreted intact in faeces was quantified using analysis of desulpho-glucosinolates.[3]

3 Results

In the absence of myrosinase (Flora –, Myro –) a small residual excretion of benzyl ITC marker could be measured suggesting a spontaneous breakdown of benzyl GLS to benzyl ITC. When only plant myrosinase was active (Flora –, Myro +) a high proportion of benzyl GLS was converted to benzyl ITC (Figure 1). Myrosinase activity derived from microflora only (Flora + , Myro –) led to a production of benzyl ITC which was not significantly different from residual breakdown. However, when both sources of myrosinase were active (Flora +, Myro +), the proportion of benzyl GLS converted into benzyl ITC was significantly lower than when plant myrosinase only was active (0.499 *vs.* 0.828, $P < 0.001$) (Figure 1).

Only trace amounts of benzyl GLS were excreted in faeces in Flora + rats suggesting a complete hydrolysis of benzyl GLS in the presence of microflora. In Flora – rats, the absence of myrosinase in the Myro – diet did not lead to a 100% excretion of benzyl GLS. When only plant myrosinase was active (Flora –, Myro +) proportion of benzyl GLS excreted intact was less than 0.100 (Figure 2).

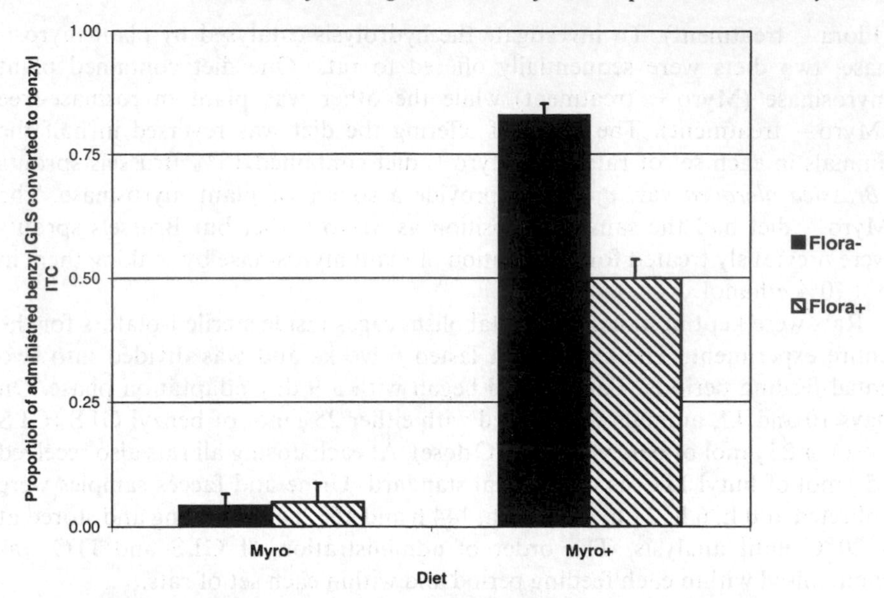

Figure 1 *Influence of the diet and bacterial status on the conversion of benzyl GLS into benzyl ITC in the intestinal tract of rats after a single oral dose of 25 µmol of benzyl ITC*

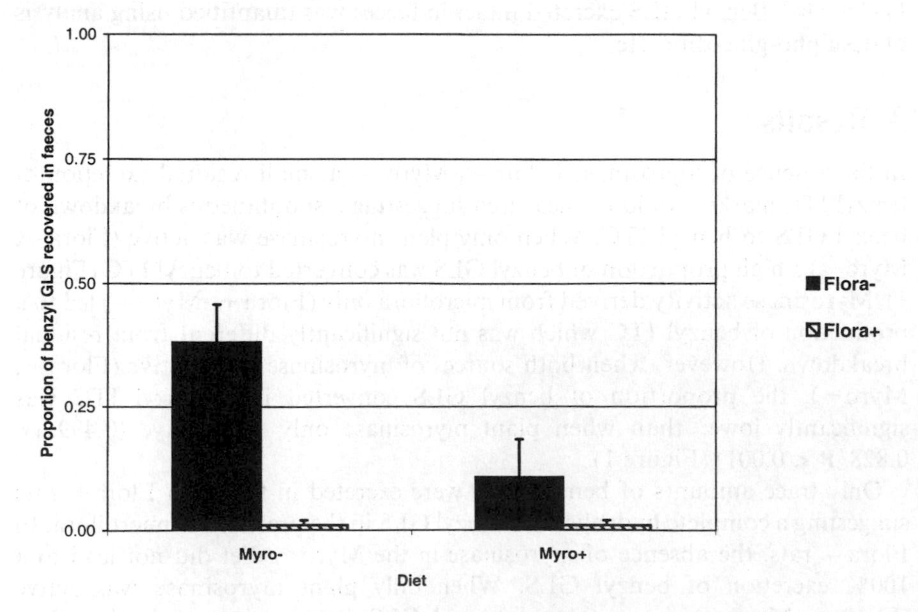

Figure 2 *Excretion of intact benzyl GLS in faeces of rats after a single oral dose of 25 µmol of benzyl GLS*

4 Discussion

The results of this experiment suggest that the bioavailability of anticarcinogenic ITCs is maximal when plant myrosinase is active. This effect was not enhanced by the activity of microflora, as shown in Flora + rats fed the Myro + diet. In fact, the microflora appeared to lower the total bioavailability of benzyl ITC, probably by consuming a part of the ITCs arising in the proximal intestine following passage to the bowel. However, when both sources of myrosinase were active, the proportion of benzyl GLS broken down into benzyl ITC was 0.499 similar to findings in conventional rats dosed with sinigrin and in human studies.[2,4]

In this study, inactivation of plant myrosinase did not enhance the contribution of the microflora to the release of benzyl ITC in the bowel. However, the microbial myrosinase activity was able to catalyse the complete breakdown of benzyl GLS. These results contrast with other findings on microbial degradation of sinigrin where a substantial release of allyl ITC was shown.[5] As the estimation of bioavailability relies here on the measurement of excretory products, it is possible that absorption in the bowel leads to a different route of excretion for benzyl ITC. Alternatively, the microflora may catalyse both the release and the subsequent breakdown of benzyl ITC resulting in minimal absorption. These aspects requires to be explored.

This study confirmed the major role played by plant myrosinase in catalysing the release of ITCs in the digestive tract. Since plant myrosinase may be inactivated by cooking, the experiment suggests that methods of preparation of the vegetables may have a major influence on the anti-cancer properties of brassica vegetables.

Acknowledgements

This research was supported by the European Community under the programme FAIR CT97 3029 entitled Effect of Food-borne Glucosinolates on Human Health (EFGLU). The authors would like to thank Novartis, The Netherlands, for providing the vegetable material.

5 References

1 A.M. Bones and J.T. Rossiter, *Physiol. Plant.*, 1996, **97**, 194–208.
2 A.J. Duncan, S. Rabot and L. Nugon-Baudon, *J. Sci. Food Agric.*, 1997, **73**, 214–220.
3 I. Minchington, J.Sang, D. Burke and R.J.W. Truscott, *J. Chromatogr.*, 1982, **247**, 141–148.
4 T.A. Shapiro, J.W. Fahey, K.L. Wade, K.K. Stephenson and P. Talalay, *Cancer Epidemiol. Biomark. Prev.*, 1998, **7**, 1091–1100.
5 L. Elfoul, S. Rabot, A.J. Duncan and L. Goddyn, in 'Agri-Food Quality II', Royal Society of Chemistry, Cambridge, 1999, pp. 294–296.

2.12

Synthesis of Isothiocyanate Mercapturic Acids

M. Vermeulen,[1] B. Zwanenburg,[2] G.J.F. Chittenden[2] and H. Verhagen[1]

[1] TNO VOEDING, FOOD ANALYSIS DEPARTMENT, PO BOX 360, 3700 AJ ZEIST, THE NETHERLANDS
[2] UNIVERSITY OF NIJMEGEN, DEPARTMENT OF ORGANIC CHEMISTRY, NSR CENTER, 6525 ED NIJMEGEN, THE NETHERLANDS

Fruits and vegetables possess anticarcinogenic properties. The responsible bioactive ingredients are currently under investigation in many laboratories worldwide. A large part of the vegetables consumed consist of cruciferous vegetables, among which are cabbages, broccoli and Brussels sprouts. One potential class of anticarcinogenic chemicals comprises isothiocyanates, breakdown products of glucosinolates that are present in all crucifers. Isothiocyanates are metabolized in the human body to their corresponding mercapturic acid derivatives and excreted into the urine. We have chemically synthesized the mercapturic acid derivatives of the most important isothiocyanates as reference standards for the determination of urinary excretion. We accomplished this by adding isothiocyanate dissolved in alcohol dropwise to a solution of *N*-acetyl-L-cysteine and sodium bicarbonate in water, whereupon the corresponding mercapturic acid was obtained. Few isothiocyanates were available commercially; others were first synthesized by adding the corresponding alkyl bromide to potassiumphthalimid. The *N*-alkylphthalimid obtained was hydrazinolysed yielding the alkyl amine, which was subsequently substituted with thiophosgene, yielding the isothiocyanate. We used TLC, GC, NMR, elemental analyses and mass spectrometry to confirm chemical identity and purity of the products. We are currently developing a method for the determination of isothiocyanate mercapturic acids in urine.

2.13

Influence of Protein Type and Digestion on Protein Benzo[*a*]pyrene Interactions

Eric Vis,* Anouk Geelen, Gerrit Alink and Tiny van Boekel

DEPARTMENT OF FOOD TECHNOLOGY AND NUTRITIONAL SCIENCES, WAGENINGEN UNIVERSITY, THE NETHERLANDS

1 Introduction

Through the years many data have been collected about dietary exposure to, for instance, micronutrients and, to a lesser extent, non-nutrients. Both the bioavailability and the influence of the diet matrix on this bioavailability are not entirely clear for many of these components. This lack of knowledge is especially true for non-nutrients such as mutagens and carcinogens. Dietary protein is known to be able to influence the bioavailability of both nutrients and medicine.[1] Protein mutagen interactions were observed in earlier *in vitro* studies with the model mutagen MNNG.[2] In those studies protein mutagen interactions were found to increase after protein digestion. Therefore, in the present study, the effects of protein and protein digestion on protein benzo[*a*]pyrene interactions will be studied.

2 Materials and Methods

In vitro binding experiments were performed by incubating protein (hydrolysates) together with 10 mg of cellulose and app. 2000 dpm (disintegrations per minute) ^{14}C labeled benzo[*a*]pyrene (Amersham, 1.85 MBq μmol^{-1}) in a test tube. Influences on protein benzo[*a*]pyrene interactions were measured by determining the distribution change of benzo[*a*]pyrene between cellulose and protein. After 30 minutes incubation at 37 °C, test tubes were centrifuged to precipitate the cellulose fraction. Radioactivity was determined in the supernatant containing both benzo[*a*]pyrene and protein using a liquid scintillation counter (Packard 1600TR). The experiment shown in Figure 4 was performed

*Corresponding author, tel. (+31-317-48 43 57), e-mail (eric.vis@algemeen.tox.wau.nl).

with small pieces of cellulose filter paper instead of plain cellulose, so no centrifugation was necessary.

Protein digestion took place in the pH-stat equipment as described by Adler-Nissen.[3] The degree of hydrolysis (DH) was expressed as the percentage of peptide bonds cleaved. Protein was digested for 30 minutes using pepsin (Merck, 2000 FIP g^{-1}) followed by a 90 minute pancreatin (Merck, 1400 FIP/g) digestion. All protein sources were at least 90% pure. Ovalbumin was heated until gelation and sonicated before use to disrupt gel structure.

3 Results and Discussion

In the first experiment, reported in Figure 1, a dose response curve was made for the amount of cellulose added and the amount of ^{14}C benzo[*a*]pyrene in the supernatant after centrifugation. Based on this experiment further incubations were performed with 10 mg of cellulose. When casein was digested, the benzo[*a*]pyrene binding capacity of the protein fraction decreased. This decrease was dependent on the degree of hydrolysis. At DH = 17% no protein benzo[*a*]-pyrene interaction could be observed (Figure 2). Similar results were obtained when ovalbumin was hydrolyzed (Figure 3). To test whether there was a difference between different dietary protein types in their interaction with benzo[*a*]pyrene several protein sources were compared (Figure 4). No obvious differences were observed between the different protein sources.

The results of above described experiments are contrary to results reported earlier with a model mutagen MNNG.[2] Apparently the protein benzo[*a*]pyrene interaction is more dependent upon intact proteins whereas MNNG interacts with much smaller protein fractions such as those present in protein hydrolysates. This difference in protein interaction may be caused by the fact that

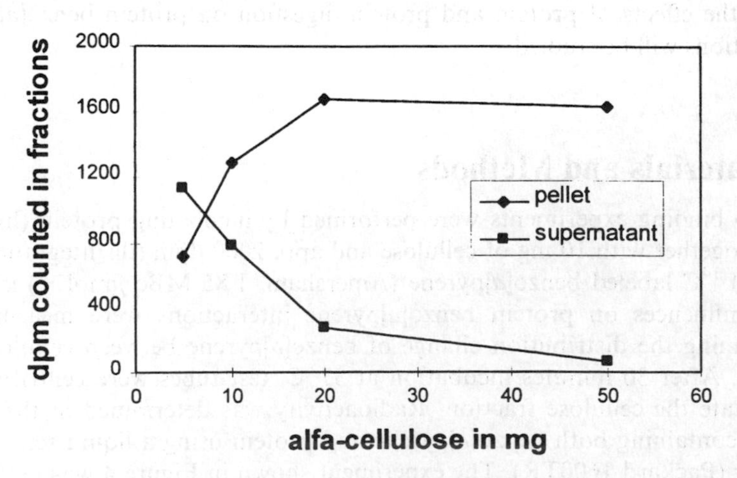

Figure 1 *Relationship between the amount of cellulose in the incubation and the amount of ^{14}C benzo[a]pyrene in both the cellulose and protein fraction*

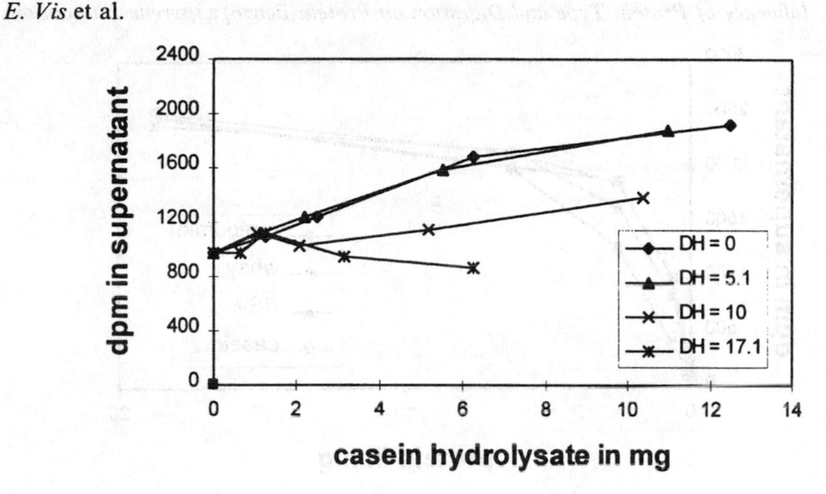

Figure 2 *The relationship between the degree of hydrolysis and the amount of ^{14}C benzo[a]pyrene in the casein fraction*

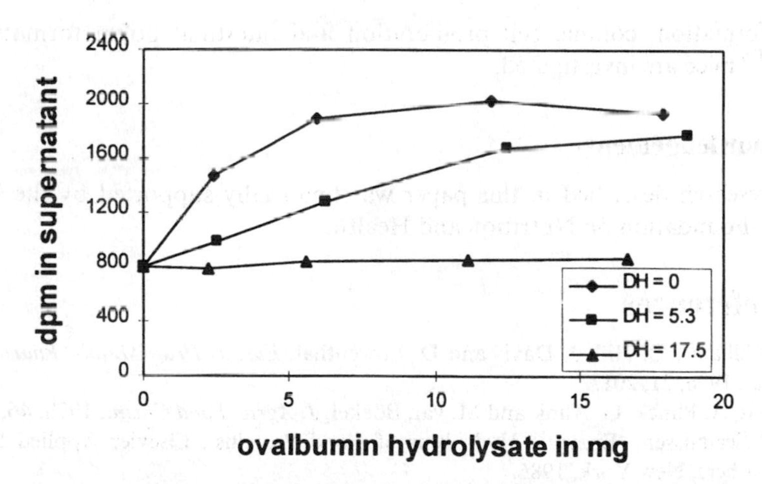

Figure 3 *The relationship between the degree of hydrolysis and the amount of ^{14}C benzo[a]pyrene in the ovalbumin fraction.*

benzo[*a*]pyrene in food is a large non-metabolized hydrophobic molecule whereas MNNG is a small and very reactive alkylating agent.

This research was performed as part of a PhD project on the influence of different dietary protein sources on their risk of (colon) cancer. From the study described in this paper it was concluded that different dietary protein sources do not seem to have a very different influence on dietary carcinogen (benzo[*a*]pyrene) bioavailability. Therefore further research was focused on the influence of different dietary protein sources on *in vivo* biomarkers for colon cancer in mice and rats. Different animal experiments are underway in which rat aberrant crypt

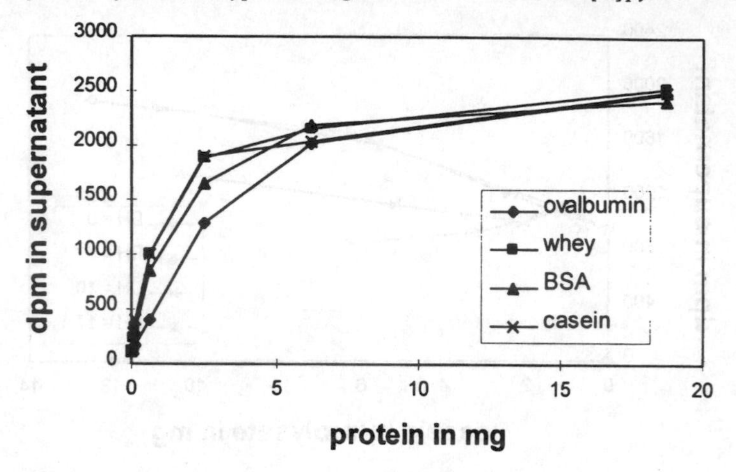

Figure 4 *The relationship between the type of dietary protein and the amount of ^{14}C benzo[a]pyrene in the protein fraction*

foci formation, colonic cell proliferation and intestinal polyp formation in Apc^{Min} mice are investigated.

Acknowledgement

The research described in this paper was financially supported by the Dutch Dairy Foundation on Nutrition and Health.

4 References

1 L. Williams, D. Hill, J. Davis and D. Lowenthal, *Eur. J. Drug Metab. Pharmacokinetics*, 1996, **21**, 201.

2 E. Vis, A. Plinck, G. Alink and M. van Boekel, *J. Agric. Food Chem.*, 1998, **46**, 3713.

3 J. Adler-Nissen, 'Enzymic Hydrolysis of Food Proteins', Elsevier Applied Science Publishers, New York, 1986.

2.14

Evaluation of Ethanol Activity in Food and Drink for Studying the Adverse Effects of Ethanol on Human Health

Carlo R. Lerici

DIPARTIMENTO DI SCIENZE DEGLI ALIMENTI, UNIVERSITY OF UDINE, VIA MARANGONI 97, 33100 UDINE, ITALY

Abstract

It is generally recognised that excessive alcohol consumption is a causative factor in several diseases. In general, risk and damage are considered a function of the amount of alcohol consumed, while 'how' alcohol is consumed is rarely considered.

In this paper, ethanol vapour pressure of model systems and of some widely consumed alcoholic beverages was measured in order to demonstrate that the effects of ethanol could be better described in terms of the physicochemical properties of ethanol, such as ethanol activity and ethanol vapour pressure, rather than of the sole ethanol content.

A gas-chromatographic method for measuring the ethanol vapour pressure and activity in food and drink, is also briefly described. In addition, vapour-liquid relationships are also discussed.

1 Introduction

The presence of ethanol in food and beverages is very common, either as a fermentation product or as an ingredient in beverages, spirits and, in the form of liqueur, in formulated foods. Small quantities of ethanol can be found in food as aroma carriers and/or additives for extending shelf life.[1] Furthermore, ethanol is always present in fruit and vegetables as a metabolite product of maturation and senescence. In terms of amount, ethanol in foods can vary from zero to more than 50% w/w in some spirits. Alcohol consumption differs within countries: intake can vary between almost zero (where drinking alcohol is prohibited) to around 10% total energy, even if individuals may consume far more. For millennia the consumption of alcoholic beverages has contributed to the pleasure of eating and drinking as well as to social interactions. Nowadays, excessive alcohol intake is generally recognised as a serious health and social risk. The epidemiological evidence that excessive alcohol consumption is a causative factor in several diseases is extensive and consistent. In particular, it increases the risk of mouth, pharyngeal, laryngeal and oesophageal cancers as well as that of primary liver cancer.[2]

Several pathways for the biological mechanisms by which alcohol might induce adverse effects on mouth and pharynx have been proposed.[2] Ethanol may damage cells, particularly cells in the upper gastrointestinal tract, and may modify the permeability of cell membranes thereby hastening the entry of carcinogens, or it may alter cell metabolism, thus leading to enhanced damage. Furthermore, alcoholic beverages may contain other toxic and carcinogenic substances. Finally, consumption may compromise nutritional status in a way that increases susceptibility to cancer.

However, in epidemiological studies, risk and damage are generally considered a function of the amount of alcohol consumed, while 'how' alcohol is consumed is rarely considered.

The 'state' of the ethanol molecule in aqueous systems, *i.e.* the ethanol 'availability' to act, can be evaluated by measuring its vapour pressure. In fact, vapour pressure of a volatile compound is defined as the pressure exerted by its vapour and can be considered an index of the overall interactions exerted by other compounds and by the environment. If the substance is within an enclosed space, the vapour pressure will reach an equilibrium value regulated by the well-known Raoult's law.

In this paper, ethanol vapour pressure of alcoholic model systems and of some widely consumed alcoholic beverages was measured in order to study how compositional and environmental factors affect the ethanol vapour pressure, and thus to measure the ethanol 'state' in the system. Furthermore, recent experimental data on the effects of ethanol on enzymatic activity in model systems are presented in order to ascertain that the action of ethanol can be best described in terms of its thermodynamic properties, such as activity and vapour pressure, rather than of the ethanol content alone.

2 Material and Methods

2.1 Sample Preparation

Two component model systems were prepared by mixing, in different weight ratios, anhydrous ethanol (Carlo Erba, Italy) and distilled water. The effects of a_w and water vapour pressures were studied in simplified ternary models in which ethanol concentration was kept constant, varying the ratio glycerol/water. In other words, while the ethanol concentration remained constant, the concentration of glycerol was progressively increased in the same quantity that water was decreased.

Beverages with different ethanol content were purchased on the local market.

In particular, non-alcoholic beers, with ethanol content less than 0.15 w/w, beers (4–8% w/w), wines (11.5–12.5% w/w), cream liqueurs (27–32% w/w), bakery liqueurs used as filling in bakery products (15–30% w/w) and spirits (33–40% w/w) were considered.

Volumes of 10 ml of each solution were put into 20 ml capacity vials that were subsequently hermetically sealed with butyl septa and metallic caps (Dani, Italy). Analyses were performed on samples stored at different temperatures for at least 24 hours in order to reach equilibrium conditions. Preliminary trials showed that the ethanol liquid-vapour equilibrium was established within 20 hours.

2.2 Ethanol Vapour Pressure Determination

The measurement of ethanol vapour pressure was carried out by means of headspace gas chromatographic (HSGC) analysis. The apparatus used was a GC mod 3700 (Varian, USA) equipped with a flame ionisation detector. A 2 m × 2 mm ID packed glass column filled with Carbowax 20M (5%) on Carbopack B (80–120 mesh) was used. The operating conditions were as follows: column temperature 80 °C, detector and injector temperature 200 °C, carrier gas (N_2) flow rate 20 ml min^{-1}. Analysis was performed using a Fisons gas-chromatograph (HRGC MEGA 2 Series, Fisons Instruments, Milano, Italy) equipped with an automatic sampler (Carlo Erba HS 250, Carlo Erba Strumentazioni, Milano, Italy) and a thermal conductivity detector (Fisons HWD Control, Fisons Instruments, Milano, Italy). Analyses were performed on samples conditioned at constant temperature in a thermostated bath. Ethanol peak areas were the average of at least three analyses; the variation coefficients were less than 5%.

The conversion of the gas chromatographic measurements into ethanol vapour pressure for non ideal systems under equilibrium conditions was performed according to the equation, derived from Raoult's law, proposed by Kolb:[3]

$$A = CX_e\gamma_e p_e^\circ \quad (T = \text{constant}) \tag{1}$$

in which, for a given temperature A is the ethanol gas chromatographic peak area, C is the response factor, X_e is the ethanol molar fraction, γ_e is the ethanol activity coefficient and p_e° is the vapour pressure of pure ethanol.

Since Raoult's law states:

$$p_e = X_e\gamma_e p_e^\circ \quad (T = \text{constant}) \tag{2}$$

combinings equation (1) and (2), the ethanol vapour pressure (p_e) is given by:

$$p_e = \frac{A}{C} \tag{3}$$

The response factor C was calculated from the pure ethanol peak area using the pure ethanol vapour pressure values at given temperatures, obtained from the literature.[4]

2.3 Water Activity Determination

Because of the interference of ethanol on the water activity (a_w) hygrometric measurement, the a_w values of water-ethanol solutions were calculated on the basis of the freezing point depression (FPD) according to the equation proposed by Lerici *et al.*:[5]

$$-\ln a_w = 722.37\frac{1}{FPD} - 2.644 \quad (T = 25\,°\mathrm{C}) \tag{4}$$

The values of p_w were calculated from a_w data:

$$p_w = a_w p_w{}^{\circ} \tag{5}$$

where $p_w{}^{\circ}$ is the pure water vapour pressure at the given temperature.

2.4 Vapour-Liquid Equilibrium Relationships

The fundamental thermodynamic equation relating activity coefficients and composition for a binary mixture is the well-known Gibbs-Duhem equation that, for ethanol(e)-water(w) system, can be written:

$$X_e\left(\frac{\partial \log \gamma_e}{\partial X_e}\right)_{T,P} + X_w\left(\frac{\partial \log \gamma_w}{\partial X_w}\right)_{T,P} = 0 \tag{6}$$

where X is the molar fraction and γ the activity coefficient.

Many solution models, which relate activity coefficients to liquid composition and satisfy the Gibbs-Duhem equation have been proposed.[6]

The third-order Margules equations for the ethanol-water binary system are:

$$\log \gamma_e = X_w^2[A_{e,w} + 2X_e(A_{w,e} - A_{e,w})] \tag{7}$$

$$\log \gamma_w = X_e^2[A_{w,e} + 2X_w(A_{e,w} - A_{w,e})] \tag{8}$$

Considering that in a binary system $X_w = 1 - X_e$, Margules equation can be written:

$$\log \gamma_e = (1 - X_e)^2[A_{e,w} + 2X_e(A_{w,e} - A_{e,w})]. \tag{9}$$

which can be developed and rearranged as:

$$\log \gamma_e = a + bX_e + cX_e^2 + dX_e^3 \tag{10}$$

where a, b, c, and d are parameters related to the constants $A_{e,w}$ and $A_{w,e}$.

Experimental data relevant to model systems were fitted using the expression obtained combining the equations (2) and (11):

$$p_e = X_e \cdot p_e^* \cdot 10^{(a+bX_e+cX_e^2+dX_e^3)} \tag{11}$$

From vapour pressure measurements, if are known ethanol and water molar fractions and vapour pressure of pure compounds at a given temperature, it is possible then to calculate other thermodynamic properties, such as the following.[4]

Activity coefficient: $$\gamma_e = \frac{p_e}{p_e{}^\circ X_e}; \quad \gamma_w = \frac{p_w}{p_w{}^\circ X_w}$$

Activity: $$a_e = \frac{p_e}{p_e^*}; \quad a_w = \frac{p_w}{p_w^*}$$

3 Results

In Figure 1 the influence of temperature on ethanol vapour pressure (p_e) of aqueous model systems having increasing ethanol content is shown. As expected, temperature strongly affected the ethanol partition from the liquid to vapour phase. The curves show p_e values calculated by fitting experimental data by equation (12) ($r^2 > 0.99$).

Besides ethanol concentration and temperature, the chemical composition of the system, and especially the water content, can also affect the liquid-vapour

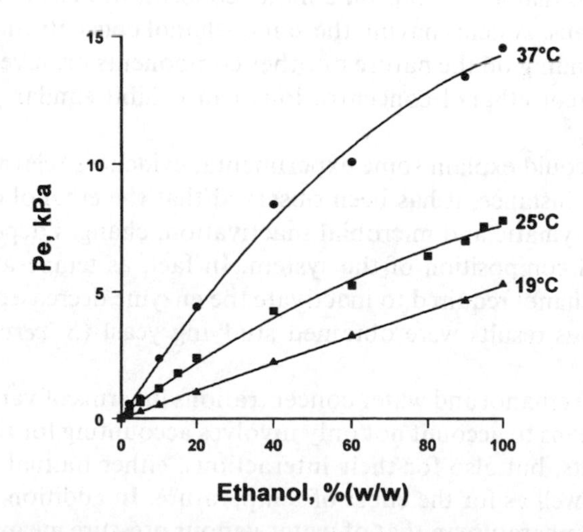

Figure 1 *Ethanol vapour pressure of water-ethanol mixtures at 19, 25 and 37°C as a function of ethanol concentration. Curves indicate ethanol vapour pressure values obtained by fitting experimental data by equation (12) ($r^2 > 0.99$, $p < 0.05$)*

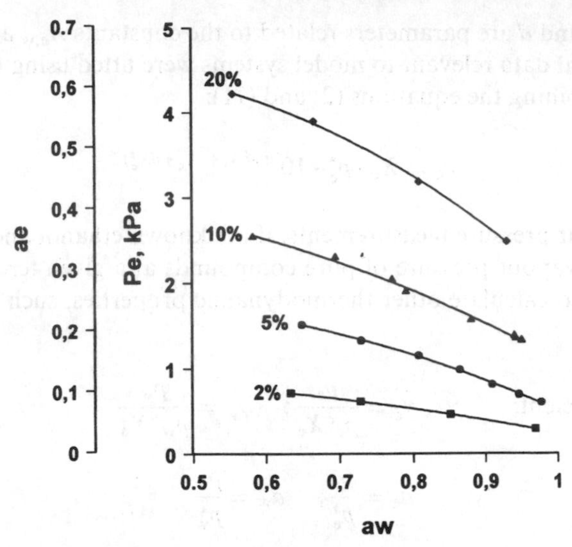

Figure 2 *Ethanol vapour pressure and ethanol activity of ternary systems having constant ethanol concentration (2, 5, 10 and 20% w/w) and different ratio glycerol/water as a function of the* a_w

partition of ethanol. In Figure 2 the changes in p_e and in ethanol activity (a_e) of mixtures containing a constant amount of ethanol and, in order to change the water activity of the systems, different quantities of water and glycerol are plotted as a function of water activity (a_w). According to previous results obtained either in model systems, or in intermediate moisture foods,[7] it can be noticed that, for constant ethanol concentrations, the p_e increased as the a_w was reduced. It means that p_e depends on ethanol concentration as well as on the a_w of the system. Thus, systems having the same ethanol concentration can exhibit different p_e depending on the nature of other components or, inversely, systems containing different ethanol concentrations can exhibit similar p_e because of their different a_w.[8]

These results could explain some experimental evidences relevant to biological systems. For instance, it has been observed that the ethanol concentration which causes enzymatic and microbial inactivation, changes depending on the temperature and composition of the system. In fact, as temperature increases the amount of ethanol required to inactivate the enzyme decreased up to zero at 70 °C.[9] Analogous results were obtained studying yeast (*S. cerevisiae*) inhibition.[10]

Expressing the ethanol and water concentrations in terms of vapour pressures (or activities) means to account not only involves accounting for the amounts of these components, but also for their interactions, either mutual or with other components, as well as for the effect of temperature. In addition, passing from the concept of temperature to that of water vapour pressure means to pass from an environmental variable, the temperature, to a compositional one, water vapour pressure.

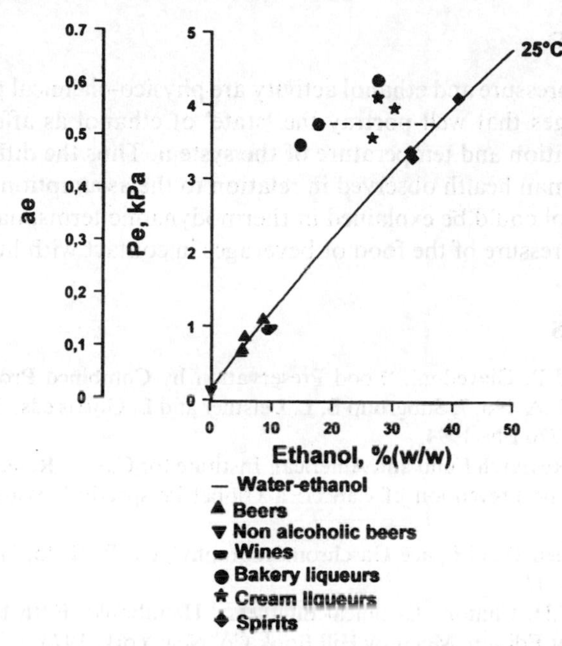

Figure 3 *Ethanol vapour pressure and ethanol activity of different alcoholic beverages as a function of ethanol concentration at 25 °C; bold line refers to water-ethanol solution.*

In Figure 3 p_e and a_e of different groups of alcoholic beverages are reported as a function of ethanol concentration. It can be noted that some beverages, despite having the same ethanol content, present strongly different p_e values. Such differences can be attributed to the different composition of the beverages, since the nature of the 'third components' (non-water and non-ethanol) affects the physico-chemical properties of the system.

It is likely that solutes contained in beverages, modifying p_e, could affect the action of ethanol. In other words, the presence of solutes such as sugars in a drink having a 20% (w/w) ethanol content would increase the p_e so that the impact could be comparable to that of a 40% (w/w) alcoholic drink (see dashed line in Figure 3). Consequently, the abrasive effect on mouth and pharynx would dramatically increase if the alcoholic beverage containing sugars is hot when consumed.

Ethanol is probably the most commonly studied and applied anti-microbial compound. It is known that ethanol interacts with biological membranes resulting in a decreased membrane integrity. As referred by Laposata,[11] fatty acid ethyl esters (FAEE) have an important role in mediating cell injury. The synthesis of FAEE in natural killer cells would be a causative factor in the inhibition of immunological activity and the promotion of tumour metastasis. Recently, Guerzoni *et al.*[12] observed that the increase of vapour pressure of the intracellular ethanol in *S. cerevisiae* under specific experimental conditions was associated to the *in vivo* formation of FAEE.

4 Conclusion

Ethanol vapour pressure and ethanol activity are physico-chemical properties in food and beverages that well portray the 'state' of ethanol as affected by the chemical composition and temperature of the system. Thus the different effects of alcohol on human health observed in relation to the assumption of the same amount of ethanol could be explained in thermodynamic terms, namely by the ethanol vapour pressure of the food or beverages in contact with human cells.

5 References

1 C.R. Lerici and P. Giavedoni, 'Food Preservation by Combined Processes', Final report FLAIR C.A. No. 7, Subgroup B, L. Leistner and L. Gorris eds., ISBN 90-900-7303-5 EUR 15776 EN, 1994.
2 World Cancer Research Fund and American Institute for Cancer Research in 'Food, Nutrition and the Prevention of Cancer: a Global Perspective', Washington, DC, 1997.
3 B. Kolb, 'Applied Head Space Gaschromatography', ed. B. Kolb, Heyden & Son Ltd, 1980, pp. 1–11.
4 H. Perry and C.H. Chilton, 'Chemical Engineers' Handbook', Fifth Edition, International Student Edition, McGraw Hill Book Co, New York, 1973.
5 C.R. Lerici, M. Piva and M. Dalla Rosa, *J. Food Sci.*, 1983, **48**, 6, 1667–1669.
6 J.H. Pemberton and C.J. Mash, *J. Chem. Thermodynamics*, 1978, **10**, 887–888.
7 M. Dalla Rosa, P. Pittia and M.C. Nicoli, *Ital. J. Food Sci.*, 1994, **4**, 421–432.
8 C.R. Lerici and M.C. Nicoli, *Adv. Food Sci.*, 1996, **18**, 5/6, 229–233.
9 C.R. Lerici and L. Manzocco, *J. Food Sci.*, 1999 (submitted).
10 M.E. Guerzoni, M.C. Nicoli, R. Massini and C.R. Lerici, *World J. Microb. Biotech.*, 1997, **13**, 11–16.
11 M. Laposata, *Prog. Lipid Res.*, 1998, **37**, 307–316.
12 M.E. Guerzoni, L. Lanciotti, F. Gardini and M. Ferruzzi, 1999, *Can. J. Microbiol.* (in press).

2.15

Application of the TNO *In vitro* Gastrointestinal Model for Research on Food (Anti)-Mutagens

Cyrille Krul,[1] Anja Luiten-Schuite,[1] Rob Baan,[2] Hans Verhagen[1] and Robert Havenaar[1]

[1] TNO NUTRITION AND FOOD RESEARCH INSTITUTE, PO BOX 360, 3700 AJ ZEIST, THE NETHERLANDS
[2] INTERNATIONAL AGENCY FOR RESEARCH ON CANCER, 150 COURS ALB. THOMAS, 69372 LYON, FRANCE

1 Introduction

On the one hand food may contain mutagenic/carcinogenic compounds (*e.g.* heterocyclic amines in meat), and on the other protective, anti-mutagenic substances (*e.g.* from vegetables and tea). To analyse the fate of food through the GI tract and to estimate the possible risks or beneficial effects of these food components an *in vitro* model which simulates the human digestion would be helpful. TNO gastro-Intestinal Model (TIM) can contribute to a reduction in the number of experimental animals in research on the efficacy and safety of (new) food products.[1]

2 Materials and Methods

TIM is a dynamic, computer-controlled system that mimics the human physiological processes of the stomach and small intestine. In the model the pH, temperature, peristaltic movements, secretion of digestion enzymes, bile and pancreatic juices and absorption of digestion products are simulated.[2] TIM was fed with green and black tea extract (with or without milk or breakfast) to investigate the antimutagenic activity of tea TIM (Table 1). Samples were taken at various time points from the jejunal and ileal dialysates and were tested for mutagenicity in the Ames test with *Salmonella* TA98 and S9-mix. MeIQx was used as model compound for food mutagens.

The potential antimutagenicity in the jejunal dialysates (200 μl/plate) of

Table 1 *Experiments performed with TIM*

Experiment	*1*	*2*	*3*	*4*	*5*	*6*	*7*
(Anti)-mutagen	Green tea[a]	Black tea[a]	Black tea[a]	Black tea[a]	Black tea[a]	MeIQx[b]	Black tea[a] + MeIQx[b]
Matrix	buffer	buffer	milk[c]	milk[d]	breakfast[e]	buffer	buffer
MeIQx in Ames assay	yes	yes	yes	yes	yes	no	no

[a] Soluble tea extract dose: 11 g/300 ml; [b] Dose: 90 g/300 ml; [c] Milk with 3.5% fat; [d] Skimmed milk 0% fat; [e] Brown and white bread (30%), cheese (10%), marmalade (7.5%), butter (5%) and water (47.5%).

Experiments 1–5 was measured against MeIQx (4 ng/plate) induced mutagenicity. In Experiments 6 and 7 MeIQx was directly introduced into TIM with or without tea. 200 μl of the dialysates were measured without addition of MeIQx.[2]

3 Results

The jejunal dialysates of the experiments with green and black tea resulted in 50–100% inhibition between 90 and 160 min. The highest inhibition was measured 2 h after the start of the digestion. No effect was found in the ileal dialysates and there were no significant differences between the dialysates with green or black tea (Figure 1).

In Figure 2 the results of jejunal dialysates with black tea, black tea with milk (0 and 4% fat) or breakfast (containing *e.g.* cheese with 40% fat and butter) are shown. Fatty milk and breakfast inhibit the antimutagenic activity of tea almost completely, unlike skimmed milk.

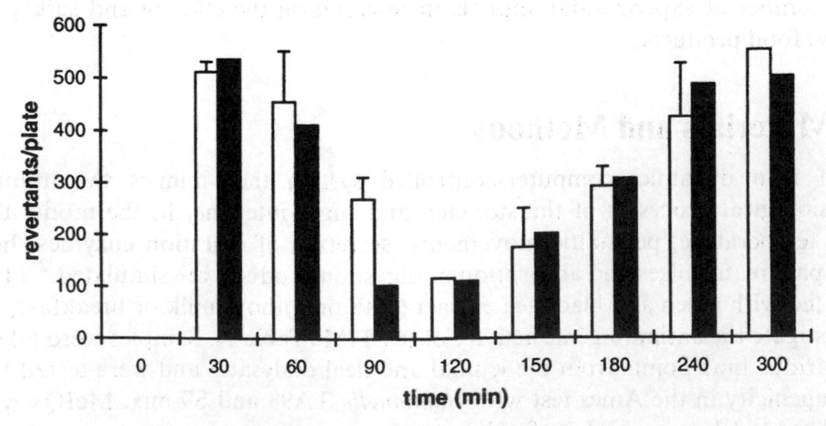

Figure 1 *Mutagenicity in jejunal dialysates (200 µl/plate) of experiments with black tea (black), green tea (white)*

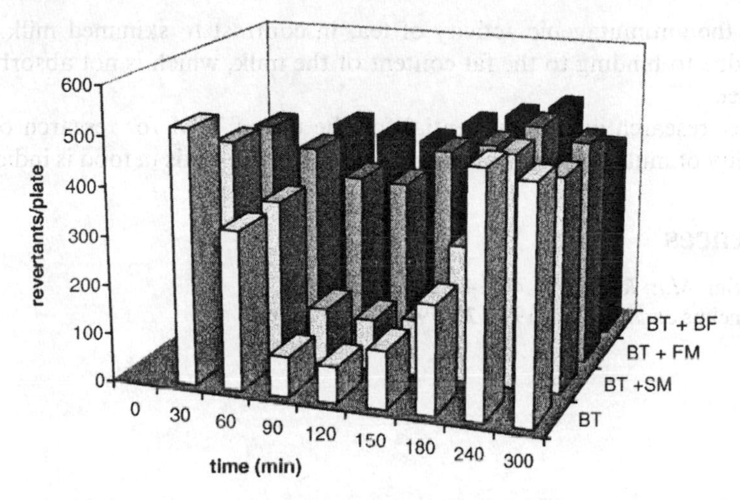

Figure 2 *Mutagenicity in jejunal dialysates (200 μl/plate) of experiments with black tea (BT), black tea and skimmed milk (BT + SM)), black tea and fatty milk (BT + FM) and with breakfast (BT + BF)*

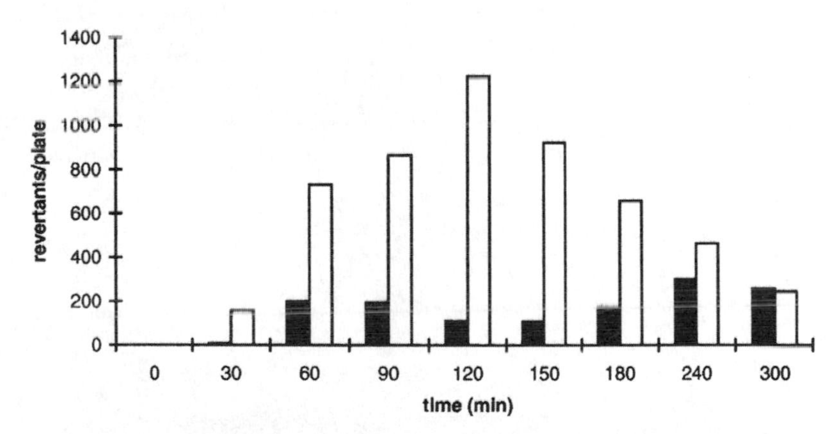

Figure 3 *Mutagenicity in jejunal dialysates (200 μl/plate) of experiments with MeIQx (white) and MeIQx with black tea (black)*

When MeIQx and tea were added together into TIM in the jejunal dialysates the mutagenicity was lower, compared to MeIQx alone (Figure 3). More than 50–90% inhibition was measured from the start of the experiment until 200 minutes.

4 Conclusion

Green as well as black tea retained its antimutagenic potential upon digestion in the model. In addition to the inhibition of activation enzymes the antimutagenic activity could be due to direct binding of mutagens. Fatty milk and breakfast reduced the antimutagenic activity of tea, in contrast to skimmed milk. This may be due to binding to the fat content of the milk, which is not absorbed in the model.

Further research into the potential for the use of TIM for research on the availability of mutagenic and/or antimutagenic compounds in food is indicated.

References

1 DeMarini, *Mut. Res.*, 1998, **400**, 457.
2 M. Minekus *et al.*, *ATLA*, 1995, **23**, 197.

Section 3

DNA Damage and Repair

3.1

Diet and DNA Damage

Siegfried Knasmüller, Fekadu Kassie, Hans Steinkellner, Gerhard Hietsch and Wolfgang Huber

INSTITUTE OF CANCER RESEARCH, UNIVERSITY OF VIENNA, AUSTRIA

1 Introduction

For more than 40 years attempts have been made to identify environmental factors which affect the stability and structure of genes. According to the multistep hypothesis of cancer formation, DNA damage in somatic cells leads to formation of initiated cells which may develop into cancer. Genotoxic effects in germ cells decrease fertility and cause heritable diseases. Initially, research focused on radiation effects and synthetic chemicals. In the last two decades, the number of studies which concern dietary factors has increased dramatically. It has been postulated that 20–60% of human cancers are related to dietary factors[1,2] and evidence is accumulating that vegetables contain a large number of DNA protective constituents. In the present contribution we will try to point out current developments and problems and discuss areas where research is needed (Figure 1).

2 DNA Reactive Compounds Formed During Food Processing

Three major groups of DNA reactive constituents are formed during cooking which cause DNA damage and its consequences, namely nitrosamines (NA), polycyclic aromatic hydrocarbons (PAH) and heterocyclic aromatic amines (HAA). These compounds have been studied very intensively and their formation as well as their biological effects are well understood. In the case of the NA and HAA it is conceivable that they contribute to human cancer risks,[3,4] the PAH concentration in Western foods is quite low and dietary exposure to these compounds does probably not pose a cancer risk relevant for man.[5]

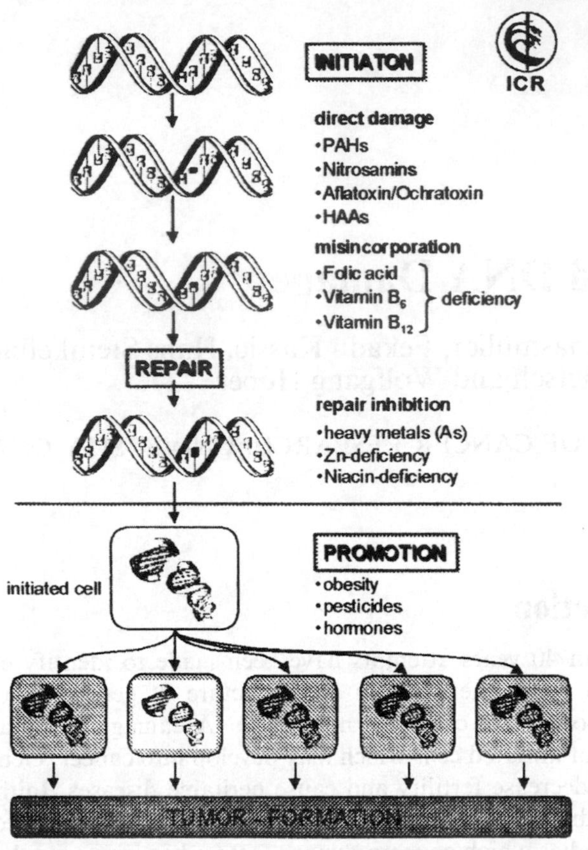

Figure 1 *Mechanism of cancer formation*

3 Mycotoxins

Molecular cloning techniques led to the detection of fingerprint mutations in the p53 tumor suppressor gene and provided evidence that contamination of foods with Aflatoxin B_1 is causally related to increased incidence of liver cancer in certain parts of Africa and in East Asia, whereas it plays apparently no role in industrialized parts of the world.[6] Ochratoxin A is widespread in foods of eastern Europe. It causes Balkan endemic nephropathy and an increase of urinary tract tumors.[7] Although this mycotoxin was initially classified as a non genotoxic carcinogen as no mutagenic effects were detected in conventional test systems, it could be shown in DNA adduct measurements that it is transformed to DNA reactive metabolites.[7] We recently tested different *Fusarium* toxins (fumonisin B1, monilifomin and vomitoxin) and found no effects in bacterial tests but a clear increase of chromosmal aberrations was seen in tests with primary rat hepatocytes at doses $< 5\,\mu g\,ml^{-1}$.[8] All these toxins have been found frequently in cereals in relative large concentrations;[6] therefore these observations merit further attempts to elucidate their potential health hazards in humans.

4 Natural Pesticides

Ames *et al.* and Gold[9,10,11] published a number of papers in which they stress that a high percentage of natural pesticides (> 50%) contained in plants which are of nutritional relevance for humans cause cancer in rodent bioassays. They also emphasize that their tumorigenic potencies equal or exceed those of synthetic compounds and that human exposure to natural carcinogens is far higher than to industrial chemicals whose toxic properties are more intensely studied. Most of the natural carcinogens they mention are DNA reactive compounds. However, numerous epidemiological studies give evidence that consumption of vegetables (which contain these genotoxic carcinogens) is associated with a reduced incidence of cancer in various organs.[12] This argues against the assumption that these natural compounds indeed have an adverse effect on human health. Carcinogens contained in common spices might be consumed in quantities which are too low to affect humans. For a number of compounds it has been shown that they even cause protection (*i.e.* antimutagenic/anticarcinogenic effects) in rodents at low doses, although they are carcinogenic and/or genotoxic at elevated doses.

5 Overestimation Due to Inadequate Testing?

Ames and Gold[9] noted that the effects of plant constituents and synthetic chemicals might be strongly overestimated by standard high dose animal cancer tests. The currently used testing strategy is that the rodents are exposed chronically to near-toxic dosed (NTD) and this may cause wounding of tissues, cell death and continous cell division of neighboring cells which is a risk factor for cancer.

Also, most of the currently used *in vitro* genotoxicity tests are not adequate for comparative quantitative risk assessments. As the indicator cells are devoid of enzymes involved in the metabolism of dietary genotoxins, exogenous enzyme homogenates are added, which catalyze phase I reactions but not conjugations (by phase II enzymes which lead in most cases to detoxification of DNA reactive metabolites). As an example, Figure 2 depicts the activation/ detoxification of HAAs in cells currently used in routine genetic toxicology tests *in vitro* (which require addition of liver enzyme homogenates) and in intact primary liver cells. Recently, successful attempts have been made to develop genotoxicity tests with metabolically competent cells (*e.g.* with the human liver cell lines Hep G2 or primary hepatocytes) which reflect the metabolism and DNA damaging potential of xenobiotics in mammals more adequately.[13,14]

A typical example for dietary genotoxins whose DNA damaging potential is overestimated on the basis of conventional *in vitro* tests are isothiocyanates (ITCs) . These are breakdown products of glucosinolates which are constituents of cruciferous vegetables. ITCs are among the most potent *in vitro* genotoxins and clastogens in bacteria and in mammalian cells ever detected.[15] The lowest effective level for allyl-ITC, which is a bladder carcinogen in rats[16] in *in vitro* tests (DNA repair tests with *E. coli* strains, chromosomal aberration assays with

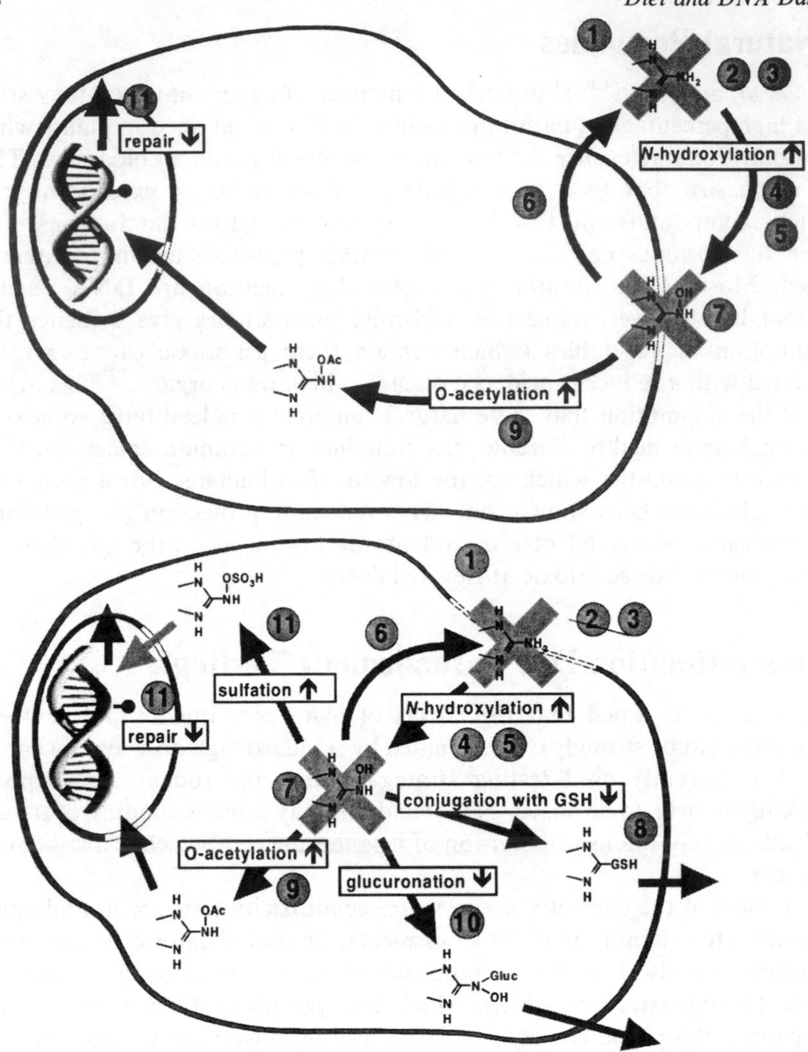

Figure 2 *Schematic representation of the metabolism of HAAs* (1a) *in cells used in conventional routine mutagenicity testing (bacteria, V79, CHO) that need addition of activation enzyme mixtures to catalyze the activation of dietary promutagens or* (1b) *in metabolically competent cells (primary hepatocytes, human Hep G2 cells) which by themselves do posses phase I and phase II enzymes. Arrows indicate reactions which either lead to formation (↑) of DNA alterations/mutations or (↓) decrease of DNA alterations. Numbers represent different antigenotoxic mechanisms:* (1) *inhibition of formation [different common spices],* (2) *direct binding [fibers, pyrroles],* (3) *enzymatic destruction [horseradish peroxidase],* (4/5) *direct or indirect inhibition of activation by cytochrome P450 enzymes [phenolics, flavonoids],* (6) *reversion to the parent compound [BHA derivative]* (7) *inactivation of N-hydroxy amines [pyrroles, consituents of beer], induction of GST [kawheol/cafestolpalmitate],* (9) *inhibition of N-hydroxylation [kawheol/cafestolpalmitate; EGCG],* (10) *induction of glucuronidation [isothiocyanates], interactions with repair/replication processes [flavorings: vanillin, cinnamaldehyde, coumarin]*

Indian muntiak cells sister chromatid exchange assays with Chinese hamster ovary cells) are between 0.2 and 2.5 $\mu g\,ml^{-1}$.[15,17,18] In human HepG2 cells and in primary rat hepatocytes which, in contrast to these cells, express several phase II enzymes, including GST that is involved in the detoxification of ITCs, the LOEL in single cell gel electrophoresis (SCGE) assays and micronucleus tests is about 2–3-fold higher. In bacterial animal mediated assays with mice (indicator cells *E. coli*) and in SCGE tests with rats (target organ liver), the corresponding values were 270 and 220 mg/kg bw. respectively.[15,19] Similar discrepancies between the *in vitro* and *in vivo* situation were seen with other ITCs,[20] (see also this volume). We could demonstrate that the genotoxic effects of ITCs are strongly reduced by non-enzymatic protein binding, by radical scavengers and also by body fluids such as gastric mucus and saliva. All these factors contribute to the detoxification of these natural genotoxins and overall it is apparent that the ITC doses required to cause DNA damage in rodents are far higher than the exposure levels of humans (overall glucosinolate intake levels are between 40 and 50 mg/P/day[21,22]).

6 DNA Reactive Carcinogens and Human Cancer Incidence

Lutz and Schlatter[23] used epidemiological data and average daily intake values of the Swiss nutrition report and attempted to attribute diet related cancers to different nutritional factors. Based on the postulate of Doll and Peto,[1] they assumed that about one third of the cancer incidences (*ca.* 80 000/10^6) are due to dietary factors. When they used the TD_{50} values of the NTP Carcinogenicity Potency Database they could attribute only a few hundred cases (200–300) to DNA reactive carcinogens such as mycotoxins, HAAs, PHAs and *N*-nitroso compounds. Uptake of arsenic (which affects DNA repair functions and thus can be classified an indirect genotoxin) accounted for several hundred cases and about 8000 incidences could be attributed to alcohol consumption. Note that ethanol is converted to the DNA reactive metabolite acetaldehyde and that alcoholics have increased rates of chromosomal aberrations in white blood cells.[24] The effects of common contaminants such as plastic monomers, pesticides, or drugs *etc.* (which are widely discussed as hazardous in the public) were classified as negligible. The top risk factor in this calculation turned out to be overnutrition, and the authors postulated that up to 60 000 cases/10^6 might be due to obesity. Again this estimate was based on animal data. It is unclear if all the assumptions of this global assessment are relevant for humans. The calculated obesity effects are not in agreement with currently available epidemiological data on cancer incidence in humans and it is possible that the estimation which is based on the addition of the effect of individual factors is too simplistic. Note that the effects caused by overweight are at least partly due to hormonal imbalances which are considered to affect the promotion of (initiated) cells formed as a consequence of DNA alterations. This situation is not reflected in the conventional carcinogenicity experiments with rats on which the risk assessment of Lutz and Schlatter is based.

7 Deficiency of Micronutrients in the Human Diet

About 40 micronutrients are required in small amounts in the human diet. It has been estimated that an unexpectedly high percentage of the USA population are deficient in certain micronutrients, *e.g.* in folic acid, vitamins C, E, B_6 and B_{12}, iron, zinc and niacin. Lack of these micronutrients may cause an increase of DNA alterations. The B vitamins and folic acid are involved in the *de novo* synthesis of nucleotides and deficiencies might cause misincorporation of uracil instead of thymidine.[25] Indeed it was possible to demonstrate that in humans folate deficiency is associated with increased micronucleus frequencies in blood cells and uracil misincorporation in spleenocytes.[26] The assumption that folate deficiency mediated uracil misincorporation in nerve cells leads to neural disorders in man and affects the integrity of germ cells in humans[25] is not experimentally verified. It has also been postulated that zinc deficiency affects the function of enzymes involved in DNA repair/replication and lack of niacin may disturb the poly-ADP ribose protective response to DNA damage,[25] but it is not known if decreased concentrations of the micronutrients in the diet may cause DNA alterations in humans. However, most research activities focused in the past on the identification of genotoxic carcinogenic dietary compounds and it is conceivable that indirect effects caused by lack of micronutrients play an important role that has been underestimated so far.

8 DNA Protective Effects of Dietary Constituents

In recent years, strong attempts have been made to identify dietary compounds which protect against DNA damage caused by environmental or intrinsic factors. Table 1 lists the most important mechanisms and gives examples for DNA protective dietary constituents.

The basic assumption has been that methods which are useful for the detection of genotoxins can be used for the identification of antimutagens as well. This paradigm is no longer valid; it became apparent that many antigenotoxic modes of action are not adequately represented in conventional *in vitro* tests which require addition of enzyme homogenates to activate dietary carcinogens (aflatoxins, PAHs, HAA or NA). As a consequence, false positive or false negative results may be obtained with these test systems. Typical examples are ethanol, which causes protective effects in bacteria *in vitro* in combination with nitrosamines but acts synergistically *in vivo*, or the food constituents quercetin, tannic acid, anthracene and anthraflavic acid that are protective in *in vitro* tests towards HAA but increase their DNA damaging effects in mammals (for details see Schwab *et al.*[27]).

Table 2 gives an overview of different methods and their predictive value for the identification of DNA protective agents. Human studies are not contained in the Table. The most frequently used human biomonitoring methods are

Table 1 *Examples for DNA protective effects of dietary constituents*

Mechanism	*Example*	*Evidence for protection in humans*
Inhibition of formation of DANN reactive compounds	Inhibition of nitrosation reactions by plant phenolics and vitamins	Reduced excretion of N-nitroso-proline
Inactivation of genotoxins by direct binding	HAA binding by fibers and pyrrols	No
Inhibition of enzymatic activation of carcinogens	Inhibition of nitrosamine activation by isothiocyanates, inhibition of aromatic amine genotoxicity via inhibition of sulfotransferases	Altered HAA excretion in humans by CYP1A inhibition after oral administration of furafylline
Induction of phase II enzymes	GST induction and UDPGT induction led to detoxification of PAHs, and HAAs	GST induction by vegetable diets in humans (see Steinkellner *et al.*, this publication)
Trapping of reactive metabolites	Inactivation of BPDE by phenolic acids, detoxification of HAAs by phenolics	No
Scavenging of reactive radicals	Prevention of radical mediated damage by phenolics and vitamins	Change of excretion of oxidized bases in the urine by vitamins
Interaction with DNA repair/replication	Antimutagenic effects of flavorings towards direct acting agents in bacterial tests	No

chromosomal analyses and MN experiments or single cell gel electrophoresis assays with peripheral blood cells or epithelial cells. These experimental models have been used mainly in intervention studies with vitamins but rarely with other dietary constituents. Urinary mutagenicity assays can be used to obtain information on the intake of compounds (*e.g.* of HAAs) by the diet and also to elucidate if dietary factors cause a change of the excretion pattern. We have recently studied the urinary bacterial mutagenicity after large hamburger meals with a *Salmonella* strain which is highly sensitive to HAAs (YG1024) and found that continuous consumption of red cabbage (300 g per day/P, 5 days) resulted in a pronounced (*ca.* 40%) reduction of free and an increase of conjugated mutagenicity. HPLC analysis of N-nitrosoproline in human urine has been used to demonstrate that the uptake of compounds that prevent nitrosation reaction indeed causes a decrease of endogenic nitrosamine formation in humans;[28] and the analyses of oxidized bases (8-hydroxguanosine) allowed to gain insight into the DNA protective effects of antioxidants in humans.[29] Another promising approach concerns investigations of the induction of DNA protective enzymes

Table 2 *Use of methods for the identification of antigenotoxic dietary constituents*

Test system	*Use*	*Comments*
In vitro tests		
Bacterial mutagenicity tests	Extensively used in antimutagenicity studies (>80%)	Do not reflect inhibition of activation and detoxification processes, predictive value limited many false positive results
Conventional mammalian cell line (CHO, V79)s	Rarely used in antimutagenicity studies	Limitations similar as for bacterial tests
Human HepG2 cells	Newly developed model	Reflect (i) induction of certain CYP450 isoenzymes and (ii) induction of GST and UDPGT13
Primary hepatocytes	Rarely used in antimutagenicity studies	Phase I and II enzymes represented (not stable), cultivation time consuming, availability of human material limited
Genetically engineered cells	Rarely used in AM studies	Useful for mechanistic studies, but only individual enzymes represented in a non inducible form
In vivo tests		
MN/CA in blood cells and bone marrow	Rarely used, negative or only weak effects with nitrosamines and HAAs	
DNA-adduct measurements	Most frequently used *in vivo* method (approximately 15% of AM studies)	Require use of radiolabelled material, adduct levels in reflect organ specific genotoxic effects, highly specific, 'side effects' of antimutagens are not detected
SCGE tests *in vivo*	New methods[20] (see also Kassie *et al.*, this publication)	Promising results with HAAs and glucosinolates reflect organ specific genotoxic effects, detection of side effects of putative antimutagens possible

by plant derived dietary constituents. The contribution of Steinkellner (this publication) describes the results of intervention studies in which induction of GST activity by cruciferous vegetable diets was found.

It has been possible to identify a number of dietary compounds which are protective towards a panel of genotoxic carcinogens such as fibers, pyrrole pigments, beverages, dairy products containing lactic acid bacteria and certain vitamins in experiments with rodents. Indeed results of epidemiological studies suggest that their consumption is associated with a decreased cancer risk in humans, but it is unclear at present if their protective role is causally related to the inactivation of DNA reactive carcinogens. To further clarify this question and also for the development of plant derived foods with high protection capacities, the development and use of improved experimental models is urgently needed.

9 References

1 R. Doll and R. Peto, *J. Natl. Canc. Inst.*, 1981, **66**, 1191.
2 R. Doll, *Canc. Res. (Suppl.)*, 1992, **52**, 2024.
3 G. Eisenbrand, 'N-Nitrosoverbindungen in Nahrung und Umwelt', Wiss. Verlags. Ges. Stuttgart, 1981.
4 R.H. Adamson, U.P. Thorgeirsson, and T. Sugimura, *Arch. Toxicol. Suppl.*, 1996, **18**, 303.
5 P.G.N. Kramers, and C.A Van der Heiden, *Toxicol. Env. Chem.*, 1988, **16**, 341.
6 IARC, 'Some Naturally occurring Substances', Monographs Vol. 56, Lyon, 1993.
7 IARC Mycotoxins, Endemic Nephropathy and Urinary Tract Tumors, Scientific Publ., Vol. 115, Lyon, 1991.
8 S. Knasmüller, N. Bresgen, F. Kassie, V. Mersch Sundermann, W. Gelderblom, and P.M. Eckl, *Mutat. Res.*, 1997, **391**, 39.
9 B.N. Ames and L.S. Gold, *FASEB J.*, 1997, **11**, 1041.
10 B.N. Ames, *Proc. Natl. Acad. Sci. USA*, 1990, **87**, 7777.
11 B.N. Ames, L.S. Gold and W.C. Willett, *Proc. Natl. Acad. Sci., USA*, 1992, **92**, 52.
12 K.A. Steinmetz and J.D. Potter, *Cancer Causes Contr.*, 1991, **2**, 325.
13 S. Knasmüller, W. Parzefall, R. Sanyal, S. Ecker, C. Schwab, M. Uhl, V. Mersch Sundermann, G. Williamson, G. Hietsch, T. Langer, F. Darroudi and A.T. Natarajan, *Mutat. Res.*, 1998, **402**, 185.
14 C. Schwab, F. Kassie, H.M. Quin, R. Sanyal, M. Uhl, G. Hietsch, S. Rabot, F. Darroudi, S. Knasmüller, *J. Env. Pathol. Toxicol. Oncol.*, 1999, **18**, 109.
15 F. Kassie, Genotoxic and antigenotoxic effects of bioactive substances contained in cruciferous vegetables with particular emphasis on isothiocyanates, Thesis, Vienna.
16 K.K. Dunnick, J.D. Prejan, J. Hasemann, R.B. Thompson, H.D. Giles and E.E. McCornell, *Fundam. Appl. Toxicol.*, 1982, **2**, 114.
17 F. Kassie, S. Musk, W. Parzefall, I. Johnson, G. Sontag, R. Schulte Hermann and S. Knasmüller, in 'Bioactive Substances in Foods of Plant Origin', ed. Kozlowska *et al.*, Polska Akademia Nauk, Vol. 2, 501–508, 1994.
18 S.R.R. Musk and I.T. Johnson. *Mutat. Res.*, 1993, **300**, 111.
19 S. Knasmüller *et al.*, unpublished.
20 F. Kassie, B. Pool-Zobel, W. Parzefall, and S. Knasmüller, *Mutagenesis*, 1999, in press.
21 K. Sones, R.K. Heany and G.R. Fenwick, *J. Sci. Food Agric.*, 1984, **35**, 712.
22 R. Lange, R. Baumgrass, M. Diedrich and M. Henschel, *Ernähr. Umsch.*, 1992, **39**, 292.
23 W.K. Lutz and J. Schlatter, *Carcinogenesis*, 1992, **13**, 2211.
24 IARC, 'Alcohol Drinking', Monographs, Vol. 44, Lyon, 1988.
25 B.N. Ames, *Toxicol. Lett.*, 1998, **102**, 5.
26 B.C. Blount, M.M. Mack, C.M. Wehr, J.T. MacGregor, R.A. Hiatt, G. Wang, S.N. Wickramasinghe, R.B. Everson and B.N. Ames, *Proc. Natl. Acad. Sci. USA*, 1997, **94**, 3290.

27 C.E. Schwab, W. Huber, W. Parzefall, G. Hietsch, F. Kassie, R. Schulte Hermann and S. Knasmüller, *Crit. Rev. Toxicol.*, submitted.
28 IARC, 'Relevance to Human Cancer of *N*-Nitroso-compounds, Tobacco Smoke and Nitrosamines', IARC Sci. Publ., 105, Lyon, 1991.
29 H.J. Helbock, K.B. Beckman, M.K. Shigenaga, P.B. Walter, A.A. Woodall, H.C. Yeo and B.N. Ames, *Proc. Natl. Acad. Sci. USA*, 1998, **95**, 288.

3.2

The Development of DNA Repair Assays Which Show That Dietary Carrots Stimulate DNA Repair Activity

Ruan Elliott, Siân Astley, Susan Southon and David Archer

INSTITUTE OF FOOD RESEARCH, NORWICH RESEARCH PARK, COLNEY, NORWICH NR4 7UA, UK

1 Introduction

Oxidative damage to DNA is proposed to play a central role in the development of some forms of cancer. There is convincing epidemiological evidence that the habitual consumption of diets rich in fruits and vegetables is associated with a reduced risk of many cancers. The 'Antioxidant Hypothesis' proposes that components of these foods, which possess antioxidant properties, are responsible for this effect by protecting cells from oxidative damage. However, studies performed recently in this laboratory suggest that at least one group of food antioxidants, the carotenoids, may directly influence DNA repair.[1] We have developed two assays for determining DNA repair activity in cell extracts. These measure the rate of radiolabelled deoxynucleotide incorporation into plasmid DNA during repair of damage introduced with either singlet oxygen or hydroxyl radicals. Singlet oxygen produces predominantly 8-hydroxydeoxyguanidine lesions.[2] Hydroxyl radicals produce a broad spectrum of damage including strand breaks and oxidised bases. The assays were used to examine the effects of dietary carotenoids on DNA repair in a lymphocyte cell line (Molt-17) *in vitro* and in human peripheral blood lymphocytes *in vivo*.

2 Methods

2.1 Repair Reactions

Plasmid pUC18 was damaged with a singlet oxygen by exposure to visible light in the presence of methylene blue.[3] Supercoiled DNA was recovered using an anion exchange HPLC method. The resultant DNA, termed MBpUC18, contained 2.4 sites per plasmid that were sensitive to FPG protein, a bacterial enzyme that generates single strand breaks at the site of oxidised purines in DNA. Plasmid pXPA (pGemT carrying a 1009 bp cDNA insert) was exposed to hydroxyl radicals produced using iron nitrilotriacetate (FeNTA) and H_2O_2.[4] Conditions were selected to produce approximately two single strand breaks per plasmid. The damaged plasmid was termed FeNTApXPA.

Repair reactions (25 μl) were performed with MBpUC18 plus pXPA or FeNTApXPA plus pUC18 (0.75 μg of each), 1–20 μg of cell protein extract, in the presence of deoxynucleotide triphosphates (dNTPs) and 500 nCi (3000 Ci $mmol^{-1}$) of one α-^{32}P-dNTP, using reaction conditions described previously.[5] DNA was repurified, linearised and electrophoresed through 1% w/v agarose gels. Plasmid recovery was determined by ethidium bromide staining. Radiolabel incorporation was measured by phosporimaging of the vacuum-dried gels.

2.2 Supplementation Studies

Molt-17 cells were supplemented with β-carotene, β-cryptoxanthin, or lutein (8 μM) for 24 hours using a liposome delivery system.[6] Cell extracts were prepared following supplementation.

For the *in vivo* study, volunteers were asked to consume a carotene capsule (4 mg of α-carotene and 8 mg β-carotene), a matched placebo, a portion of cooked, mashed carrots (200 g), or a portion of mandarin oranges (298 g), daily for 3 weeks in addition to their normal diet. A fifth group was given 60 mg vitamin C tablets to take every other day to match the vitamin C content of the mandarins. Peripheral blood lymphocytes were isolated from fasting blood samples taken before, and at the end of the intervention period (weeks 0 and 3), and after a 6 week washout period (week 9).

3 Results

For both assays, radiolabel incorporation was approximately linear for 4 hours after which it tailed off reaching a maximum by 8 hours. All 4 dNTPs were incorporated into FeNTApXPA to the same extent. Incorporation of dGTP into MBpUC18 was 2–3 times greater than for each of the other dNTPs, reflecting the preferential reaction of singlet oxygen with guanidines in DNA. Over 1 hour, incorporation of dGTP was directly proportional to added protein in the range 0–10 μg for both assays. A 1 hour incubation and 2 μg of protein extract per reaction were selected as standard conditions for repair assay analyses.

Carotenoid supplementation did not alter DNA repair activity in Molt-17 cells. However, *in vivo*, dietary supplementation with carrots provoked a moderate but significant ($p < 0.05$) increase in repair activity measured with both assays (Table 1). No effect was observed for the volunteers who were given the carotene capsule, mandarins or vitamin C.

Table 1 *Values indicate mean ± standard error with n denoting the number of volunteers in each group. *denotes values significantly greater than at baseline (week 0, p < 0.05).*

Dietary intervention	*Repair activity (fmol dGMP incorporated per hour per μg protein)*					
	FeNTApXPA Assay			*MBpUC18 Assay*		
	Week 0	*Week 3*	*Week 9*	*Week 0*	*Week 3*	*Week 9*
Carotene Capsules (n=11)	10.6±1.5	10.7±1.5	10.4±2.3	19.1±2.2	18.9±2.1	17.0±2.8
Carrots (n=11)	9.8±1.1	12.6±1.5*	10.7±1.3	15.9±1.4	19.8±1.9*	18.1±2.1
Mandarins (n=9)	9.4±1.7	8.7±1.1	9.4±1.5	20.2±2.8	20.1±2.3	20.2±2.8
Placebo Capsules (n=9)	9.8±2.2	11.9±2.1	9.4±1.4	20.4±2.9	22.7±2.4	19.8±2.6
Vitamin C (n=9)	10.0±0.8	10.0±0.8	9.7±1.4	18.9±3.4	20.8±2.5	22.6±3.9

4 Discussion

The assays described here represent sensitive tools for assessing the cellular capacity for repair of oxidative DNA damage. Although dietary supplementation with cooked carrots increased DNA repair activity, no equivalent effect was seen with the mixed carotene supplement, suggesting that either the form in which the carotenoids are presented or another component of carrots may be important. Taken together with other results obtained using different markers of DNA damage and repair,[1] it is clear that carotenoids can regulate DNA repair processes but that the mechanisms involved are complex.

Acknowledgements

This work was funded by the Ministry of Agriculture, Fisheries and Food, UK.

5 References

1 S. Astley, R. Elliott, D. Archer and S. Southon, Royal Society of Chemistry meeting 'Food and Cancer Prevention III', Norwich, UK, 5–8 September 1999, paper 3.5.
2 M. Pflaum, S. Boiteux and B. Epe, *Carcinogenesis*, 1994, **2**, 297.
3 S. Boiteux, E. Gajewski, J. Laval and M. Dizdaroglu, *Biochemistry*, 1992, **31**, 106.

4 A.H. Sarker, S. Watanabe, S. Seki, T. Akiyama and S. Okada, *Mut. Res.*, 1995, **337**, 85.
5 R.D. Wood, P. Robins and T. Lindahl, *Cell*, 1988, **53**, 97.
6 P. Grolier, V. Azais-Braesco, L. Zelmire and H. Fessi, *Biochim Biophys Acta*, 1992, **1111**, 135.

3.3

All-*trans*-β-Carotene Reduces Genetic Damage *In vitro* in Human Lymphocytes

M. Glei,[1] G. Rechkemmer,[2] B. Spänkuch[2] and B. L. Pool-Zobel[1]

[1]FRIEDRICH SCHILLER UNIVERSITY, INSTITUTE FOR NUTRITION, DORNBURGER STRAẞE 25, D-07743 JENA, GERMANY
[2]FEDERAL RESEARCH CENTRE FOR NUTRITION, HAID-UND-NAU-STRAẞE 9, D-76131 KARLSRUHE, GERMANY

1 Introduction

A previous dietary intervention study with 23 healthy, non-smoking male subjects aged 27–40 had shown that the supplementation of the diet with tomato, carrot or spinach products resulted in a significant reduction in endogenous levels of strand breaks in lymphocyte DNA. Oxidative DNA damage in peripheral lymphocytes was significantly reduced during a carrot juice intervention.[1] These reductions may be due to scavenging of reactive oxygen species (ROS) and to induction of chemopreventive enzymes such as glutathione S-transferase (GST), which also may deactivate ROS.[2] It was the aim of this study to determine whether major carotenoid ingredients of the protective juices (namely lycopene of tomato juice and all-*trans*-β-carotene of carrot juice) have similar modes of activity and could therefore be regarded as responsible for the *in vivo* effectiveness of the juices in the intervention study. For this, the compounds were assessed *in vitro* in human lymphocytes for their capacity to reduce DNA damage, modulate repair of endogenous and bleomycine-induced damage, and to induce GST π expression.

2 Materials and Methods

Peripheral lymphocytes were isolated from healthy volunteers by gradient centrifugation with HISTOPAQUE®–1077 from Sigma. The blood donors either consumed their usual unrestricted diet ('undepleted' subjects [Figures 3–5]) or they had refrained from consuming carotenoid-containing foods for two weeks ('depleted' subjects [Figures 1, 2, 6]). The lymphocytes were incubated

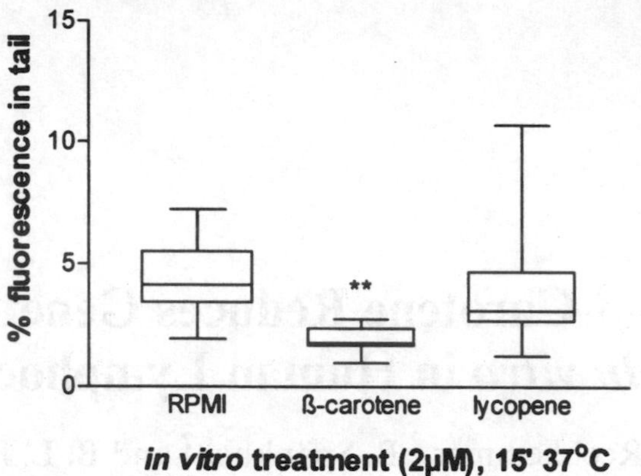

Figure 1 *DNA strand breaks in lymphocytes (**$p < 0.01$; two sided unpaired t-test; $n = 8$)*

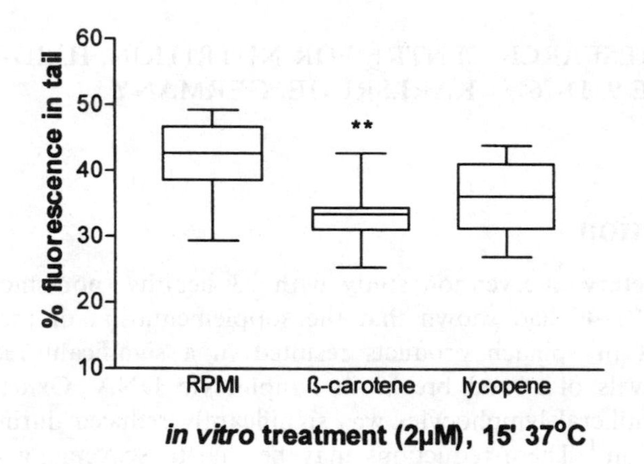

Figure 2 *Oxidised pyrimidine bases in lymphocytes (**$p < 0.01$; paired t-test; $n = 8$)*

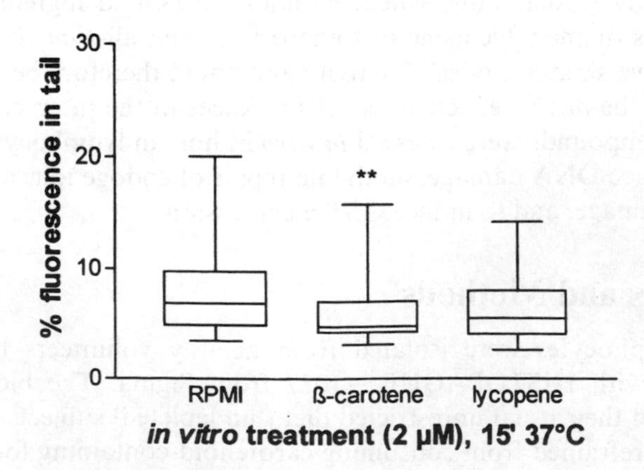

Figure 3 *DNA strand breaks in lymphocytes (**$p < 0.01$; paired t-test; $n = 16$)*

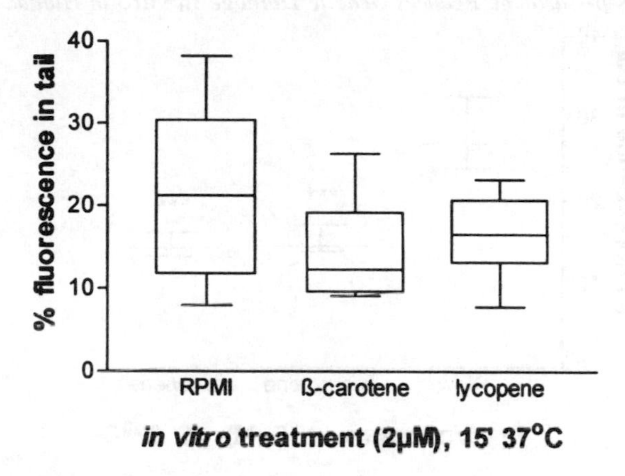

Figure 4 *Oxidised pyrimidine bases in lymphocytes (n=8)*

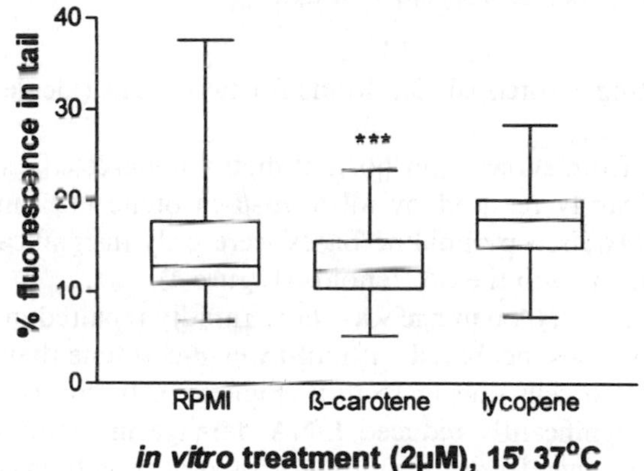

Figure 5 *Effect of carotenoids on DNA damage induced by bleomycin in lymphocytes (***$p < 0.001$; paired t-test; $n=6$)*

(15 min, 37 °C) with or without 2 μM of water soluble carotenoids (kindly provided by L.E. Schlipalius, Betatene, Australia), a concentration, near the physiological dose found after the tomato and carrot juice intervention.[3] The cells were then worked up in the comet assay with or without using the repair specific enzyme endonuclease III to reveal oxidised DNA bases.[4–6] Modulation of DNA repair was assessed by inducing damage with 20 μg bleomycin/ml cell suspension (30 min, 37 °C) (from Mack) and observing kinetics of damage-persistence after incubation with the carotenoids. Proteins were isolated from aliquots of the treated lymphocytes and analysed by ELISA (HEPKIT-Pi, Biotrin) to determine GST π.[2]

3 Results

In vitro treatment with all-*trans*-β-carotene lead to a significant reduction of DNA breaks and oxidised DNA bases in lymphocytes from 8 subjects who had

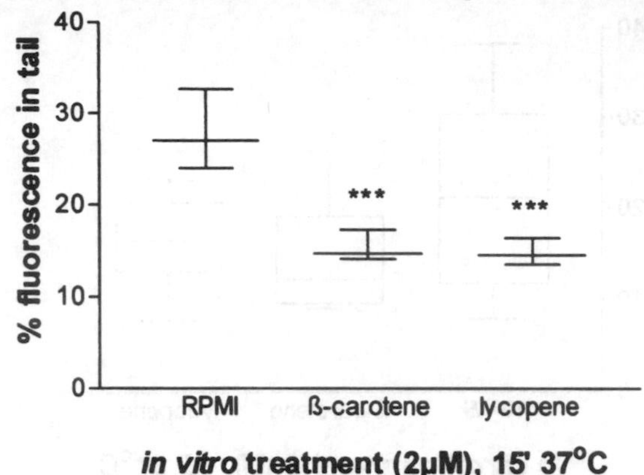

Figure 6 *Effect of carotenoids on DNA damage induced by bleomycin in lymphocytes (***p < 0.001; two sided unpaired t-test; n = 5)*

refrained from eating carotenoid-rich foods for two weeks ('depleted' subjects) (Figures 1, 2).

In lymphocytes from subjects on normal diets ('undepleted') strand breaks were again significantly reduced by all-*trans-β*-carotene (15 min incubation time) (Figure 3). Oxidised pyrimidine bases were only insignificantly reduced after *in vitro* treatment with the carotenoids (Figure 4).

Bleomycin induced DNA damage was more rapidly repaired in lymphocytes from undepleted persons incubated with all-*trans-β*-carotene than in untreated or in lycopene-treated cells during 15 min (Figure 5). In contrast to Figure 5 both carotenoids significantly reduced DNA damage in lymphocytes of one subject who had refrained from consuming carotenoid-rich foods (Figure 6).

Finally, the mean GST π protein levels of nine depleted probands were significantly ($p < 0.05$) increased from 3.9 ± 1.5 to 7.1 ± 2.0 ng GST $\pi/10^6$ cells (means ± SEM) after incubation with 2 μM all-*trans-β*-carotene. Lycopene increased the level only to 5.5 ± 1.5 ng/10^6 cells ($p > 0.05$). In both cases, lymphocytes from only six of the nine investigated persons showed an GST π induction. In the cells of the other subjects there was a slight decrease of GST π content.

4 Conclusion

All-*trans-β*-carotene but not lycopene reduced DNA damage in peripheral lymphocytes. This effect together with the preliminary findings on potential to GST π levels may explain the marked reduction of oxidised bases in subjects consuming carotenoid-containing juices, especially during the carrot juice intervention phase.[1,2] Further studies are being performed to elucidate the relative roles of antioxidative and GST π inducing activities of these two individual carotenoids for reducting genetic damage. Altogether this type of mechanism – directed *in vivo/in vitro* approach – should help increase our knowledge on the nature of protective properties by complex plant foods.

5 References

1 B.L. Pool-Zobel, A. Bub, H. Müller, I. Wollowski and G. Rechkemmer, *Carcinogenesis*, 1997, **18**, 1847.

2 B.L. Pool-Zobel, A. Bub, U. M. Liegibel, A. Treptow-van Lishaut and G. Rechkemmer, *Cancer Epidemiol., Biomarkers and Prevention*, 1998, **7**, 891.

3 H. Müller, A. Bub, B. Watzel and G. Rechkemmer, *Eur. J. Nutr.*, 1999, **38**, 35.

4 N.P. Singh, M.T. McCoy, R.R. Tice and E.L. Schneider, *Exp. Cell Res.*, 1988, **175**, 184.

5 B.L. Pool-Zobel and U. Leucht, *Mutat. Res.*, 1997, **375**, 105.

6 B.L. Pool-Zobel, N. Lotzmann, M. Knoll, F. Kuchenmeister, R. Lambertz, U. Leucht, H.G. Schröder and P. Schmetzer, *Environ. Mol. Mutagen.*, 1994, **24**, 23.

3.4

Relationship Between Diet and Oxidative DNA Damage

Emily R. Beatty,[1] Timothy G. England,[2] Barry Halliwell[2,3] and Catherine A. Geissler[1]

[1]NUTRITION FOOD & HEALTH RESEARCH CENTRE, DEPARTMENT OF NUTRITION AND DIETETICS, KING'S COLLEGE LONDON, FRANKLIN-WILKINS BUILDING, 150 STAMFORD STREET, LONDON SE1 8WA, UK
[2]INTERNATIONAL ANTIOXIDANT RESEARCH CENTRE, KING'S GUY'S AND ST THOMAS'S SCHOOL OF BIOMEDICAL SCIENCES, ST THOMAS STREET, LONDON SE1 9RT, UK
[3]DEPARTMENT OF BIOCHEMISTRY, NATIONAL UNIVERSITY OF SINGAPORE, KENT RIDGE CRESCENT, SINGAPORE 119260

1 Introduction

Consumption of fruit and vegetables is associated with reduced risk of many cancers.[1] This has been attributed in part to the antioxidants and other phytochemicals they contain which may protect DNA from oxidative damage or regulate antioxidant defences and DNA repair mechanisms. The present study aimed to examine the relationship between diet and steady state levels of oxidative DNA damage as measured by GC-MS.

2 Methods

Fasting blood samples were taken and seven-day diet records[2] were completed by forty-eight healthy, non-smoking, non-supplement taking volunteers. DNA was extracted from whole blood and damaged bases measured by gas chromatography–mass spectrometry (GC-MS).[3] Estimated nutrient intakes were calculated using Integrated Dietary Analysis version 3.[4] Subjects provided written informed consent and the study was approved by King's College Research Ethics committee. Relationships between dietary intake and DNA damage were examined using two-tailed Pearson's correlations for normally distributed data and Spearman's rank correlations for non-normally distributed data, using SPSS version 8.

3 Results

All subjects (37 female, 11 males) completed 7 d diet diaries satisfactorily as determined by energy intake: basal metabolic rate ratios (Table 1). Significant negative correlations were found between total measured products of DNA base damage and intake of dietary fibre ($R = -0.41$, $P = 0.005$); NSP ($R = -0.38$, $P = 0.011$); potassium ($R = -0.40$, $P = 0.007$); magnesium ($R = -0.40$, $P = 0.006$); iron ($R = -0.40$, $P = 0.007$); zinc ($R = -0.34$, $P = 0.022$); manganese ($R = -0.34$, $P = 0.021$); riboflavin ($R = -0.32$, $P = 0.032$); niacin ($R = -0.35$, $P = 0.020$); folic acid ($R = -0.40$, $P = 0.006$), and percentage energy from protein ($R = -0.30$, $P = 0.044$). 8-OH Guanine, the base product most frequently measured in other studies, was negatively correlated only with intake of NSP ($R = -0.30$, $P = 0.043$) and copper ($R = -0.34$, $P = 0.02$) and positively with percentage energy from fat ($R = 0.32$, $P = 0.032$). The only associations between antioxidant intake and DNA damage were negative correlations between 8-OH adenine and dietary vitamin C ($R = -0.31$, $P < 0.043$), carotene equivalents ($R_{sp} = -0.47$, $P = 0.001$), and retinol equivalents ($R_{sp} = -0.40$, $P = 0.007$). A number of other significant negative correlations between individual DNA base oxidation products and folic acid, niacin, and the dietary minerals iron, copper, zinc, sodium and chloride were seen (Table 1).

4 Discussion

Levels of DNA damage measured in this study have been reported elsewhere,[3] and are similar to other reported levels reported in humans.[5] Estimated mean intake of nutrients met Reference Nutrient Intakes[6] (RNI), with those of vitamin C and folic acid being 3.3 and 1.6 × RNI respectively and vitamin E intake 2.9 × the minimum safe intake.[6] The high intake of vitamin C, E, folic acid and NSP, and relatively low energy from fat (35%) reflects the characteristics of the volunteers who, in general, were interested in healthy eating. In this group, levels of only one of the eleven base products measured (8-OH-adenine) were associated with lower intake of antioxidants (vitamin C and carotene equivalents). However, folic acid (perhaps a marker of leafy vegetable intake), riboflavin, niacin and a number of dietary minerals were associated with lower total measured products of DNA base damage and lower levels of other measured base products. Folic acid deficiency has been implicated in uracil misincorporation and chromosome breakage,[7] and the minerals iron, copper, zinc and manganese are all important in antioxidant or detoxifying (cytochrome P450) enzyme systems, while riboflavin is important for glutathione reductase activity. All of these may influence steady-state levels of DNA damage. These results suggest that in subjects with adequate intakes of antioxidants, antioxidant intake is not an important determinant of steady-state levels of oxidative

Table 1 *Estimated daily nutrient intakes of healthy subjects (n = 48), and significant (P < 0.05) correlations with DNA base oxidation products.*

	Mean	*S.E.M.*	*Base product correlations*
Energy: Basal metabolic rate	1.53	0.03	
Energy (MJ)	9.3	0.4	
Protein (g)	77	2.6	5-OH Me, Hydantoin (−0.37)
Total fat (g)	86	3.2	
Saturates (g)	29	1.5	
Monounsaturates (g)	28	1.1	
Polyunsaturates (g)	16	1.0	
Carbohydrate (g)	266	8.5	
Starch (g)	147	5.7	5-OH Me, Uracil (−0.32); 5-OH-Cytosine (−0.31)
Sugars (g)	113	5.2	
Energy from protein (%MJ)	14.0	0.33	T.B.P.* (−0.30)
Energy from fat (%MJ)	35.2	0.9	8-OH Guanine (0.32)
Energy from carbohydrate (%MJ)	45.6	1.0	
Energy from alcohol (% MJ)	5.2	0.64	
Non-starch polysaccharides (g)	17.9	1.2	8-OH Guanine (−0.30); T.B.P. (−0.38)
Dietary fibre (g)	24.2	1.5	5-OH Hydantoin (−0.31); T.B.P. (−0.41)
Alcohol (g)	17.1	2.4	
Cholesterol (mg)	222	14.5	
Retinol (μg)	405	46.7	FAPy Guanine (0.45)
Retinol equivalents (μg)	942	83.4	8-OH Adenine (−0.40)
α-Tocopherol equivalent (mg)	9.2	0.5	
Carotene equivalent (mg)	3.12	0.45	8-OH Adenine (−0.47)
Vitamin C (mg)	115	9.2	8-OH Adenine (−0.31)
Folic acid (μg)	312	13.1	5-OH Hydantoin (−0.33); FAPy Adenine (−0.29); T.B.P. (−0.40)
Sodium (g)	3.1	0.14	5-OH Me, Hydantoin (−0.39)
Calcium (mg)	910	38	8-OH Adenine (0.31)
Iron (mg)	14.2	0.6	5-OH Me, Hydantoin (−0.31); T.B.P. (−0.40)
Selenium (μg)	55	3.3	
Copper (μg)	1.6	0.08	FAPy Guanine (0.31); 8-OH Guanine (−0.34);
Magnesium (mg)	338	12.7	T.B.P. (−0.40)
Manganese (mg)	3.92	0.24	T.B.P. (−0.34)
Potassium (g)	3.35	0.10	T.B.P. (−0.40)
Zinc (mg)	9.3	0.4	5-OH Me, Hydantoin (−0.35); T.B.P. (−0.34)
Iodine (μg)	99.4	5.4	8-OH Adenine (0.42)
Chloride (g)	4.7	0.2	5-OH Me, Hydantoin (−0.39)
Riboflavin (mg)	2.0	0.1	T.B.P. (−0.32)
Niacin (mg)	22.1	1.1	5-OH Hydantoin (−0.32); T.B.P. (−0.35)
Vitamin B12	4.2	0.5	Thymine glycol (0.331)

*T.B.P., Total Base Products.

DNA damage, and other factors, including rates of repair, as well as other dietary constituents may be more important.

Acknowledgements

This work was funded by World Cancer Research Fund. We are grateful to MRC Dunn Human Resource Centre for assistance with vitamin C analysis.

5 References

1 World Cancer Research Fund and American Institute for Cancer Research, 'Food nutrition and the prevention of cancer: a global perspective' American Institute for Cancer Research, Washington, D.C. 1997.

2. S.A. Bingham, A. Cassidy, T.J. Cole, A. Welch, S.A. Runswick, A.E. Black, D. Thurnham, C. Bates, K.T. Khaw, T.J.A. Key and N.E. Day, *Brit. J. Nutr.*, 1995, **73**, 531.

3 T.G. England, E.R. Beatty, C.A. Geissler and B. Halliwell, *Am. J. Clin. Nutr.*, submitted.

4 'Integrated Dietary Analysis', IDA Publications, London, 1997.

5 S. Senturker and M. Dizdaroglu, *Free Radical Biol. Med.*, 1999, **27**, 370.

6 Department of Health, 'Dietary Reference Values for Food Energy and Nutrients for the United Kingdom' 1991, London.

7 B.C. Blount and BH.N. Ames, *Anal. Biochem.*, 1994, **219**, 195.

3.5

DNA Damage and Repair: Relative Responses to Antioxidant Nutrients in the Diet

Siân Astley, Ruan Elliott, David Archer and Susan Southon

INSTITUTE OF FOOD RESEARCH, NORWICH RESEARCH PARK, COLNEY, NORWICH NR4 7UA, UK

1 Introduction

Single cell gel electrophoresis (Comet assay, SCGE), in its original form, is a simple visual method for measuring single-strand breaks (SSBs) in eukaryotic cells[1], which may arise from damage and repair. However, being able to measure SSBs does not provide sufficient information. There are other forms of damage, specifically oxidised bases, which do not create SSBs, and are more significant to cells. Numbers of oxidised purines or pyrimidines are estimated by incorporating an enzyme modification step to the basic comet procedure[2]. The enzymes used are endonuclease III, which nicks DNA at sites where there are oxidised pyrimidines, and formamidopyrimidine glycosylase (FPG-protein), which demonstrates endonuclease activity towards oxidised purines. Neither enzyme is 100% efficient or specific but their inclusion means DNA damage can be measure more effectively.

Arguably, repair may be the more important factor in the damage:repair balance since, provided DNA damage is rapidly and efficiently repaired, there are no further consequences for the cell. However, DNA repair systems are not faithful and, while it would be advantageous to increase the rate and fidelity of repair mechanisms, it is also vital to keep damage to a minimum. We combined the basic Comet assay technique and the enzyme adapted method, over a time course, in order to measure both damage and repair. More specifically, nett baseline SSBs (damage and repair); baseline numbers of oxidised bases (purines and pyrimidines separately); resistance to oxidative damage, following treatment with hydrogen peroxide (H_2O_2); stress induced base oxidation; rate and repair of SSBs and rate and repair of oxidised bases.

Oxidant stress can cause DNA damage, and the Comet assay has been used to investigate the antioxidant capacity of various compounds including dietary components such as the carotenoids.[3,4] The hypothesis being, unless the stress is sufficient to kill the cell, oxidative damage to DNA may increase cancer risk.[5] Compounds with antioxidant capacity may reduce this risk, which would explain the epidemiological association between higher consumption of fruits and vegetables and lower incidence of cancer.[6] The Comet is one useful way of testing this hypothesis but results can be ambiguous and interpretation difficult.

We have investigated the response of human lymphocytes (as accessible nucleated cells) to varying concentrations of carotenoids, using the different versions of the Comet assay. Initially, a human secondary cell line[7] (Molt-17) was used, and the carotenoids delivered as enriched liposomes rather than solvents.[8] Later, the response of human primary lymphocytes was investigated following manipulation of volunteer's diets with carotenoid isolates or carotenoid-rich foods. These results are an amalgamation of our finding, and their interpretation so far.

2 Single-strand Breaks *In vitro*

Molt-17 lymphocytes were cultured under normal humidified conditions ($175\,cm^2/800$ ml flasks, 37 °C, 5% CO_2) in RPMI 1640 containing 100 $mmol\,l^{-1}$ glutamine, 100 IU penicillin ml^{-1}, 100 $mg\,ml^{-1}$ streptomycin, and 10% (v/v) foetal calf serum (all from Life Technologies). Supplementation was between 0.5 and at 8 $\mu mol\,l^{-1}$ β-carotene, β-cryptoxanthin, lutein or lycopene.

- No change in levels of baseline SSBs following supplementation was observed
- Cells supplemented with 1 to 8 $\mu mol\,l^{-1}$ β carotene had significantly lower numbers of SSBs following treatment with H_2O_2 suggesting an increased resistance to oxidative damage.
- Protection against oxidative attack decreased around 8 $\mu mol\,l^{-1}$ although no additional damage, above that observed in control cells, was noted.
- Similar responses occurred for the other major carotenoids – lutein, lycopene and β-cryptoxanthin

3 Single-strand Breaks *In vivo*

Two separate studies were examined. In **Study 1**, the volunteers received 15 mg synthetic β-carotene per day for four weeks, which resulted in up to a ten-fold increase in plasma concentrations (0.2–2 $\mu mol\,l^{-1}$). In the **Study 2**, volunteers received 8 mg of β-carotene from palm fruit oil for three weeks, which is more akin to dietary manipulation by food than direct supplementation. This resulted in a four-fold increase in plasma concentration (0.2–0.8 $\mu mol\,l^{-1}$). Primary lymphocytes from **Study 1** were examined using the basic Comet Assay only, while those from **Study 2** underwent the adapted enzyme method.

There were no differences in numbers SSBs between the test and placebo groups at baseline. In **Study 1**, higher intake of β-carotene increased DNA resistance to oxidation *ex vivo*. The same effect was not seen with the other carotenoids lutein and lycopene but their plasma levels were much lower ($<1\ \text{mmol}\ l^{-1}$) following supplementation. In the placebo group from **Study 1**, there was a significant increase in SSBs following treatment with H_2O_2 that was not present in the group that had received β-carotene. In **Study 2**, there was no difference in numbers of SSBs in lymphocytes from placebo and test groups following H_2O_2 treatment *ex vivo*.

- Increased resistance to oxidation is dose responsive, with the threshold lying somewhere between 0.8 and 2 $\mu\text{mol}\ l^{-1}$.

4 Oxidised Bases *In vitro* and *In vivo*

Molt-17 lymphocytes supplemented with β-carotene (8 $\mu\text{mol}\ l^{-1}$) were assayed for oxidised purines and pyrimidines as were primary lymphocytes from **Study 2**.

- H_2O_2 treatment increased numbers of oxidised bases but supplementation with the carotenoids had no effect on baseline numbers of oxidised bases or the numbers resulting from oxidative attack.

5 Repair Activity *In vitro*

Molt-17 lymphocytes were treated as described above then divided between two flasks. Cells in one flask from each pair were treated with H_2O_2 (100 μmol in HBSS, 5 min, 4 °C), and the reaction stopped with dimethylsulphoxide (DNase and RNase-free, 50 μl, Sigma). HBSS was added to cells in the paired flask. Lymphocytes (400 μl) were removed, from both flasks, at times 0, 30, 60, 90 and 120 minutes. Each aliquot was stored on ice, and in the dark, until all the samples had been collected, and then subjected to SCGE.

There were no differences in any of the groups at baseline, at any time. There were no differences between the groups at time zero following H_2O_2 treatment. There was, however, a significant decrease in numbers of SSBs in lymphocytes that had been pre-incubated with β-carotene by 60 minutes, which continued to decline over a further 60 minutes. A similar pattern was observed for the other carotenoids (β-cryptoxanthin and lutein) except the decrease was not significant until 90 minutes.

- Increased cellular carotenoids stimulate SSBs repair.

6 Repair *In vivo*

Lymphocytes isolated from whole blood were made up to volume in HBSS (10 ml, 1.5×10^6 cells ml^{-1}), and the volume split between two culture flasks ($25\ \text{cm}^2$/50 ml). After one hour, the cells were treated, or not, with H_2O_2 as previously described. Lymphocytes (400 μl) were removed, from both flasks, at times 0, 120, 180 and 240 minutes. Each aliquot was stored on ice and in the dark until all the samples had been collected.

There were no differences between the placebo and β-carotene supplemented group at baseline or at time zero, following H_2O_2 treatment. However, lymphocytes from the volunteers in the β-carotene group showed a significant decrease in numbers of SSBs over four hours. At four hours, there were no differences between the pre- and post-H_2O_2 treated cells. No significant differences between the groups at baseline, or over time at baseline, were detected.

7 Repair of Oxidised Bases *In vivo*

Repair of oxidised bases in primary lymphocytes from **Study 2** was measured. In H_2O_2-treated cells from the placebo group, SSBs solely attributable to oxidised bases decreased, returning to baseline by four hours. There was no significant difference between the groups at baseline or in baseline numbers over time, and no differences between the groups at time zero following H_2O_2 treatment. However, oxidised bases remained elevated at four hours in cells from β-carotene supplemented individuals.

- Increased intakes of β-carotene delays repair of oxidised bases.

8 Conclusion

The apparent delay in repair of oxidised bases has considerable implications to the safety of the cell while repair of SSBs is of much less importance. Whatever is happening with the repair process it is worth noting that it is apparently more sensitive than the response of protection against damage. It is also important to emphasise that our results are not consistent with the carotenoids acting as pro-oxidants and that there is no evidence of on-going damage.

The apparent discrepancy between increased repair of SSBs and slower repair of oxidised bases appears to be contradictory but the excision and repair of oxidised bases results in the appearance of a SSB. If the repair of oxidised bases is delayed by β-carotene then there would be less SSBs created, which would in turn give the impression that SSBs were decreasing. Less SSBs would be seen in the β-carotene group using the basic Comet method, which only measures SSBs while more strand breaks would be seen in the β-carotene group using enzyme adapted method because of the remaining oxidised bases. The converse would hold true for the placebo group, which is concordant with our results. The difficulty lies in why this might happen.

One hypothesis that appears to fit the data available is that increased β-carotene down-regulates expression of the glycosylases – those enzymes responsible for removing oxidised bases. Ligation enzymes – those that close the SSBs – would still be at normal levels since they have a role in repair of non-oxidative lesions. Antioxidant protection from the carotenes may not reduce the overall levels of damage but could reduce its frequency, causing the cell to switch off non-essential genes. When oxidatively attacked, following H_2O_2 treatment, the cell must up-regulate those proteins but is unable to do so quickly enough to prevent accumulation of damaged bases.

The area of DNA damage:repair in human cells is very complex. The Comet assay is used by many different laboratories with different aims, and it has shown us that the carotenoids have a role beside their antioxidant capacity. However, the interpretation is troublesome and may lead to different conclusions depending on the level of understanding or information available from other methods. A more mechanistic approach is needed, which is based on molecular biology techniques if the true role of the carotenoids is ever to be elucidated.

Acknowledgement

This work was supported by the Ministry of Agriculture, Fisheries and Food.

9 References

1 McKelvey-Martin VJ, Green MHL, Schmezer P, Pool-Zobel BL, de Meo MP and Collins AR (1993) The Single Cell Gel Electrophoresis (SCGE) Assay (Comet Assay): A European Review. *Mutat Res*, **288**, 47–64

2 Collins AR, Duthie SJ and Dobson VL (1993) Direct Enzymatic Detection of Endogenous Oxidative Base Damage in Human Lymphocyte DNA. *Carcinogenesis* **14**: 1733–1735.

3 Astley SB, Hughes DA, Wright AJA, Peerless ACJ and Southon S (1996) Effect of beta-carotene supplementation on DNA damage in human blood lymphocytes *Biochem Soc Trans* **24**: 526S.

4 Astley SB, Hughes DA, Wright AJA, Peerless ACJ and Southon S (1997) Supplementation of the diet with β-carotene or lycopene: Comparison of the effects on DNA damage in primary T-lymphocytes using the 'Comet Assay'. *Proc Nutr Soc* **56**: 105A.

5 Halliwell B and Gutteridge JMC (1999) Free radicals in biology and medicine. 3rd Edition Oxford: Oxford University Press.

6 WCRFund (1997) Food, nutrition and the prevention of cancer: A global prospective. Washington DC: American Institute for Cancer Research.

7 Drexler HG and Minowada J (1989) Morphological, Immunological and Isoenzymatic Profiles of Human Leukaemia Cells and Derived T-cell Lines. *Haematol Oncol* **7**: 115–125.

8 Grolier P, Azais-Braesco V, Zelmire L and Fessi H (1992) Incorporation of Carotenoids in Aqueous Systems: Uptake by Cultured Rat Hepatocytes. *Biochim Biophys Acta* **1111**: 135–138.

3.6

The Influence of Folic Acid on DNA Stability in Human Cells

Susan J. Duthie and Sabrina Narayanan

ROWETT RESEARCH INSTITUTE, ABERDEEN AB21 9SB, UK

1 Introduction

Folate deficiency has been implicated in the development of cancer, particularly cancer of the colorectum.[1] There appear to be two mechanisms through which folate deficiency may increase the risk of cancer. Folate deficiency may induce an imbalance in DNA precursors, leading to uracil mis-incorporation into DNA in place of thymine. 'Catastrophic' DNA repair of this lesion may lead to DNA instability *via* DNA double strand breakage, chromosome damage and malignant transformation.[2] Alternatively, folate deficiency, by inhibiting synthesis of the primary methyl donor, *S*-adenosylmethionine (SAM), may decrease DNA methylation, leading to proto-oncogene activation.[3]

In this study we have investigated the effects of folate deficiency *in vitro*, both on DNA stability and DNA methylation in normal human lymphocytes and in cultured human colonocytes (Caco-2).

2 Methods

2.1 Cell Culture

Caco-2 cells (1×10^5 cells per flask) were grown for up to 14 days either in folate-containing GMEM ($2\,mg\,ml^{-1}$; F+) or folate-deficient GMEM ($0\,mg\,ml^{-1}$; F−). Isolated human lymphocytes[4] were grown in F+ or F− RPMI 1640 media for up to 10 days. In certain experiments, cells were exposed to hydrogen peroxide (200 μM) for 5 min on ice before incubation in culture media at 37 °C for up to 8 hours to stimulate DNA repair. Samples were taken for comet analysis, cell growth (haemocytometer) and DNA methylation.[5]

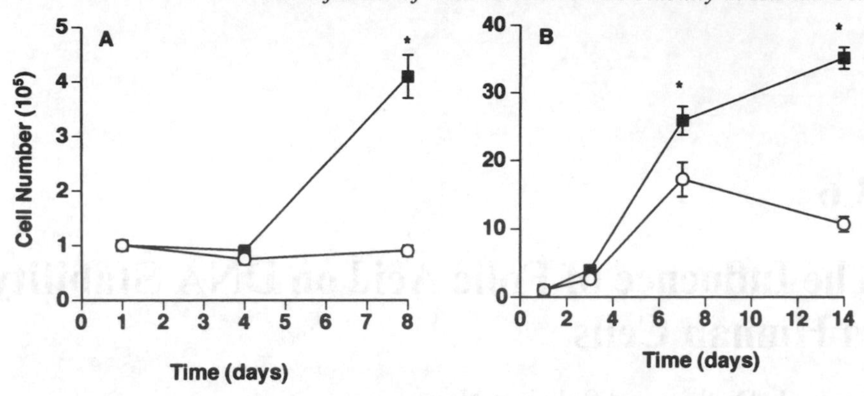

Figure 1 *The effect of folate depletion of lymphocyte* (A) *and Caco-2* (B) *growth. Results are mean +/− SEM for n=8. *P<0.001, where P values refer to differences between cells grown in the presence (squares) or absence (circles) of folic acid*

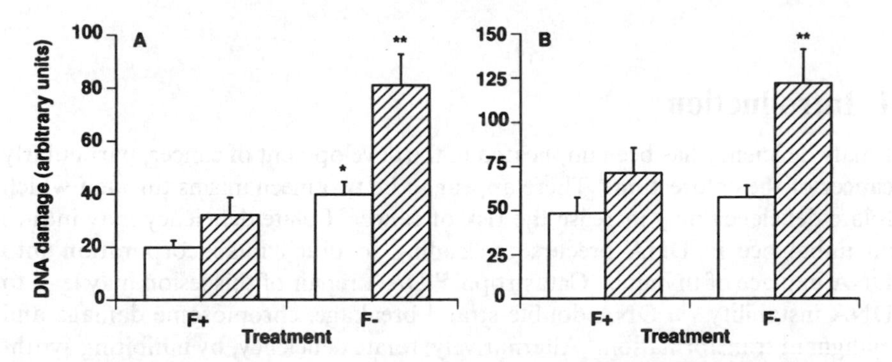

Figure 2 *The effect of folate depletion on DNA stability in lymphocytes* (A) *and Caco-2* (B) *cells. DNA strand breaks (open columns) and mis-incorporated uracil (hatched columns). Results are mean +/− SEM for n=8. *P<0.02, where P values refer to differences in strand breaks, and **P<0.02 where P values refer to differences in uracil mis-incorporation, in cells either grown in F+ or F− media for 10–14 days*

2.2 DNA Stability Measured Using the Comet Assay

Endogenous DNA strand breakage, mis-incorporated uracil and DNA repair were determined using a modified comet assay.[6] Comets were classified visually depending on the amount of fluorescence in the tail. DNA damage can range from 0 to 400 arbitrary units.[4]

3 Results

Folate deficiency inhibits cell proliferation (Figure 1). DNA instability (strand breakage and uracil mis-incorporation) is significantly increased by folate depletion in human lymphocytes (Figure 2A). However, while uracil levels are increased in Caco-2 cells, folate deficiency does not induce strand breakage (Figure 2B).

Table 1 *The effect of folate deficiency* in vitro *on DNA repair in normal human lymphocytes and Caco-2 cells*

	Human lymphocytes		*Caco-2 Cells*	
	F+	*F−*	*F+*	*F−*
Untreated	7.0 ± 3.8	115.3 ± 7.2	11.0 ± 0.9	9.0 ± 5.1
0 Hours	256.3 ± 24.9	394.0 ± 2.7	326.5 ± 8.6	310.8 ± 7.4
4 Hours	89.5 ± 13.2	332.7 ± 11.0	68.3 ± 14.1	67.5 ± 11.4
8 Hours	28.3 ± 6.5	323.3 ± 17.7	47.3 ± 4.3	47.0 ± 4.1

DNA damage (strand breakage) is in arbitrary units. Cells were grown in F+ or F− media for 10–14 days. Results are mean ± SEM for n – 4.

Moreover, DNA repair in response to oxidative damage is identical in Caco-2 cells grown either in F− or F+ media, but is severely compromised in lymphocytes cultured under conditions of folate deficiency (Table 1). Similarly, folate depletion reduces genomic DNA methylation in human lymphocytes but does not effect Caco-2 DNA methylation status (Figure 3).

4 Discussion

Folate deficiency is associated with an increased risk of colorectal cancer.[1] In this study we have investigated the effect of folate deficiency *in vitro* on DNA stability in isolated human lymphocytes and malignantly transformed human colonocytes. Normal human lymphocytes and cultured human colonocytes contain detectable levels of DNA strand breaks and mis-incorporated uracil. Moreover folate deficiency destabilises lymphocyte DNA further (inability to

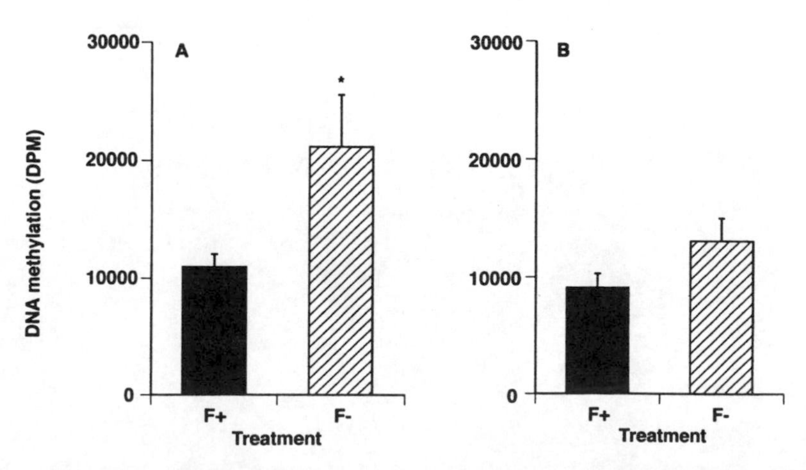

Figure 3 *Genomic DNA methylation status in lymphocytes* (A) *and Caco-2* (B) *cells. Results are mean DPM +/− SEM for n=6. *P<0.04 where P values refer to differences in DNA methylation in cells grown in F+ or F− media for 10–14 days. A high DPM count reflects a lower degree of DNA methylation*

proliferate, increased strand breakage and mis-incorporated uracil, altered DNA repair and DNA hypomethylation). Conversely, despite folate deficiency inhibiting cell growth, DNA stability in Caco-2 cells is maintained. It remains to be determined whether or not lymphocytes are unusually sensitive to folate depletion or if folate status can alter DNA stability in normal human colonocytes.

Acknowledgements

World Cancer Research Fund and the Scottish Executive Rural Affairs Department.

5 References

1 Y-I. Kim, *J. Nutr. Biochem.*, 1999, **10**, 66.
2 B.C. Blount, M.M. Mack, C.M. Wehar, J.T. MacGregor, R.A. Hiatt, G.Wang, S.N. Wickramasinghe, R.B. Everson and B.N. Ames, *Proc. Natl. Acad. Sci. USA*, 1997, **94**, 3290.
3 E. Giovannucci, M.J. Stampfer and G.A. Colditz, *J. Natl. Cancer Inst.*, 1995, **85**, 875.
4 S.J. Duthie and A. Hawdon, *FASEB J.*, 1998, **12**, 1491.
5 M. Balaghi and C. Wagner, *Biochem. Biophys. Res. Commun.*, 1993, **193**, 1184.
6 S.J. Duthie and P. McMillan, *Carcinogenesis*, 1997, **18**, 1709.

3.7

N-Nitroso Compounds from Dietary Bacon Do not Initiate or Promote Aberrant Crypt Foci in the Colon of F344 Rats

G. Parnaud,[1] B. Pignatelli,[2] S. Taché,[1] and D.E. Corpet[1]

[1]ECOLE NATIONALE VÉTÉRINAIRE, INRA, 23 CAPELLES, 31076 TOULOUSE, FRANCE
[2]INTERNATIONAL AGENCY FOR RESEARCH ON CANCER, 69372 LYON CDX8, FRANCE

1 Introduction

The intake of meat, specifically red or processed meat, is associated with increased risk of colon cancer in many case-control studies, and in some cohort studies.[1] The endogenous formation of *N*-nitroso compounds (NOC) from meat constituents may explain this association.[2] The mutations that are common in human colorectal cancers are consistent with the effect of alkylating agents such as NOC. In human volunteers, a diet high in red meat increases NOC level in stools.[2] We speculated (i) that a diet high in beef meat or bacon would increase fecal NOC level in rats, (ii) that the presence of NOC in the gut might initiate aberrant crypt foci (ACF) in the colon of rats, and (iii) that fecal NOC might promote the growth of ACF. Since the fat content of diet may modify the effect of diet on colon carcinogenesis,[3] the effect of meat and bacon was tested in both low-fat and high-fat contexts.

2 Methods

Study 1 – Low-fat diet post-initiation: Fifty Fisher-344 rats were given a single i.p. injection of azoxymethane (20 mg kg^{-1}), then randomized to 5 different diets, given *ad libitum* for 100 days. Control rats were fed an AIN-76-based low-fat control diet, containing 2% corn oil, 5% lard, and 25% casein. Four experimental groups were given similar diets containing 30% meat: beef, chicken, pork or bacon. Protein and fat contents were adjusted and identical in the five diets. Dietary and fecal NOC were assayed by thermal energy analysis:

samples were denitrosated and the released nitric oxide was detected by chemiluminescence.[4] Fecal nitrate and nitrite were also measured (data not shown). Colons were fixed flat in formalin, then stained for 10 min with methylene blue and scored blindly for ACF by a single observer, according to Bird's procedure.[5] Promotion was assessed by the multiplicity of ACF (number of crypts per ACF) at day 100.

Study 2 – High-fat diet post-initiation: As previously reported,[6] 100 Fisher-344 rats were given a single i.p. injection of azoxymethane ($20\,mg\,kg^{-1}$), then randomized to 10 different AIN-76-based high-fat diets, given *ad libitum* for 100 days. Five diets were adjusted to 14% fat and 23% protein, and 5 other diets to 28% fat and 40% protein. Fat and protein were supplied by (1) lard and casein, (2) olive oil and casein, (3) beef, (4) chicken with skin, and (5) bacon. Colonic ACF and fecal NOC were assayed as described for Study 1.

Study 3 – Initiation: Twenty-five Fisher-344 rats were randomized into 3 groups: ten control rats were fed an high-fat AIN-76 diet containing 28% corn oil and 40% casein; five positive control rats were fed the same diet and given azoxymethane ($5\,mg\,kg^{-1}$ BW, i.p. injection); ten experimental rats were fed a diet containing 60% bacon. This bacon-based diet was the same as in the promotion study 2 described above, and was given for 30 days to the rats. Experimental rats were then given the control diet for 15 days. Initiation was assessed by the number ACF per rat 45 days after the beginning of the bacon-based diet. NOC were assayed as described above, on fecal samples collected before day 30.

3 Results

Study 1 – Low-fat diet post-initiation: Compared with controls, the ACF multiplicity was reduced by 17% in rats fed the bacon-based diet (low-fat diet containing 30% bacon). The number of crypts per ACF was 2.9 ± 0.2 and 2.4 ± 0.2 respectively ($p < 0.01$). In contrast, the ACF multiplicity was the same in rats fed beef, chicken and pork meat and in control rats fed casein and lard (*e.g.*, control-diet: 2.9 ± 0.2 and beef-diet: 2.9 ± 0.3). The total number of ACF per rat was not changed by the diet. The bacon-based diet brought 13.3 nmol NOC daily to the rats, which excreted $13.1\,nmol\,NOC\,d^{-1}$. Rats given bacon had 10.5 times more fecal NOC than controls: 14.4 ± 5.4 and $1.4 \pm 0.6\,nmol\,g^{-1}$ respectively ($p < 0.0001$). Rats fed with beef, pork or chicken meat had less fecal NOC than controls (beef: 0.3 ± 0.1, pork: 0.5 ± 0.5, chicken: 0.9 ± 0.8, and controls 1.4 ± 0.6 nmol/g feces). Thus bacon-fed rats has more NOC in feces, but smaller ACF than control rats.

Study 2 – High-fat diet post-initiation: Compared with their respective casein and lard fed controls, the ACF multiplicity was reduced by 12% in rats fed the 30% bacon diet, and by 20% in rats fed the 60% bacon diet.[6] This protective effect of bacon was very significant ($p < 0.001$). In contrast, the ACF multiplicity was the same in other groups, with no difference between rats fed beef, chicken, or casein, as previously published ($p=0.7$).[6] Diets containing 30% and 60% bacon brought 4.3 and 12.2 nmol NOC daily to the rats, which excreted 8.6 and 12.3 nmol NOC d^{-1} respectively. Rats given 30% bacon had 20 times more NOC in feces than controls fed 14% lard: 9.3 ± 2.1 and 0.5 ± 0.2 nmol/g respectively ($p<0.0001$). Rats given 60% bacon had 12 times more fecal NOC than controls fed 28% lard: 13.7 ± 3.1 and 1.10 ± 0.49 nmol/g respectively ($p<0.0001$). Rats fed with beef or chicken meat had less fecal NOC than controls (beef: 0.3 ± 0.2, chicken: 0.4 ± 0.2 nmol/g feces). Thus, bacon fed rats had more NOC in feces, but smaller ACF than control rats.

Study 3 – Initiation: No ACF were detected in saline injected rats fed with a bacon-bacon diet or a control diet. A mean number of 28 ACF was detected in azoxymethane-injected rats. Fecal NOC concentration was 1.1 ± 0.1, 1.4 ± 0.3 and 22.0 ± 2.8 nmol/g feces in control rats, AOM-injected rats, and bacon-fed rats respectively ($p<0.001$). The NOC concentration was thus 21-fold higher in feces of bacon-fed rats than in feces of controls, but no ACF was detected in bacon-fed rats.

4 Discussion

The feeding of a diet containing 30 or 60% bacon to rats increased the level of NOC in the gut. Compared with control rats fed with casein and lard, the bacon-based diets led to a 10 to 20-fold increase in fecal NOC in three independent studies. In human volunteers, a high red meat diet increases the fecal NOC 4-fold, to an estimated concentration of 2 μg/g feces.[2] In this study, assuming that the median MW of fecal NOC is 150, the fecal concentration might be 1.5 to 2.5 μg/g feces. The concentration of NOC in feces of bacon-fed rats is thus similar to the concentration found in stools of volunteers eating a high meat diet. In contrast, a diet containing 30% or 60% beef meat did not increase the fecal NOC in rats. However, in humans the origin of NOC in feces is likely to be endogenous.[2] This may not be true in rats, where fecal NOC seems to come from ingested NOC occurring in bacon.

In this study, a high level of fecal NOC was not associated with ACF initiation or ACF promotion in rats. Hasegawa could show the initiation of few colon tumors in rats given a total dose of 100 μmol of pure genotoxic NOC.[7] He also showed a weak promotion of colon tumors in rats given a total dose of 1 mmol of non-genotoxic NOC.[7] Here, each rat had received a total dose of 0.4 μmol of NOC (initiation study), or of 1.3 μmol of NOC (promotion studies). Hence, the NOC doses from bacon might have been too small to initiate or promote ACF in the colon of rats. It is also possible that the specific NOC in bacon and feces are not genotoxic or tumor-promoting for the rat colon mucosa.

Rats given the high-bacon diets consistently had smaller ACF than controls, as previously shown.[6] This protective effect was found in two independent studies, using bacon from different suppliers. Bacon-based diets containing 7, 14 and 18% fat decreased significantly the ACF growth in rats colon, when compared with matched controls. A bacon-based diet thus appears to protect against carcinogenesis, perhaps because bacon contains 5% NaCl and increased the rats' water intake.[6]

The present results suggest that (i) a high-bacon diet can increase 10- to 20-fold the fecal NOC level in rats, leading to fecal concentrations similar to what is found in human volunteers eating a high red meat diet, (ii) in contrast, a diet high in beef meat does not increase fecal NOC in rats, (iii) the intestinal concentration of NOC due to a bacon-based diet does not initiate or promote colon carcinogenesis, assessed by the ACF assay in rats. Taken together, the data do not support the hypothesis that NOC can explain the association of meat intake with colon cancer risk.

Acknowledgements

This work was supported by INRA and DGER of French Ministry of Agriculture, and by the Comité du Gers de la Ligue Nationale contre le Cancer, France.

5 References

1 G. Parnaud and D.E. Corpet, *Bull. Cancer*, 1997, **84**, 899.
2 S.A. Bingham, B. Pignatelli, J.R.A. Pollock, A. Ellul, C. Malaveille, G. Gross, S. Runswick, J.H. Cummings and I.K. Oneill, *Carcinogenesis*, 1996, **17**, 515.
3 B.C. Pence, M. Landers, D.M. Dunn, C.L. Shen and M.F. Miller, *Nutr. Cancer*, 1998, **30**, 220.
4 B. Pignatelli, I. Richard, M. Bourgade and H. Bartsch, *Analyst*, 1987, **112**, 945.
5 R.P. Bird, *Cancer Lett.*, 1987 **37**, 147.
6 G. Parnaud, G. Peiffer, S. Tache and D.E. Corpet, *Nutr. Cancer*, 1998, **32**, 165.
7 R. Hasegawa, M. Futakuchi, Y. Mizoguchi, T. Yamaguchi, T. Shirai, N. Ito and W. Lijinsky, 1998, *Cancer Lett.*, **123**, 185.

3.8

Cellular Effects of Lignans: Modulation of Growth, Oxidative DNA Damage and Cell Metabolism in Human Colon Cancer Cells

T.W. Becker,[1] M.G. Peter[2] and B.L. Pool-Zobel[1]

[1]INSTITUTE FOR NUTRITION, FRIEDRICH SCHILLER UNIVERSITY, JENA, GERMANY
[2]INST. ORG. CHEMIE UND STRUKTURANALYTIK, UNIVERSITY OF POTSDAM, GERMANY

1 Introduction

It is generally well accepted that individual dietary factors have potential to reduce risks of cancer development in many tissues. Dietary ingredients that are implicated as being the actual chemopreventive factors include phytoprotectants (micronutrients, secondary plant ingredients) and products, formed during the fermentation of non-digestible components in the gut. Also gut metabolites arising from plant lignan precursors, like enterolacton and enterodiol, have been implicated to exert chemopreventive potentials (Figure 1).

These putative effects include antiestrogenicity in hormone-dependent cells expressing the estrogen receptor (ER) and other diverse intracellular effects in ER-non expressing cells. Specifically, it was the purpose of this study to compare the effects of enterolacton and enterodiol on oxidative stress, cell metabolism and cell growth in colon cells (ER^-). The experiments were performed with the human colon cancer cell line HT 29 and its redifferentiated subclone HT 29 clone 19a, which resembles primary human colon cells.[1] These cells were used as models to assess lignans in two different stages of malignancy.

2 Materials and Methods

Oxidative DNA damage was detected using single-cell-microgel-electrophoresis with repair enzymes.[2]

Cell metabolism was determined with a Cytosensor Microphysiometer (Molecular Devices) which allows the determination of pH values in perfused

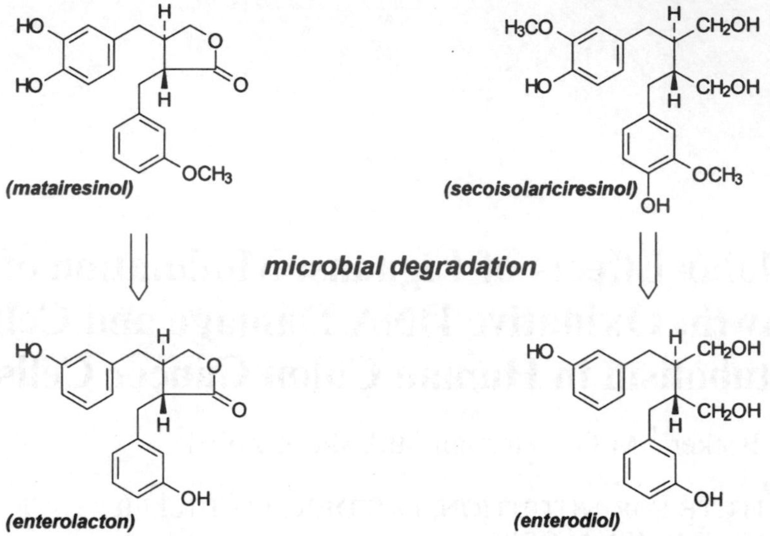

Figure 1 *Microbial transformation of the plant lignans matairesinol or secoisolariciresinol to the corresponding mammalian lignans enterolacton or enterodiol, respectively*

cell cultures within a detection-volume of less than 2 μl. pH changes were recorded before, during and after repeated treatments with various concentrations of lignans.[3]

Growth of colon cells was determined in microtiter plates. 48 h after seeding, the cells were treated with various concentrations of lignans in culture-medium. The cell number was detected 24 and 72 h after treatment by fixation and permeabilisation cells with methanol for 5 min, followed by adding DAPI. After 30 min, the DNA content was detected fluorometrically with Ex/Em 360/465 nm.

3 Results

a. The mammalian lignans enterolacton and enterodiol reduce oxidative DNA damage (Table 1).
b. Lignans induce a fast cellular acidification response in human colon cancer cells (Figure 2).

Table 1 *DNA damage in HT 29 clone 19a cells detected with the 'comet-assay'. (Tail intensity [%]; means ± SEM; n = 6)*

	Untreated	*Enterolacton*	*Enterodiol*
DNA-strand-breaks	2.1 ± 0.3	2.3 ± 0.4	2.3 ± 0.4
Oxidised pyrimidine bases	29.1 ± 7	10.3 ± 3	12.7 ± 3.4
Oxidised purine bases	47.7 ± 7	23.2 ± 3.4	36.2 ± 4.4

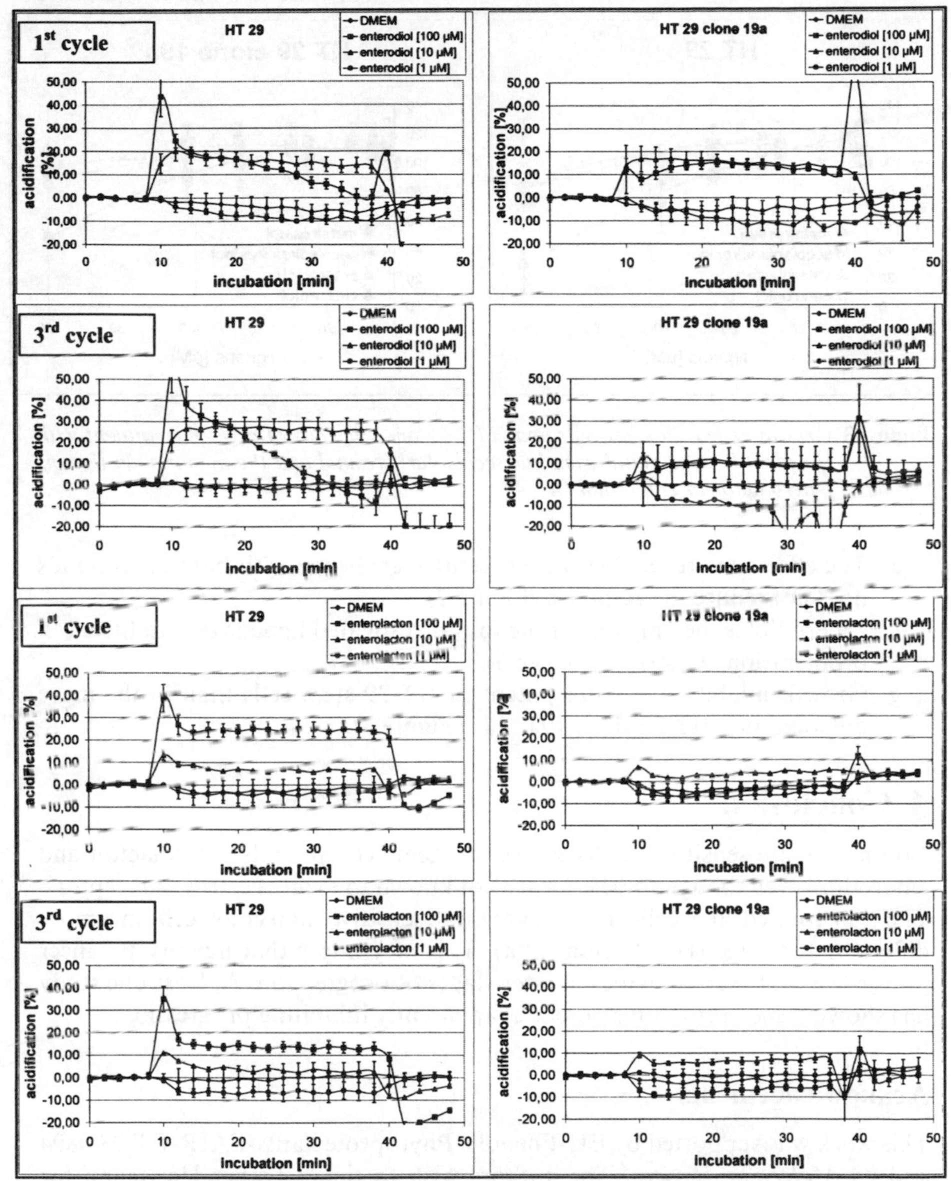

Figure 2 *Cellular acidification response following repeated treatments with enterodiol or enterolacton in the two human colon cancer cell lines HT 29 and HT 29 subclone 19a. Lignan treatment started after 8 min incubation with DMEM and was stopped after 40 min*

c. HT 29 stem cells are more sensitive for both enterolacton and enterodiol than the more differentiated HT 29 clone 19a cells (Figure 2).

d. Enterodiol is more potent in eliciting the cellular response than enterolacton (Figure 2).

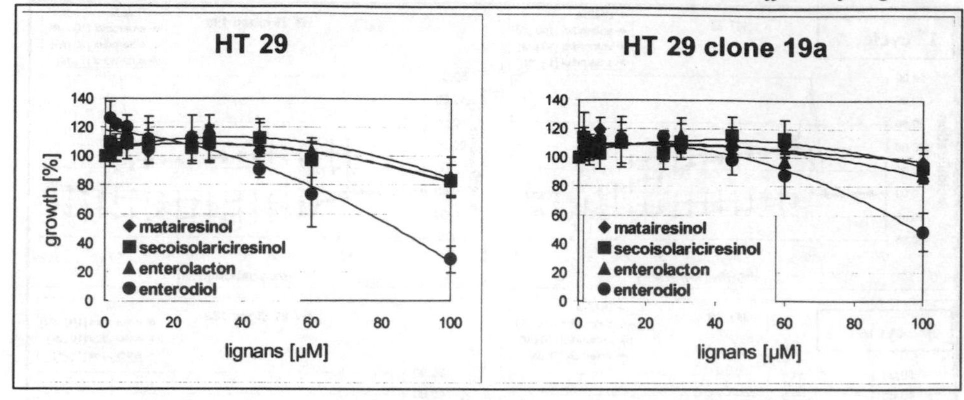

Figure 3 *Growth of HT 29 stem cells and HT 29 clone 19a cells after 72 h treatment with the plant lignans matairesinol or secoisolariciresinol and the mammalian lignans enterolacton or enterodiol (n = 4)*

e. The effects are reversible since repetitive challenge with both compounds does not reduce the response (Figure 2).
f. Enterodiol is the only one of the four investigated lignans that reduces cell proliferation at 100 μM (Figure 3).
g. Growth inhibition is more potent in HT 29 stem cells than in the more differentiated HT 29 clone 19a cells (Figure 3).

4 Conclusion

Colon cells are sensitive to the phytoestrogenic compounds enterolacton and enterodiol, even though these cells are not known to express estrogen receptors. The transformed stem cells (HT 29) are more sensitive than colon cells in a more differentiated stage (HT 29 clone 19a). This may imply that lignans, if cancer preventive, will exert activities in later stages of progression. At least one study has shown tumor preventive potential apparently inhibiting progression.[4]

Acknowledgements

This work was supported by EU-Phenolic Phytoprotectants-FAIR-CT-98 0894 and by *Molecular Devices*. The authors wish to thank Esther Hartmann for technical assistance.

5 References

1 M. Rousset, *Biochimie*, 1986, **68**, 1035.
2 A.R. Collins, S.J. Duthie and V.L. Dobson, *Carcinogenesis*, 1993, **14**, 1733.
3 H.M. McConnell, J.C. Owicki, J.W. Parce, D.L. Miller, G.T. Baxter, H.G. Wada and S. Pitchford, *Science*, 1992, **257**, 1906.
4 M.J. Davies, E.A. Bowey, H. Adlercreutz, I.R. Rowland and P.C. Rumsby, *Carcinogenesis*, 1999, **20**, 927.

3.9

Genotoxic and Antigenotoxic Effects of Isothiocyanates

Fekadu Kassie,[1] Qin Hong-Min,[1] Sylvie Rabot,[2] Beatrice Pool-Zobel,[3] Maria Uhl,[1] Wolfgang Huber[1] and Siegfried Knasmüller[1]

[1]INSTITUTE OF TUMOR BIOLOGY-CANCER RESEARCH, UNIVERSITY OF VIENNA, VIENNA, AUSTRIA
[2]INSTITUT NATIONAL DE LA RECHERCHE AGRONOMIQUE, JOUY-EN-JOSAS, FRANCE
[3]INSTITUTE OF HYGIENE AND TOXICOLOGY, FEDERAL RESEARCH CENTER OF NUTRITION, KARLSRUHE, GERMANY

1 Introduction

Isothiocyanates (ITCs) are enzymatic breakdown products of glucosinolates, natural thioglycosides present in cruciferous plants. Epidemiological evidence indicates that consumption of ITC-containing *Brassica* vegetables is inversely associated with the incidence of cancer.[1] Moreover, experimental results from laboratory studies indicate that ITCs are able to block the tumorigenic effect of several environmental carcinogens.[2] On the other hand, some ITCs have been reported to be clastogens[3] and allyl isothiocyanate was found to induce transitional papilloma in the urinary bladder of rats. Therefore, in view of the presence of abundant quantities of ITCs in the diet of human beings, a thorough investigation on the health risks as well as the benefits of these compounds is highly required.

2 Results

2.1 Genotoxic Effects of Selected ITCs

We have tested the genotoxic effects of four ITCs [benzyl isothiocyanate (BITC), allyl isothiocyanate (AITC), methyl isothiocyanate (MITC) and phenethyl isothiocyanate (PEITC)] commonly found in cruciferous plants using different *in vitro* and *in vivo* test sytems. The results are summarized in

Table 1 *Genotoxic effects of ITCs[a] in different* in vitro *and* in vivo *test systems[b]*

Test system[1]	*Amount of ITCs used*	*End point*	*Result*
Salmonella/microsome assay with TA98 and TA100 ± S9[1]	10–150 $\mu g\ ml^{-1}$	His^+ revertants	Moderate effect (80–150% increase in the number of his^+ revertants with all four ITCs compared with the spontaneous background level); addition of S9 attenuated the effect completely.
In vitro differential DNA repair assay using two *E.coli* strains differing in DNA repair capacity (±S9)	1–25 $\mu g\ ml^{-1}$	repairable DNA damage (decrease in *uvr*B/*rec*A colonies compared to *uvr*B^+/*rec*$^+$)	Survival rates of *uvr*B/*rec*A colonies decreased by 50% due to repairable DNA damage at a concentration of 2, 11, 12 and 25 µg/ml by BITC, MITC, AITC and PEITC, respectively; almost complete attenuation of genotoxic effect upon addition of S9 mix.
Micronucleus induction assay with human Hep G2 cells	0.25–5 $\mu g\ ml^{-1}$	micronuclei (MN)	BITC and PEITC caused a 2-fold increase in the MN frequency (compared to the background level) at concentrations of 2 and 3 $\mu g\ ml^{-1}$, respectively. With AITC and MITC, concentrations more than 5 $\mu g\ ml^{-1}$ were required.
SCGE assay with rat gastric mucosa cells and primary hepatocytes	1.25–5 $\mu g\ ml^{-1}$	comet formation as a consequence of DNA strand breaks	75% of the cells exhibited severe DNA strand breakage at the highest concentration
In vivo differential DNA repair assay in mice using two *E. coli* strains differing in DNA repair capacity	90 and 270 $mg\ kg^{-1}$	repairable DNA damage (decrease in *uvr*B/*rec*A colonies compared to *uvr*B^+/*rec*$^+$)	Survival rates of *uvr*B/*rec*A colonies decreased by more than 50% due to repairable DNA damage at the higher BITC and AITC dose level; with MITC and PEITC, survival rate decreased by 20% only even at the higher concentration.
In vivo SCGE assay with gastric mucosa cells and primary hepatocytes	220 mg/kg^{-1}	comet formation as a consequence of DNA strand breaks	The DNA was severely damaged only in 18% of the cells

[a] ITCs used in the assays were BITC, AITC, MITC, and PEITC; SCGE assay was performed with BITC only.
[b] The *Salmonella*/microsome assay was carried out according to the protocol of Maron and Ames[5] with *Salmonella typhimurium*. TA98 and TA100 strains. Differential DNA repair assay was performed as described by Kassie *et al.*,[6] and relative survival was determined by comparing the viability of *uvr*B/*rec*A colonies (repair deficient) with their *uvr*B^+/*rec*$^+$ (repair proficient) counterparts. The micronuleus assay protocol was done as described by Natarajan and Darroudi.[7] Comet assay was carried out as described by Pool-Zobel *et al.*[8]

Table 1. All the ITCs induced marked genotoxic effects in all *in vitro* test systems used except in the *Salmonella*/microsome assay. In comparison to the results obtained under *in vitro* test conditions, the *in vivo* genotoxicity of ITCs was by far weaker. We have also recently reported the genotoxicity of crude juices prepared from ITC-containing *Brassica* vegetables in bacterial and mammalian cells.[4] The most pronounced effects were seen with juices prepared from Brussels sprouts, white cabbage and cauliflower.

Table 2 *Reduction of the genotoxic effects of BITC (5 $\mu g\,ml^{-1}$) by human saliva, human gastric juice, bovine serum albumin and various antioxidants (vitamin E, β-carotene, sodium benzoate and ascorbic acid)*[a]

Putative antigenotoxin	*Concentration range*	*Result* (ID_{50}[b])
Human saliva	70–300 $\mu g\,ml^{-1}$	180 $\mu g\,ml^{-1}$
Human gastric juice	70–300 $\mu g\,ml^{-1}$	120 $\mu g\,ml^{-1}$
Bovine serum albumin	1–9 $mg\,ml^{-1}$	4 $mg\,ml^{-1}$
Vitamin E	10–180 $\mu g\,ml^{-1}$	30 $\mu g\,ml^{-1}$
β-Carotene	10–180 $\mu g\,ml^{-1}$	25 $\mu g\,ml^{-1}$
Sodium benzoate	10–180 $\mu g\,ml^{-1}$	25 $\mu g\,ml^{-1}$
Vitamin C	10–180 $\mu g\,ml^{-1}$	No effect

[a] The assay used was the *in vitro* differential DNA repair assay with *E. coli* strains and performed as described by Kassie *et al.*[6] Relative survival was determined by comparing the viability of *uvr*B/*rec*A colonies (repair deficient) to that of *uvr*B^+/*rec*$^+$ (repair proficient) colonies.
[b] ID_{50} is the concentration at which BITC-induced repairable DNA damage was decreased by 50% upon adding the various substances to the incubation mixture containing the indicator bacteria and BITC (5.6 $\mu g\,ml^{-1}$).

2.2 Detoxification of ITCs

The pronounced differences between the *in vitro* and *in vivo* genotoxicity of ITCs prompted us to test the role of human body fluids, proteins and antioxidants (vitamins E and C, β-carotene and sodim benzoate) in the attenuation of the genotoxic effects of BITC *in vitro*. Exposure of a mixture of *E. coli* strains to BITC for 2 h in the presence of the above substances resulted in a marked attenuation of the repairable DNA damage induced by this compound (Table 2). While low concentrations of the antioxidants were protective, higher concentrations augmented the effect of BITC. Moreover, we studied the genotoxicity of BITC in single cell gel electrophoresis (SCGE) assay with rat gastric mucosa cells in the presence of gastric mucus or after washing the mucus with phosphate buffered saline. As with human gastric mucus, a reduction in the effect of BITC was observed with rat gastric mucus (cells with DNA strand breaks was reduced by 80% in the presence of the mucus, data not shown).

2.3 Possible Mechanisms of Genotoxicity of ITCs

The protective effects observed with the different radical scavengers as well as results of lipid peroxidation experiments with Hep G2 cells (Figure 1) suggest that reactive oxygen species might be involved in the genotoxicity of ITCs. However, a discrepancy of three orders of magnitude or more was found between the ITC concentrations required to cause marked micronuclei induction and malondialdehyde formation in Hep G2 cells. Thus, the observed genotoxic effects of ITCs might be associated with the direct effect of free radicals and/or other intermediates and final products of lipid peroxidation rather than malondialdehyde.

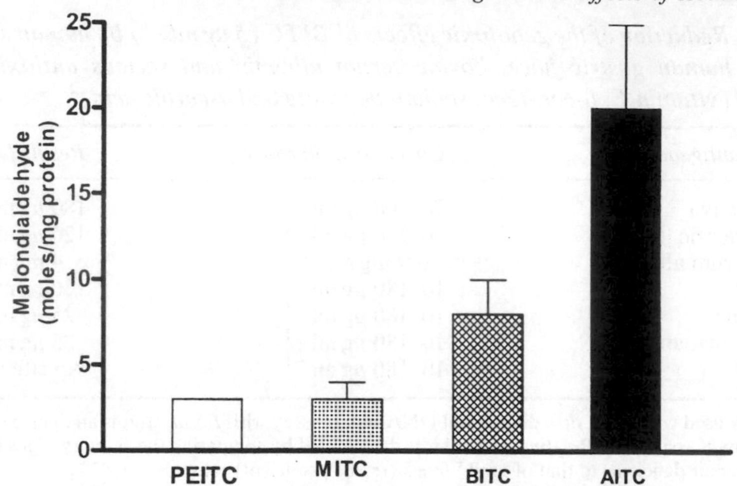

Figure 1 *Induction of malondialdehyde formation in Hep G2 cells treated with different ITCs (30 mg ml^{-1}). Exposure time was 3 h and the assay was performed according to Yagi.[7] The values showed here are mean ± SD of three assays made in paralllel in a single experiment*

2.4 Antigenotoxic Effects of ITCs

We investigated the potential protective effects of ITCs towards a variety of mutagens/carcinogens and these results are summarized in Table 3. Protection was found in both *in vitro* and *in vivo* studies. Concentrations of ITCs required to reduce the effects of the different genotoxins under *in vitro* test conditions were < 5 μg ml^{-1} whereas relatively high doses of the compounds were required (*ca.* 200 mg kg^{-1}) in experiments with animals. The protective effects observed here are in line with previous reports in which ITCs were found to block the tumorigenic effects of many carcinogens (see a review by Zhang and Talalay[2]).

2.5 Mechanisms of Antigenotoxic Effects of ITCs

At present two mechanisms are known by which ITCs may block the tumorigenic effects of chemical carcinogens: suppression of carcinogen activation by cytochrome P450 and induction of enzymes that detoxify carcinogens. Th first mechanism accounts for the protective effect of ITCs towards nitrosamines[10] while protection towards polycyclic aromatic hydrocarbons was attributed to the induction of glutathione S-transferase.[12,13] Recent results from our laboratory suggest that the reduction of the genotoxic effects of heterocyclic aromatic amines is due to the induction of UDP-glucuronosyl transferase.[14]

3 Conclusions

ITCs are among the most biologically active compounds. In addition to their genotoxic and antigenotoxic effects they were also reported to exert cytotoxic,

Table 3 *Antigenotoxic effects of ITCs against different environmental mutagens/carcinogens in different test systems*

Mutagen/carcinogen	*ITC*	*Test system*[a]	*Result*
IQ (0.5 μg ml^{-1})	PEITC (1 μg ml^{-1})	*In vitro* differential DNA repair asay with *E. coli*	98% decrease in repairable DNA damage
IQ (40 mg kg^{-1} bw, orally)	PEITC (220 mg kg^{-1} orally)	Host mediated assay in mice with *E. coli*	50% decrease in repairable DNA damage
IQ (90 mg kg^{-1} bw, orally)	BITC (70 mg kg^{-1} orally for three days)	*In vivo* SCGE assay with rat liver and colon cells	60% and 40% decrease in DNA damage in liver and colon cells, respectively.
PhIP (5 μg ml^{-1})	PEITC (1 μg ml^{-1})	*In vitro* differential DNA repair assay with *E. coli*	90% reduction in repairable DNA damage
PhIP (120 μg ml^{-1})	PEITC (1 μg ml^{-1})	Micronucleus induction in Hep G2 cells	no protective effect
PhIP (50 mg kg^{-1} bw, orally)	BITC (70 mg kg^{-1} bw, orally)	DNA adduct measurement in different organs of rat	reduced by 30–100%
PhIP (175 mg kg^{-1} bw, orally)	PEITC (210 mg kg^{-1}, orally)	Host mediated assay in mice with *E. coli*	20–40% decrease in repairable DNA damage in different organs of mice
Benzo[a]pyrene (4 μg ml^{-1})	BITC (0.8 μg ml^{-1})	*In vitro* SCGE assay with Hep G2 cells	50% reduction in comet tail moment
N-Nitrosodimethyl-amine (2.2 μg ml^{-1})	PEITC (1 μg ml^{-1})	*In vitro* MN assay with Hep G2 cells	60% decrease in MN frequency
N-Nitrosodimethyl-amine (60 mg kg^{-1} bw)	PEITC (210 mg kg^{-1}, orally)	Host mediated assay in mice with *E. coli*	54% decrease in repairable DNA damage in liver

[a]Differential DNA repair assay was performed as described by Kassie *et al.*[6] The micronuleus assay test was carried out as described by Natarajan and Darroudi.[8] SCGE asay was done following the protocol of Pool-Zobel *et al.*[9] PhIP-DNA adducts were quantified by the ^{32}P-post-labelling method as described by Gupta *et al.*,[19] and Kaderlik *et al.*[20]

carcinogenic, fungicidal, and bactericidal activities.[15] The present data indicate that ITCs are 'janus' compounds in that they possess both genotoxic and antigenotoxic properties. Although these compounds are among the most potent *in vitro* genotoxins, such effects were seen under *in vivo* conditions only when very high doses of ITCs (>200 mg kg^{-1} bw) were administered. This seems to be due to the *in vivo* attenuation of the activity of ITCs as a result of direct binding to proteins, scavenging by antioxidants or enzymatic detoxification. As the daily intake of glucosinolates by human beings was found to be only 46 mg/p/day[16] or less,[17] dietary exposure of humans to ITCs does not seem to be associated with a risk of genetic damage.

In our investigations, we found ITCs to be protective towards the genotoxic effects of different groups of mutagenic/carcinogenic compounds. Strong antitumorigenic effects were also observed in long-term carcinogenicity experiments.[12,13] However, it remains to be further elucidated whether increased intake of ITCs is associated with lower cancer rates in humans. Currently available epidemiological data show an inverse association between consumption of cruciferous vegetables and cancer incidence and the most significant negative association was found between *Brassica* vegetables and colon cancer.[18]

Acknowledgement

This work was supported by EU grant to S.K.

4 References

1 G. Block, B. Patterson and A. Subar, *Nutr. Cancer*, 1992, **18**, 1.
2 Y. Zhang and P. Talalay, *Cancer Res.*, 1994, **54**, 1976.
3 S.R.R. Musk and I.T. Johnson, *Mutat Res.*, 1993, **300**, 111.
4 F. Kassie, W. Parzefall, S. Musk, I. Johnson, G. Lamprecht, G. Sontag and S. Knasmüller, *Chem. Biol. Interact.*, 1996, **102**, 1.
5 D.M. Maron and B.N. Ames. 'Handbook of Mutagenicity Test Procedures', Elsevier, Amsterdam, 1984, p. 93.
6 F. Kassie, W. Parzefall, L. Kronberg, R. Franzen, R. Schulte-Hermann and S. Knasmüller, *Environ. Mol. Mutagen.*, 1994, **24**, 317.
7 K. Yagi. 'Lipid Peroxides and Its Clinical Significance', New York, 1982, p. 223.
8 A.T. Natarajan and F. Darroudi, *Mutagenesis*, 1991, **6**, 399.
9 B.L. Pool-Zobel, N. Lkatzmann, M. Knoll, F. Kuchenmeister, L. Lambertz, U. Leucht, H. G. Schröder and P Schmezer, *Env. Mol. Mutagen.*, 1994, **24**, 23.
10 M.A. Morse, S.G. Amin, S.S. Hecht and F.L. Chung, *Cancer Res.*, 1989, **49**, 2894.
11 S. Barcelo, J.K. Chipman, A. Gescher, K. Mace and A. Pfeifer, '5th International Conference on Mechanisms of Antimutagenesis and Anticarcinogenesis', Okayama, Japan, 1996, 41.
12 L. Wattenberg, *J. Natl. Cancer Inst.*, 1977, **58**, 395.
13 L. Wattenberg, *Cancer Res.*, 1981, **41**, 2991.
14 F. Kassie, H.M. Qin, S. Rabot and S. Knasmüller, *Neoplasma*; in press.
15 B.N. Ames, M. Profet and L.S. Gold, *Proc. Natl. Acad. Sci. USA*, 1990, **87**, 7777.
16 R.K. Heaney and G.R. Fenwick, 'Natural Toxicants in Food, Progress and Prospects', Wiley, Chichester, UK, 1987, 76.
17 R. Lange, *Ernährungs-Umschau*, 1992, **39**, 292.
18 K.A. Steinmetz and J.D. Potter, *Cancer Causes Control*, 1991, **2**, 427.
19 R.C. Gupta, M.V. Reddy and K. Randerath, *Carcinogenesis*, 1982, **3**, 1081.
20 K.R. Kaderlik, R.F. Minchin, G.J. Mulder, K.F. Ilet, M. Daugaard Jenson, C.H. Teitel and F.F. Kadlubar, *Carcinogenesis*, 1994, **15**, 1703.

3.10

The Effect of Cooking on the Protective Effect of Broccoli Against Damage to DNA in Colonocytes

Brian Ratcliffe,[1] Andrew R. Collins,[2] Helen J. Glass,[1,2,3] Kevin Hillman[3] and Rebecca J.T. Kemble[1,3]

THE BOYD ORR RESEARCH CENTRE – ABERDEEN RESEARCH CONSORTIUM AT
[1]THE ROBERT GORDON UNIVERSITY, KEPPLESTONE, ABERDEEN AB15 4PH, UK
[2]THE ROWETT RESEARCH INSTITUTE, BUCKSBURN, ABERDEEN AB21 9SB, UK
[3]THE SCOTTISH AGRICULTURAL COLLEGE, CRAIBSTONE, ABERDEEN AB21 9YA, UK

1 Introduction

The consumption of fruits and vegetables is associated with reduced risk of cancer.[1] Such associations have led to investigations of individual plant components and extracts including vitamins and non-nutrient antioxidants. While there is some evidence that supplements of antioxidant vitamins reduce the risk of cancer development, this has not been found consistently. The supplementation trial of Greenberg *et al.*[2] failed to prevent colorectal adenomas. Studies with plant extracts may fail to elucidate the mechanisms whereby the consumption of raw vegetables and salads seems to protect against colorectal cancer.[3]

The authors have hypothesized that intact plant cells act as 'vehicles' for the delivery of antioxidants and anticancer agents to the colon where they are 'unpacked' by bacterial fermentation.[4] In a study that supports the hypothesis, the authors showed that whole broccoli protects against endogenous damage to the DNA in coloncytes and that this protection is removed if the vegetable is physically processed (homogenized) immediately prior to consumption.[5] The aim of this experiment was to examine the effect of cooking on this protective effect of broccoli. Again, oxidative DNA damage was used as an index of carcinogenic potential. Microwave heating was used as the cooking method to minimize nutritional losses and differences compared with whole raw broccoli and because it could be more precisely standardized for each subject.

2 Materials and Method

To obtain appropriate populations of colonocytes for the assessment of endogenous damage to DNA, it was necessary to use experimental animal models. Fifteen, male Landrace × Large White pigs were divided into three groups that were matched for age and starting weight. Each group received subsequently a standard, high quality, cereal-based diet (control group, C) or this same diet supplemented with whole, raw broccoli (treatment group, W) or with similar amounts of broccoli which had been cooked by microwaves (treatment group, M). The latter group's vegetable supplement was cooked using a domestic microwave cooker. The cooking was standardized using similar weights of whole broccoli on each occasion. The pigs were 57 d of age at the start of the experiment and they were maintained on the diets for twelve days. The animals received their feed daily in two meals, to simulate human eating patterns, and the amounts were controlled to maintain similar dry matter intakes in all groups. Feed consumption was monitored on a daily basis.

At the end of the experimental period, the pigs were weighed then anaesthetized and killed. *Post mortem,* the colon was excised from caecum to rectum. In addition, samples of faeces, colonic contents, and blood were obtained for further analyses. Enumeration of selected microbial groups (coliforms, lactobacilli and bifidobacteria) was performed immediately on the samples of colonic contents and faeces. The anatomical mid-point of the colon was located and a section of 200 mm length was taken for the isolation of colonocytes by a modification of the method of Brendler-Schwaab *et al.*[6] Isolated colonic cells were suspended in freezing medium (90% FCS, 10% DMSO) at a density of 3×10^6 ml^{-1} and were stored at −80 °C. The thawed colonocytes were assayed subsequently for DNA strand breaks using a modification of the 'comet' method of Collins *et al.*[7] With this technique, 'comets' arising from colonic cells were allocated a score from 0–4 indicating the increasing degree of damage. A total of 100 colonocytes was examined blind and at random for each pig. Plasma vitamin C concentrations were determined by reversed phase HPLC using ion-pairing reagent with UV detection.[8] Plasma α-tocopherol was also determined by HPLC.[9]

3 Results

Food intake over the experimental period (mean 860 g dry matter d^{-1}) and final body weights (mean 27.9 kg) were similar across the groups. Groups W and M consumed similar amounts of broccoli (approximately 600 g d^{-1}).

Examination of the microflora revealed that there was no significant difference between the treatment groups in the bacterial numbers of coliforms, lactobacilli and bifidobacteria within the colonic contents and faeces.

There was no significant difference between plasma vitamin C levels which were 38.9 (SD 7.94), 27.7 (SD 12.59), and 21.7 (SD 14.33) μM, for W, M, and C respectively. Similarly, plasma α-tocopherol levels were not significantly different with a mean value 1.3 (pooled SD 0.34) μM.

However, the mean pooled scores (arbitrary units) for 'comets' were respectively 120 (SD 11.6) for C, 147 (SD 54.8) for M and 61 (SD 15.7) for W ($P = 0.005$) indicating substantially less endogenous DNA damage in the pigs consuming whole, raw broccoli.

4 Discussion and Conclusions

Pigs were used because rats are inappropriate models for this work since their gnawing habit reduces plant material to small particles that are unlike the larger food particles detectable in the stomachs of pigs and humans. The vegetable supplement produced no effect on the levels of plasma antioxidants but there was clearly a very marked protective effect against DNA strand breaks as measured by 'comet' analysis from the consumption of whole, raw broccoli. This has been demonstrated in earlier work [5] Cooking of the whole broccoli removed this protection. The protective effect could not be explained by antioxidant status or changes in bacterial populations since these factors were not affected.

It is accepted that DNA damage is not necessarily an indicator of carcinogenic processes. Nevertheless, it can act as an indicator of undesirable conditions that may be conducive to mutagenesis or carcinogenesis.

While it is accepted that there may be other explanations, the results support the hypothesis that intact plant cells are important for the health of the large bowel. It is not yet clear which cellular component of these intact plant cells is the principal protective agent. Nevertheless, the consumption of raw vegetables may be important for reducing the risk of colonic cancer.

Acknowledgement

This work was supported by a grant from the World Cancer Research Fund.

3 References

1 World Cancer Research Fund, American Institute for Cancer Research, 'Food, nutrition, and the prevention of cancer: a global perspective', American Institute for Cancer Research, Washington, DC,1997.
2 E.R. Greenberg, J.A. Baron, T.D. Tosteson, D.H. Freeman, G.J. Beck, J.H. Bond, T.A. Colacchio, J.A. Coller, H.D. Frankl, R.W. Haile, J.S. Mandel, D.W. Nierenberg, R. Rothstein, D.C. Snover, M.M. Stevens, R.W. Summers and R.U. van Stolk. *New England J. Med.*, 1994, **331**, 141.
3 K.A. Steinmetz and J.D. Potter, *Cancer Causes Control*, 1991, **2**, 325.
4 B. Ratcliffe, A.R. Collins, H.J. Glass, K. Hillman and R.J.T. Kemble, *Cancer Lett.*, 1997, **114**, 57.

5 B. Ratcliffe, A.R. Collins, H.J. Glass, and K. Hillman, in 'Natural Antioxidants and Anticarcinogens in Nutrition, Health and Disease', ed. J.T. Kumpulainen and J.T. Salonen, Royal Society of Chemistry, Cambridge, 1999, p. 440.
6 S.Y. Brendler-Schwaab, P. Schmezer, U. Liegibel, S. Weber, K Michalek, A Tompa and B.L. Pool-Zobel, *Toxicol. in Vitro*, 1994, **8**, 1285.
7 A.R. Collins, S.J. Duthie and V.L. Dobson, *Carcinogenesis*, 1993, **14**, 1733.
8 M. Ross, *J. Chromatogr.*, 1994, **675**, 197.
9 D. Hess, H.E Keller, B. Oberlin, R. Bonfanti and W. Schuep, *Int. J. Vit. Nutr. Res*, 1991, **61**, 232.

3.11

Processed Wheat Aleurone Is a Rich Source of Bioavailable Folate

Michael Fenech,* Manny Noakes, Peter Clifton and David Topping

CSIRO DIVISION OF HUMAN NUTRITION, PO BOX 10041, GOUGER STREET, ADELAIDE SA AUSTRALIA 5000

1 Background

The important role of folic acid in the prevention of neural tube defects[1] has increased the importance of identifying natural dietary sources of folate that can make a useful contribution to intake of this vitamin in the general population. Using novel milling technology it has become possible to isolate aleurone cells with sheared cell walls from wheat grain and to prepare a novel cereal product that has a natural folate level of approximately $500\,\mu g\,(100\,g)^{-2}$. We have therefore performed intervention trials in humans to assess the relative bioavailability of natural folate in cereal made from aleurone flour (ALF) and to determine whether such a cereal could produce a significant increment in blood folate status when consumed.

2 Methods

Using a randomised series of short-term intervention trials with a cross-over involving 8 men and 8 women aged between 29 and 50 years we have compared the increment of plasma folate following ingestion of (a) 100 g wheat bran (WB) cereal (low folate control), (b) 100 g aleurone cereal (ALF) and (c) a tablet containing 500 µg folic acid taken together with 100 g wheat bran cereal (high folate control). The extent of folate absorption was measured by estimating the area under the plasma folate concentration *versus* time curve.

3 Findings

The results for the change in plasma folate over a seven hour period following ingestion of the above products have shown that the extent of increase in plasma

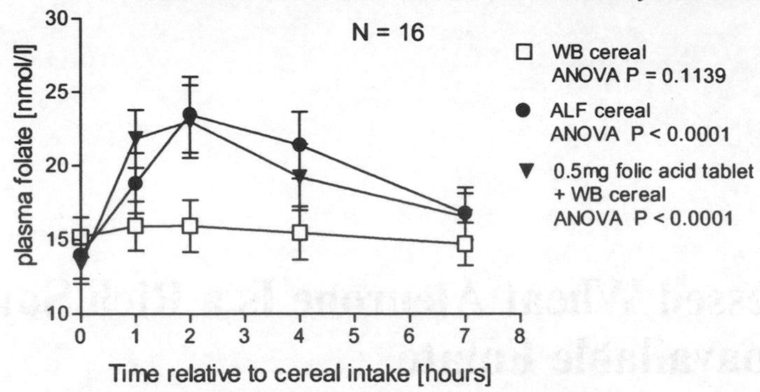

Figure 1 *Change in plasma folate following ingestion of WB cereal, ALF cereal and 0.5 mg folic acid with WB cereal. Results represent the mean ± SEM, n = 16, males and females combined. The ANOVA P values for the change in plasma folate with time for the WB cereal, ALF cereal and 0.5 mg folic acid with WB cereal were 0.1139, < 0.0001, < 0.0001 respectively*

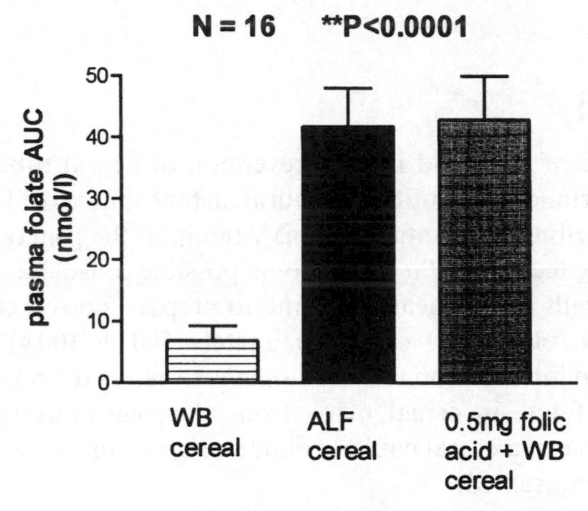

Figure 2 *Relative levels of plasma folate AUC during 7 hours following ingestion of WB cereal, ALF cereal and 0.5 mg folic acid + WB cereal. Results represent the mean ± SEM, n = 16, data for males and females combined. Column bars indicated with different letters show values that are significantly different from each other (P < 0.0001). AUC, area under plasma folate concentration vs time curve*

folate following ingestion of aleurone cereal was more than four-fold greater than that observed following the wheat bran cereal (P < 0.0001) and equivalent to that observed following the 500 μg folic acid tablet taken with wheat bran cereal (Figures 1 and 2). These results were also statistically significant when the results for males and females were analysed separately. The kinetics of folate increment in the blood (Figure 1) was more gradual following ingestion of aleurone cereal relative to folic acid from the tablet and the observed folate increments following ingestion of aleurone cereal and folic acid tablet were not significantly related. The extent of folate absorption was apparently significantly and positively related to an individual's base-line folate status.

4 Interpretation

This study has shown that cereal made from wheat aleurone flour is a significant source of natural bioavailable folate that can, when eaten in moderate amounts, make an effective contribution to increasing blood folate status. This effect is of a similar magnitude to that observed following ingestion of 500 μg folic acid. These results suggest that inclusion of foods made from wheat aleurone flour in the diet can be considered as an alternative important strategy for increasing folate intake in the general population.

5 References

1 L.B. Bailey, Folate requirements and dietary recommendations, in 'Folate in Health and Disease', ed. L.B. Bailey, Marcel Dekker Inc., New York, 1995, pp. 123–151.
2 N. Stenvert, Novel natural products from grain fractionation, in 'Cereals – Novel Uses and Processes', ed. G.M. Cambell, C. Webb and S.L. McKee, Plenum Press, New York, 1997, pp. 241–245.

3.12

Folate, Vitamin B12, Homocysteine Status and DNA Damage in Young Australian Adults

Michael Fenech,* Clare Aitken and Josephine Rinaldi

CSIRO DIVISION OF HUMAN NUTRITION, PO BOX 10041, GOUGER STREET, ADELAIDE SA AUSTRALIA 5000

Elevated chromosome damage rates are associated with an increased risk for cancer and accelerated ageing.[1,2] Our main objective is to determine optimal vitamin B intake for genomic stability in humans. In previous cross-sectional studies we had observed a significant negative correlation between chromosome damage (measured using the cytokinesis-block micronucleus assay) in lymphocytes and plasma vitamin B12 in subjects who were healthy and not vitamin B12 or folate deficient.[3] Furthermore the micronucleus index was positively correlated with plasma homocysteine.[3] These data suggested that above average intakes of folate and vitamin B12 may be required to minimise DNA damage rates.

We therefore performed a cross-sectional study (N = 49 males, 57 females) and a randomised double-blind placebo-controlled dietary intervention study (N = 31, 32 per group) to determine the effect of folate and vitamin B12 (B12) supplementation on DNA damage (micronucleus formation and DNA methylation) and plasma homocysteine (HC) in young Australian adults aged 18–32 years. None of the volunteers were folate deficient (*i.e.* red cell folate $< 136\,\mathrm{nmol\,l^{-1}}$) and only 4.4% (all females) were vitamin B12 deficient (*i.e.* serum B12 $< 150\,\mathrm{pmol\,l^{-1}}$). The cross-sectional study showed that (a) the frequency of micronucleated (MNed) cells was positively correlated with plasma HC in males ($R = 0.293$, $P < 0.05$) and (b) in females MNed cell frequency was negatively correlated with serum B12 ($R = -0.359$, $P < 0.01$) but (c) there was no significant correlation between micronucleus index and folate status. The results also showed that the level of unmethylated CpG (DNA) was not significantly related to vitamin B12 or folate status.

The dietary intervention involved supplementation with 3.5 times the RDI of folate and vitamin B12 in wheat bran cereal for three months followed by ten

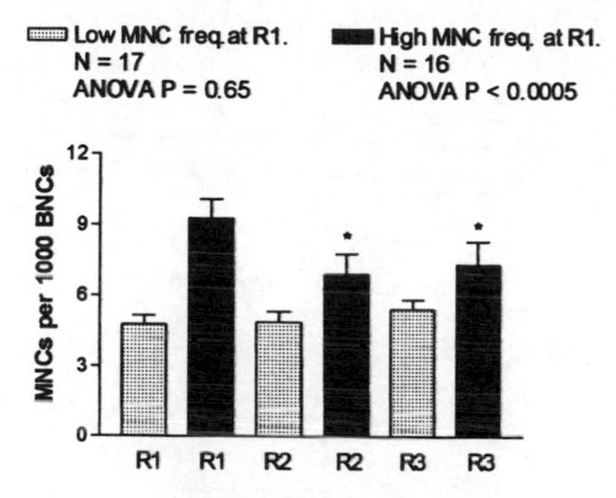

Figure 1 *Comparison of changes in MNed cell (MNC) frequency during the intervention in folate and B12 supplemented subjects who were either in the high or low 50th percentile of age- and gender-adjusted MNed cell frequency at the beginning of the intervention. R1 = blood samples just before the intervention started, R2 = blood samples after three months on cereal supplemented with 3.5 times the RDI of folate and B12, R3 = blood samples collected after a further three months on tablets containing 10 times the RDI of folate and B12. * P < 0.01 for comparisons to corresponding data at R1*

times the RDI of these vitamins via tablets for a further three months. In the supplemented group MNed cell frequency was significantly reduced during the intervention by 25.4% in those subjects with initial MNed cell frequency in the high 50th percentile but there was no change in those subjects in the low 50th percentile for initial MNed cell frequency (Figure 1) . The reduction in MNed cell frequency was significantly correlated with serum B12 ($R = -0.49$, $P < 0.0005$) and plasma HC ($R = 0.39$, $P < 0.006$), but was not significantly related to red cell folate. DNA methylation status was not altered in the supplemented group. The greatest decrease in plasma HC (by 37%) during the intervention was observed in those subjects in the supplemented group with initial plasma HC in the high 50th percentile, and correlated significantly with increases in red cell folate ($R = -0.64$, $P < 0.0001$) but not with serum B12. The results from this study suggest that (a) MNed cell frequency is minimised when plasma HC is below 7.5 μmol l^{-1} and serum B12 is above 300 pmol l^{-1} and (b) dietary supplement intake of 700 μg folic acid and 7ug vitamin B12 is sufficient to minimise MNed cell frequency and plasma homocysteine. Thus it appears that elevated plasma HC, a risk factor for cardiovascular disease, may also be a risk factor for chromosome damage.

References

1 B.N. Ames, *Toxicol. Lett.*, 1998, **102–103**, 5.

2 L. Hagmar, S. Bonassi, U. Stromberg, A. Brogger, L.E. Knudsen, H. Norrpa, C. Reuterwall and European Study Group on Cytogenetic Biomarkers and Health, *Cancer Res.*, 1998, **58**, 4117.

3 M. Fenech, I.E. Dreosti and J.R. Rinaldi, *Carcinogenesis*, 1997, **18**, 1329.

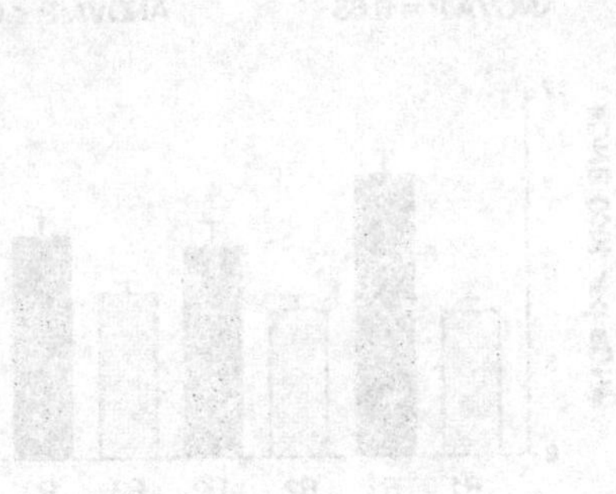

Section 4

Defence Systems: Enzyme Induction

4.1

Enzyme Induction in Laboratory Liver Models by Tea Ingredients

Lu Qi, Chi Han and Junshi Chen

INSTITUTE OF NUTRITION AND FOOD HYGIENE, CAPM,
BEIJING 100050, CHINA

1 Introduction

The cancer chemopreventive effects of polyphenols in food and beverages have received much attention recently.[1–5] Tea is one of the most widely consumed beverages in the world. Evidence from a number of experimental and epidemiologic studies has shown the anticarcinogenic properties of tea. The antioxidant activity of tea was regarded as one of the main mechanisms involved in its protective effects.[4–8] The free radical scavenging activity of tea and tea ingredients has been reported in many previous studies. However, the role of tea treatment on the levels of antioxidant defense and phase II enzyme activities has not yet been extensively studied. Khan *et al.*[9] reported that oral administration of green tea polyphenols and green tea extracts enhanced the activities of antioxidant and phase II enzymes. But inconsistent results also existed.[10] So, further studies on the effects of tea on the antioxidant defense and metabolic enzyme systems during animal carcinogenesis are needed. Herein, we report the induction effects of tea and tea ingredients on these defense enzyme systems.

2 Materials and Methods

2.1 Animals

Male weaning Wistar rats (90–120 g) were purchased from the Animal Center of Chinese Academy of Medicine Sciences, Beijing. Rats were fed with regular chow.

2.2 Tea Ingredients

All tea ingredients, including tea polyphenols (99%), tea pigments (oxidized product of 40% green tea polyphenols), thearubinens (95%), theaflavins (95%), (−)-epigallocatechin-3-gallate (EGCG, 95%), (−)-epicatechin-3-gallate (ECG, 95%), (−)-epigallocatechin (EGC, 95%), and (−)-epicatechin (EC, 65%), were provided by the Institute of Tea Science and Research, Chinese Academy of Agricultural Sciences. The water extract of tea was prepared freshly every day as follows: 1.5 g of fresh green tea leaves (Long Jin brand) was infused in 100 ml boiling water, allowed to stand at room temperature for 30 minutes, and then filtered. Mixed tea was prepared by mixing freeze-dried green tea water extract, tea polyphenols (40% purity) and tea pigments in the ratio of 4:1:1 based on the results of short-term screening on various tea ingredients[11] and dissolved in tap water at a concentration of 0.5%. The solutions were prepared freshly each day.

2.3 Methods

Hepa G2 cells were plated on 96-well microtiter plates at a density of 10,000 cells/well in 200 μl of DMEM with 10% fetal calf serum, 100 U ml^{-1} of penicillin G, 100 $\mu g\, ml^{-1}$ of streptomycin, and 0.1% DMSO. The cells were grown for 24 h in a humidified incubator with 5% CO_2 at 37 °C. The medium was decanted, then each well was re-fed with 100 μl of culture medium and test tea samples (2 and 10 mg l^{-1} for each sample); control cells were fed with media only. After the plates were exposed to test samples for 24 h, the medium was decanted, and the cells were lysed by incubation at 37 °C for 10 min with 50 μl in each well of 0.8% digitonin. The plates were then shaked for an additional 10 min at 25 °C. After this, 200 μl of the complete reaction mixture was added to each well, and the reaction was arrested after 5 min by addition 50 μl of 0.3 mM dicoumarol. The plates were then scanned at 610 nm.[12] The number of cells or the amount of protein in each well was determined based on the method reported by Delong *et al.*[13]

The precancerous liver cancer model in rats was developed by CCl_4 treatment followed by DEN and partial hepatomy. Sixty rats were randomly divided into five groups. The animals in the negative control group (6 rats) was not subjected to DEN treatment and hepatomy, and drank tap water; the animals in the CCl_4 control group (9 rats) were given 0.5ml/rat 20% CCl_4 i.p. once a day at day 40 and 41 and drank tap water; the positive control group (15 rats) was given DEN (100 mg kg^{-1} b.w.) i.p. once a day from day 7 to 17 and drank tap water only; the green tea group (15 rats) was also treated with DEN, but drank 2 green tea water extract as the sole source of drinking fluid from week 1 to 6, and drank tap water only from the 7th week; the mixed tea group (15 rats) was also treated with DEN, but drank 0.5 mixed tea water extract as the sole source of drinking fluid from week 1 to 6, and drank water only from the 7th week. Except the negative control group, all the animals in the other 4 groups were subjected to 2/3 hepatomy under *ether ana* at the end of the sixth week. The liver samples were frozen in liquid nitrogen and stored at −80 °C. All the animals were sacrificed at the end of the eighth week, and the livers were removed and cleaned.[14]

Gamma glutamyl transpeptidaose (r-GGT) positive foci in liver were measured as discribed by Rutenburg *et al.*[15] Briefly, the liver samples were sliced with cryoultramicrotome (−30 °C), and the slices were fixed in acetone for 12 h (4 °C). After the acetone was dried, the liver slices were then immersed in 0.05 mol l^{-1} Tris-HCl (pH 7.4) for 10 min and incubated under normal temperature for 60 min. After this procedure, the slices were rinsed with physiological saline and then immersed in 0.1 mol l^{-1} cupric sulfate for seconds and 0.01 mol l^{-1} cupric sulfate for 2 min. The slices were rinsed again with physiological saline for 2 min and with water for two times (1 min once). The r-GGT positive foci was examined under microscope (4 × 15).

The determination of antioxidant enzyme activities: The livers were homogenized in 0.1 mol l^{-1} Tris-HCl buffer (pH 7.4), and then centrifuged at 9000 g (4 °C) for 15 min. The supernatant fractions were centrifuged again at 10,500 g (4 °C) for 60 min. The supernatant fraction (liver cytosol) and liver microsomes were used for analysis immediately or stored at −80 °C.

The activities of catalase (CAT),[16] superoxide dismutase (SOD),[17] glutathione peroxidase (GSH-Px),[18] glutatione reductase(GR),[19] quinone reductase (QR)[12,13] and glutatione S-transferase (GST)[20] in liver cytosol were measured.

The determination of peroxidation products in rat liver microsome: 0.5 ml liver microsome preparation and 200 μl of freshly prepared DNPH (3 mg DNPH dissolved in 10 ml 2M HCl) were shaked in a closed system for 15 min and then extracted with 1 ml ethyl acetate, and the solvents were removed under reduced pressure. The residues were redissolved in 250 μl acetonitrile for HPLC analysis. The hydrazone derivatives of the lipid peroxidation products were separated by HYPERSIL ODS column (5 μm particle size, 125A 4.6 mm × 250 mm) eluted isocratically with 45% acetonitrile in water at 1 ml min^{-1}. Lipid peroxidation products such as malondiadehyde (MDA), acetaldehyde (ACT), propionaldehyde (PP) and acetone (ACON), were measured at 340 nm by HPLC.[21]

2.4 Statistical Analysis

The Rank test was used in the comparison of the number of glutamyl transpeptidaose-positive foci among different groups. The Student's *t*-test was used in comparing the concentration of lipid peroxidation products and cytosol enzyme activities among different groups.

3 Results

The data in Table 1 show that, at the concentrations tested, tea polyphenols, tea pigments and mixed tea significantly induced QR activity in Hepa G2 cell line. At 2 mg l^{-1}, the QR activity increased 35.8%, 58.8%, and 74.6% respectively; at 10 mg l^{-1}, the QR activity increased 73.0%, 60.4%, and 60.6% respectively. Among the 4 individual catechins, only EGCE ($p < 0.01$) and ECG ($p < 0.05$) induced QR activity significantly at 10 mg l^{-1}, but no significant induction was observed with EC and EGC treatment. Thearubigens and theaflavins, the major polyphenols in black tea, increased QR activity 30.3% and 18.5% respectively

Table 1 *Induction of QR activity in Hepa G2 cells by tea components (means ± SD)*

Groups	*Dose* $(mg\ l^{-1})$	*QR activity* $(mmol\ g^{-1})$	*Enhancement rate (%)*
Control	–	248.6 ± 62.6	–
Tea polyphenols	2	336.8 ± 43.8	35.8
	10	430.4 ± 71.8[a]	73.0
Tea pigments	2	394.8 ± 97.4[b]	58.8
	10	399.0 ± 57.6[b]	60.4
Mixed tea	2	434.2 ± 64.2[a]	74.6
	10	399.4 ± 14.6[a]	60.6
EGCG	2	291.2 ± 49.0	17.1
	10	380.8 ± 55.0[a]	53.1
ECG	2	292.4 ± 52.2	17.6
	10	374.8 ± 68.4[b]	50.7
EGC	2	272.2 ± 17.0	9.4
	10	280.2 ± 50.2	12.6
EC	2	288.6 ± 15.0	16.0
	10	275.0 ± 60.4	10.6
Thearubigens	2	298.2 ± 43.6	19.9
	10	324.2 ± 29.0[b]	30.3
Theaflavins	2	244.4 ± 47.0	–
	10	294.8 ± 51.8	18.5

[a] Compare with control group, $P < 0.01$. [b] Compare with control group, $P < 0.05$.

at 10 $mg\ l^{-1}$. However, only thearubigens showed statistically significance ($p < 0.05$). In general, among those tea ingredients studied, the multi-component mixtures were more effective than the single components in inducing QR activity.

Exposure of rats to green tea extracts at a concentration of 2% (w/v) and mixed tea at a concentration of 0.5% (w/v) as the sole source of water fluid for 8 weeks did not affect body weight gain (results not shown). The results in Table 2 showed that the numbers of r-GGT positive foci in the green tea and mixed tea

Table 2 *Effects of drinking green tea and mixed tea on r-GGT positive foci formation induced by DEN (numbers/10 × 9.62 mm²)*

Groups	*Numbers of r-GGT positive foci*
Negative control	0, 0, 0, 0, 0, 0
CCl_4	0, 0, 0, 1, 0, 0, 1, 0
DEN[b]	20, 4, 10, 8, 7, 6, 10, 7, 8, 9, 1
Green tea[a]	4, 1, 4, 3, 2, 1, 4, 6, 1, 7
Mixed tea[a]	1, 4, 2, 0, 0, 0, 1, 1, 1, 7

[a] Compare with group DEN, $P < 0.01$. [b] Compare with group negative control, $P < 0.01$.

Table 3 *Effects of drinking green tea and mixed tea on the formation of peroxidation products in the rat liver microsomes induced by DEN (means ± SD)*

Groups	*MDA* ($\mu mol\ g^{-1}$)	*ACT* ($\mu mol\ g^{-1}$)	*ACON* ($\mu mol\ g^{-1}$)	*PP* ($\mu mol\ g^{-1}$)
Negative control	1.70 ± 0.10	2.05 ± 0.77	2.68 ± 0.40	0 ± 0
CCl_4	2.34 ± 0.62	2.14 ± 0.80	3.02 ± 0.21	0.40 ± 0.19
DEN	5.73 ± 0.54[b]	3.26 ± 1.02[b]	5.01 ± 0.73[b]	2.18 ± 0.23[b]
Green tea	4.12 ± 0.90[a]	1.96 ± 0.44[a]	4.73 ± 1.04	1.68 ± 0.16[a]
Mixed tea	4.03 ± 0.65[a]	2.88 ± 0.32	4.34 ± 0.87	1.75 ± 0.21[a]

[a] Compare with group DEN, P < 0.01. [b] Compare with group negative control, P < 0.01.

groups were significantly lower than in the positive control group ($p < 0.01$) at the end of the eighth week. A greater inhibition of r-GGT positive foci formation occurred in the mixed tea group when compared with the green tea group ($p < 0.05$). The effect of green tea and mixed tea drinking on the formation of peroxidation products in the rat liver microsome is illustrated in Table 3. The levels of peroxidation products (MDA, ACT, ACON, PP) in the positive control group were significantly higher than those of the negative control group ($p < 0.01$). Comparing with the positive control group, the overall levels of peroxidation products in the green tea and mixed tea groups were significantly reduced. Among them, the levels of MDA, ACT and PP in the green tea group, as well as the levels of MDA and PP in the mixed tea group, were significantly lower than in the positive control group ($p < 0.01$).

The data in Table 4 show that drinking of green tea resulted in a significant increase of liver SOD, GST and QR activities ($p < 0.01$), while the activities of CAT, GSH-Px and GR did not change dramatically. In the mixed tea group, the activities of CAT, SOD, GST and QR were elevated significantly as compared with the positive control group ($p < 0.01$). But the mixed tea group showed no significant effects on the activities of GSH-Px and GR.

Table 4 *Effects of drinking green tea and mixed tea on activities of cytosolic antioxidant enzymes and phase II metabolism enzymes in rat liver (means ± SD)*

Groups	*CAT* $U\ g^{-1}\ s^{-1}$	*SOD* *U mg*	*GSH-Px* $U\ g^{-1}\ min^{-1}$	*GR* $U\ g^{-1}\ min^{-1}$	*QR* $\mu mol\ mg\ min^{-1}$	*GST* $U\ g^{-1}\ min^{-1}$
Negative control	214.2 ± 86.2	15.8 ± 1.2	48.3 ± 14.4	77.5 ± 19.7	10.9 ± 1.9	172.9 ± 33.2
CCl_4	191.4 ± 65.1	15.3 ± 3.8	47.3 ± 12.8	51.7 ± 3.3	9.3 ± 1.2	173.4 ± 24.4
DEN	201.2 ± 50.0	14.9 ± 2.7	32.9 ± 6.8	48.5 ± 15.4	9.3 ± 0.9	168.7 ± 26.6
Green tea	219.8 ± 65.1	19.9 ± 4.7[a]	43.9 ± 12.4	53.3 ± 21.9	11.2 ± 1.4[a]	219.9 ± 15.7[a]
Mixed tea	302.0 ± 88.1[a]	21.8 ± 3.5[a]	41.1 ± 15.1	63.6 ± 8.8	14.1 ± 1.8[a]	244.3 ± 41.4

[a] Compare with group DEN, P < 0.01.

4 Discussion

Significant increase of QR activity occurred in the human liver cancer Hepa G2 cell line treated by a variety of tea components in the present study, including tea polyphenols, tea pigments and mixed tea, as well as individual components such as EGCG and ECG, and thearubigens. Our data suggested that the induction ability of tea depended on the collective effects of several kinds of active ingredients, which mainly include tea catechins and pigments.

Quinones are among the toxic products of the oxidative metabolism of aromatic hydrocarbons. Quinone reductase protects cells against the toxicity of quinones and their metabolic precursors by promoting the obligatory two-electron reduction of quinones to hydroquinones. On the other hand, the one-electron reduction of quinones to semiquinone radical anions are capable of damaging DNA, RNA, and other essential cellular macromolecules. In addition, quinone reductase is induced coordinately with other electrophile-processing phase II enzymes (glutathione S-transferases and UDP-glucuronosyltransferases) by a variety of compounds that have anticarcinogenic effects.[12,22]

In the present animal study, drinking green and mixed tea inhibited the precancerous hepatic lesions r-GGT positive foci in rats, and also reduced the formation of peroxidation products in liver microsome and enhanced the activities of some antioxidant and phase II enzymes such as CAT, SOD, QR and GST in rat liver. It is concluded that the anticarcinogenic effects of green tea and mixed tea might be caused by their antioxidant properties, although other mechanisms may also exist.

Tea, including black tea, green tea and olong tea, contains a wide variety of polyphenols that can function as scavengers of reactive oxygen species.[1–6,23] Our data suggested an alternative mechanism, *i.e.* induction of the enzyme systems that deactivate reactive oxygen species and metabolites of precarcinogens. The activities of such enzymes generally result in the production of less deleterious molecular species from a wide variety of mutagenic and carcinogenic substances, and inhibiting oxdative injury, specifically damage to macromolecules, such as DNA and protein, hence inhibiting carcinogenesis.

It is also worthy to notice that tea pigments, as well as thearubigens, showed similar antioxidant properties as catechins, which are regarded as the active constituents in preventing cancer.[24] Tea pigments, as the oxidized products of catechins, are the major constituent of black tea. Recent studies demonstrated that theaflavins and thearubigens, the major constituents of tea pigments, also have antioxidant activities *in vitro*.[25,26] Our observation indicated that tea pigments might also possess anticarcinogenic effects, due to their antioxidant properties.

In the previous short-term screening tests on tea ingredients, we found that the inhibitory effects of any single tea ingredient on the initiation, promotion or progress phase of carcinogenesis were not as strong as those of the whole tea water extracts.[11] Based on this concept, a mixed tea was developed in our laboratory, which is composed of water extracts of green tea, tea polyphenols and tea pigments. It is clear that this mixture is stronger than any other kinds of tea ingredients in inducing enzyme activities, and thus may aid in preventing cancer.

Acknowledgements

Supported by the Chinese National Natural Science Foundation research grant No 39330189. The author would like to thank Prof. Qikun Chen of the Institute of Tea Science, Chinese Academy of Agricultural Science for supplying green tea and tea ingredients; Dr. Peter Fu of NCTR (USA) for helping in measuring the peroxidation products in rat liver microsomes.

5 References

1 KT Chung, TY Wong, CI Wei, *et al.*, Tannins and human health: a review. *Crit Rev Food Sci Nutr*. 1998; **38**, 421–64.

2 S Ren and EJ Lien, Nature products and their derivatives as cancer hemopreventive agents. *Prog Drug Res*. 1997; **48**, 147–71.

3 A Challa, N Ahmad and H Mukhtar, Cancer prevention through sensible nutrition. *Int J Oncol*. 1997; **11**, 1387–92.

4 LB Joan, Green tea and cancer in humans: a review of the literature. *Nutr Cancer*. 1998; **31**(3), 151–9.

5 C Han, Tea and cancer. *J Hygiene Res*. 1995; **24**, 29–33.

6 K Santosh, Katiyar and M Hasan, Tea in chemoprevention of cancer: epidemiologic and experimental studies (Review). *Int J Oncol*. 1996; **8**, 221–238.

7 L Kohlmier, KGC Weterings, S Steck and JK Frans, Tea and cancer prevention: an evaluation of the epidemiologic literature. *Nutr Cancer*. 1997; **27**(1), 1–13.

8 JH Weisburger, Tea and health: the underlying mechanisms. *Proc Soc Exp Biol Med*. 1999; **220**(4), 271–5.

9 SG Khan, SK Katiyar, R Agarwal and H Mukhtar, Enhancement of antioxidant and phase II enzymes by oral feeding green tea polyphenols in drinking water to SKH-1 hairless mice: possible role in cancer chemoprevention. *Cancer Res*. 1992; **52**, 4050–4052.

10 A Bu-Abbas, MN Clifford, C Ioannides and R Walker, Stimulation of rat hepatic UDP-glucuronosyl transferase activity following treatment with green tea. *Food Chem Toxic*. 1995; **33**, 27–30.

11 LJ Liu, C Han and JS Chen, Short-term screening of anticarcinogeneic ingredients of tea by cell biology assays. *J Hyg Res*. 1998; **27**, 53–56 (in chinese).

12 HJ Prochaska, MJ Delong and P Talalay, On the mechanisms of induction of cancer-protective enzymes: a unifying proposal. *Proc Natl Acad Sci*. 1985; **82**, 8232–8236.

13 MJ Delong, HJ Prochaska and P Talalay, Induction of NAD(P)H-quinone reductase in murine hepatoma cells by phenolic antioxidants, azo dyes, and other chemoprotectors: a model system for the study of anticarcinogens. *Proc Natl Acad Sci*. 1986; **83**, 787–791.

14 R Mao, L Ding and JY Chen, The inhibitory effects of epicatechin complex on diethylnitrosamine induced initiation of hepatocarcinogenesis in rats. *Chung-Hua-Yu-Fang-I-Hsueh-Tsa-Chih (Chinese J Prev Med)*. 1993; **27**, 201–204.

15 AM Rutenburg, H Kim, JW Fischbein, *et al.*, Histochenical and ultrastructural

demonstration of gamma-glut amyl-transpeptidase activity. *J Histochem Cytochem.* 1969; **17**, 517–523.

16 H Aebi, Catalase *in vitro*. *Methods Enzymol.* 1984; **105**, 121–126.

17 KW Lan, JM Huang, ZR Xie, BF Deng, YZ Fang and ZF Liu, Measurement of superoxide dismutase: Comparision of pyrogallol-NBT coloromatometry and chemistry luminescence method. *Prog Biochem Biophy*. 1988; **15**, 138–140.

18 J Mohandas, JJ Marshall, GG Duggin, JS Horvath JS and D Tiller, Differential distribution of glutathione and glutathione-related enzymes in rabbit kindey:possible implication in analgesic nephropathy. *Cancer Res.* 1984; **44**, 5086–5091.

19 DJ Worthington and MA Rosemeyer, Human glutathione reductase: Purification of the crystalline enzyme from erythrocyte. *Eur J Biochem.* 1974; **48**, 167–171.

20 WH Habig, MJ Pabst and Jokoby, Glutathione S-transferase. The first enzymatic step in merapturic acid formation. *J Biol Chem.* 1974; **249**, 7130–7139.

21 YC Ni, TY Wong, FF Kablubar and PP Fu, Hepatic metabolism of chloral hydrate to free radicals,and induction of lipid peroxidation. *Biochem Biophys Res Commun.* 1994; **204**, 937–943.

22 AM Benson, MJ Hunkeler and P Talalay, Increase of NADPH: quinone reductase by dietary antioxidants: possible role in protection against carcinogenesis and toxicity. *Proc Natl Acad Sci USA*. 1980; **77**, 5216–5220.

23 HN Graham, Green tea composition, consumption, and polyphenol chemistry. *Prev Med.* 1992; **21**, 334–350.

24 GD Stoner and H Mukhtar, Polyphenols as cancer chemopreventive agents. *J Cell Biochem.* 1995; Suppl 22, 169–180.

25 NJ Miller, C Castelluccio, Tijburg and C Rice-Evans, The antioxidant properties of theaflavins and their gallate esters–radical scavengers or metal chelators? *FEBS Lett.* 1996; **392**, 40–44.

26 K Yoshino, Y Hara, M Sano and L Tomita, Antioxidative effects of black tea theaflavins and thearubigin on lipid peroxidation of rat liver homogenates induced by tert-butylhydroperoxide. *Biol Pharm Bull.* 1994; **17**, 146–149.

4.2

Modulation of Hepatic and Renal Glutathione-S-Transferase by Naturally Occurring Plant Phenolics in Rat and Mouse

Violetta Krajka-Kuzniak, Jerzy Gnojkowski and Wanda Baer-Dubowska

DEPARTMENT OF PHARMACEUTICAL BIOCHEMISTRY, K.MARCINKOWSKI UNIVERSITY OF MEDICAL SCIENCES, POZNAN, POLAND

1 Introduction

Plant phenolics are abundant in nature and are present in human diet in representative amounts. Protocatechuic acid, a simple phenolic acid and tannic acid, ester of glucose and digallic acid are inhibitors of chemical carcinogenesis in rodents and may be considered as potential chemopreventive agents in humans.[1,2] The glutathione-S-transferases (GSTs) (EC 2.5.1.18) are phase II biotransformation enzymes that catalyse the conjugation of many electrophilic compounds with glutathione. The cytosolic GSTs have been divided into four different classes: alpha, mu, pi and theta. Their expression varied depending on tissue and species. Mannervik *et al.*[3] demonstrated that GST isozymes from rat, mouse and man contains at least one isozyme of each class alpha, mu or pi. The induction of GSTs is usually assumed to result in a decrease of cancer risk and certain anticarcinogens like oltipraz are able to induce specific GST isozyme subunits.[4] In this study we examined the effect of protocatechuic acid and tannic acid on GSTs in liver and kidney of male Wistar rats and female Swiss mice. Rats received both compounds at the dose of 50 mg kg^{-1} twice a week for 14 days. Mice were treated with a single dose of 80 mg kg^{-1} or 800 mg kg^{-1} of tannic acid and protocatechuic acid respectively. Control groups of animals received only vehiculum. GSTs from cytosol of liver and kidney of both species were purified using S-hexylglutathione affinity chromatography. The protein content was determined by the method of Lowry[5] and enzyme activities were determined using 1-chloro-2,4-dinitrobenzene as a second substrate.[5] GST subunits were separated by HPLC using a LKB Bromma model 2150 liquid chromatograph, equipped with a LiChrospher WP300 RP-18 column (250 × 4 mm, 5 μm particle, Merck), Spectral Detector (LKB) operating at 214 nm and gradient of 0.1% trifluoroacetic acid (TFA) in water (solvent A) and 0.1% TFA in acetonitrile (solvent B). The gradient system was as follows: (a) 50–60% solvent B (linear, 20 min); (b) 60–100% solvent B (linear, 5 min).

2 Results and Discussion

Figure 1, panel A shows hepatic and renal GST activities after treatment with protocatechuic acid at a single dose of 800 mg kg^{-1} and tannic acid at a dose of a 80 mg kg^{-1}. These doses were chosen based on the results of *in vitro* experiments, which showed that protocatechuic acid was less potent modulator of xenobiotics metabolising enzymes than tannic acid.[7] Protocatechuic acid increased the GST activity in liver and tannic acid in mouse kidney. In mouse liver tannic acid reduced the total activity of this enzyme in comparison with control. The same effect was observed after treatment of rats for the period of two weeks with 4 doses of 50 mg kg^{-1} of both compounds (Figure 1, panel B). The earlier observations of Athar *et al.*[1] and Baer-Dubowska *et al.*[8] showed the induction of GST by tannic acid in mouse stomach and epidermis. The levels of induction in these tissues were significantly higher than those found in mouse and rat livers and kidneys in the present study. The analysis of GST subunits by HPLC (Figure 2) showed that GST induction was primarily caused by significant increase of GST class mu with simultaneous decrease of class pi or alpha. Protocatechuic acid and tannic acid represent structurally diversified phenolics; however, no clear structure-activity relationships were found. The human mu class GSTM1 deficiency has been associated with increased risk of lung and bladder cancer and increased polyaromatic hydrocarbons (PAH)-DNA adducts.[9] In our earlier studies we demonstrated decrease of PAH binding to DNA by tannic acid in Swiss mice epidermis.[8] Same effect was observed also by Das *et al.* in Sencar mice lungs and epidermis.[10] The induction of GST class mu by tannic acid may explain in part the reduced DNA adducts formation after treatment with this compound. It is generally believed that the anti-

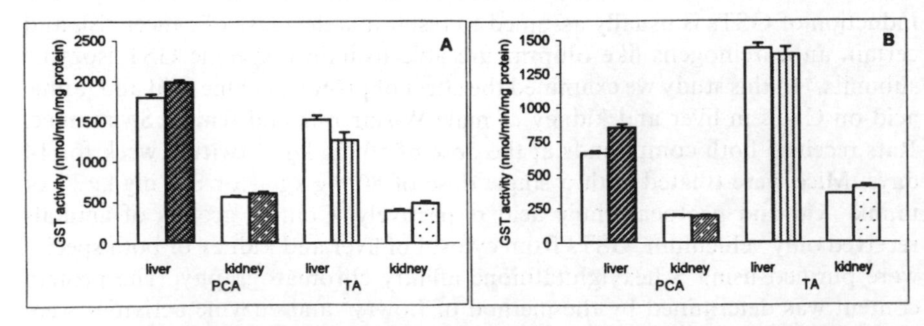

Figure 1 *GST activity in liver and kidney of mice* (A) *and rats* (B) *treated with protocatechuic acid (PCA) and tannic acid (TA) vs control (open columns)*

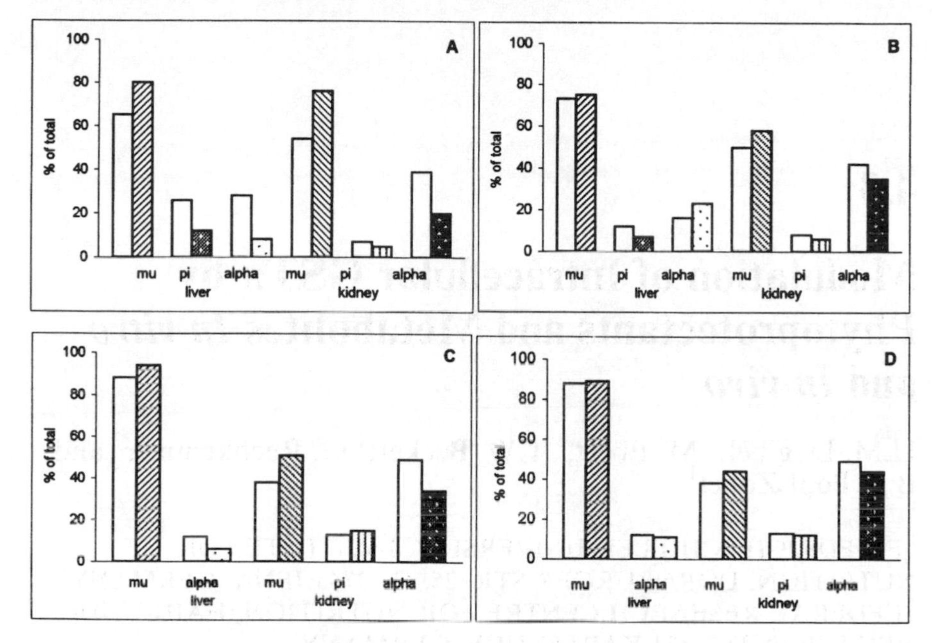

Figure 2 *Relative proportion of GST classes in liver and kidney of mouse treated PCA (*A*) and TA (*B*) and rats (*C,D*) vs control (open columns)*

mutagenic and anticarcinogenic activity of plant phenolics is the result of scavenging the ultimate carcinogenic metabolites.[11] The results of the present study indicate that, although moderate, their effect on GST can contribute also to overall anticarcinogenic effect of these compounds.

3 References

1 M. Athar, W.A. Wasiuddin, A. Khan and H. Mukhtar, *Cancer Res.*, 1988, **49**, 5784.
2 T. Tanaka, T. Kojima, T. Kawamori and H. Mori, *Cancer*, 1995, **75**, 1433.
3 B. Mannervik, P. Alin, C. Guthenberg, H. Jensson, M.K. Tahir, M. Warholm and H. Jornvall, *Proc. Natl. Acad. Sci. USA*, 1985, **82**, 7202.
4 T. Primiano, P.A. Egner, T.R. Sutter, G.J. Kelloff, B.D. Roebuck and T.W. Kensler, *Cancer Res.*, 1995, **55**, 4319.
5 O.H. Lowry, N.J. Rosebrough, A.L. Farrand and R.J. Randall, *J. Biol. Chem.*, 1951, **193**, 265.
6 W.H. Habig, M.J. Pabst and W.B. Jakoby, *J. Biol. Chem.*, 1974, **249**, 7130.
7 W. Baer-Dubowska, H. Szaefer, V. Krajka-Kuzniak, *Xenobiotica*, 1997, **28**, 735.
8 W. Baer-Dubowska, J. Gnojkowski and W. Fenrych, *Nutrition Cancer*, 1997, **29**, 42.
9 R.A. Grinberg-Funes, V.N. Singh, F.P. Perera, D.A. Bell, T. Lan Young, Ch. Dickey, L.W. Wang and R.M. Santella, *Carcinogenesis*, 1994, **15**, 2449.
10 M. Das, W.A. Khan, P. Asokan, D.R. Bickers and H. Mukhtar, *Cancer Res.*, 1987, **47**, 760
11 H.L. Newmark, *Can. J. Physiol. Pharmacol.*, 1987, **65**, 461.

4.3

Modulation of Intracellular GSTπ by Phytoprotectants and Metabolites *In vitro* and *In vivo*

U.M. Liegibel,[1] M. Ebert,[1] T.W. Becker,[1] G. Rechkemmer[2] and B.L. Pool-Zobel[1]

[1] FRIEDRICH-SCHILLER-UNIVERSITY, INSTITUTE FOR NUTRITION, DORNBURGER STR. 25, D-07743 JENA, GERMANY
[2] FEDERAL RESEARCH CENTRE FOR NUTRITION, HAID-UND-NEU-STR. 9, D-76131 KARLSRUHE, GERMANY

1 Introduction and Aim

Cytosolic glutathione S-transferases (GSTs) are an important class of phase II enzymes which may detoxify exogenous and endogenous compounds. The GSTπ class, represented by a single enzyme, GSTP1, is involved in the cellular response to oxidative stress. Induction of intracellular GSTπ by phytoprotectants is thought to protect the cells from oxidative and genotoxic damage.

We have performed *in vitro*-studies to determine GSTπ in lymphocytes after incubation with carotenoids, and in human colon tumour cells (HT29 clone 19A) after treatment with butyrate. The results were compared to two *in vivo*-studies. It was our aim to determine the modulation of cytosolic GSTπ, total protein, and/or GST activity *in vitro* and *in vivo*.

2 Materials and Methods

2.1 *In vitro* Studies

Human lymphocytes were incubated with 2 μM β-carotene or lycopene (water-soluble compounds, Betatene, Victoria, Australia) in RPMI for 24 h, HT29 clone 19A cells with 4 mM butyrate in DMEM for 72 h.

2.2 Carotenoid Intervention Study

23 healthy, male non-smokers daily received for two consecutive weeks (1) 330 ml tomato juice (40 mg lycopene), (2) for another two weeks 330 ml carrot juice (15.7 mg α-carotene + 21.6 mg β-carotene) and (3) finally 10 g spinach powder with 11.5 mg lutein (kindly provided by Fa. Schoenenberger, Magstadt and Fa. Voelpel, Königmoos, Germany).[1]

2.3 'Crystalean' Study

Groups of 3 female F344-HFA-rats received 0%, 10% and 15% resistant starch ('Crystalean') with their standard diet for 29 days. Colon cells were isolated using proteinase K (50 U ml^{-1}) in HBSS.[2]

2.4 GST and Total Protein

Cytosols were prepared by ultracentrifugation. Human GSTπ was quantified using a commercially available ELISA test (Biotrin, Sinsheim, Germany), rat GSTπ with reversed-phase HPLC.[3] Protein was measured using the method of Bradford. The GST activity was measured by a procedure of Habig *et al.* using glutathione and 1-chloro-2,4-dinitrobenzene as substrates ($A_{340\ nm}$).[4]

3 Results

GSTπ as well as the GST activity and total protein were slightly increased in lymphocytes following *high-cis-β*-carotene incubation, but were not affected by *all-trans-β*-carotene or lycopene (Figure 1).

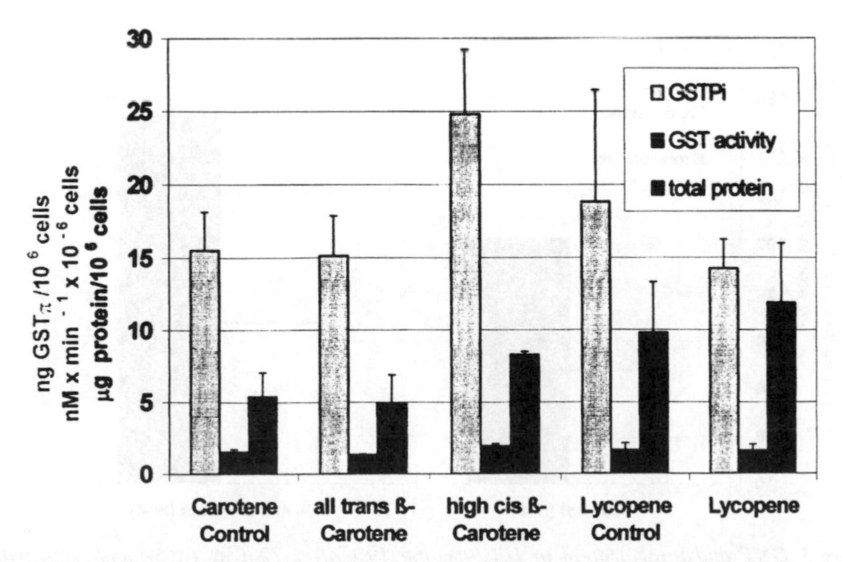

Figure 1 *GSTπ, GST activity, and total protein in lymphocytes after 24 h* in vitro *incubation with different carotenoids (2 µM); mean ± SEM; n = 3*

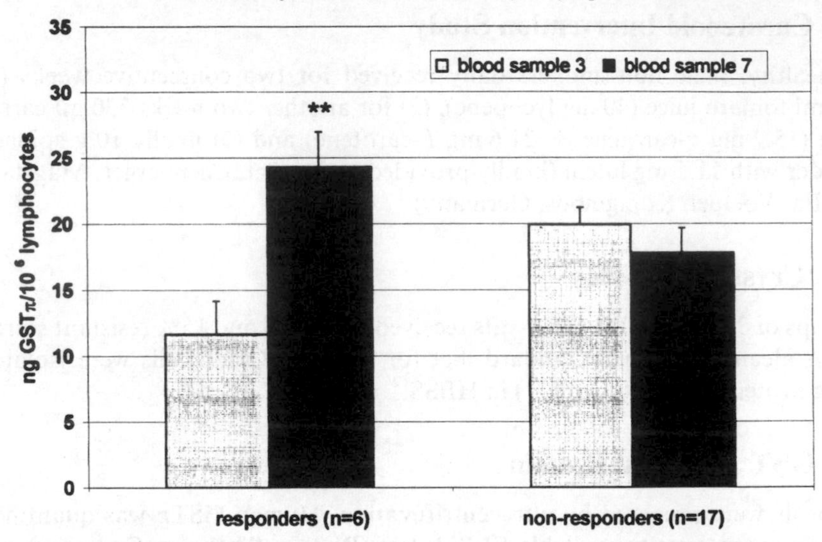

Figure 2 *Modulation of GSTπ in 'responders' and 'non-responders'; sample 3 (depletion) versus sample 7 (carrot juice intervention); mean ± SEM; unpaired, two-tailed t-test; **p > 0.01*

Intervention with carotenoid-rich juices, leading to enhanced plasma levels of α- and β-carotene, increased cytosolic GSTπ in 6/23 participants ('responders'; Figure 2).[5]

In HT29 clone 19A, GST activity and total protein were significantly increased by 4 mM butyrate (Figure 3).

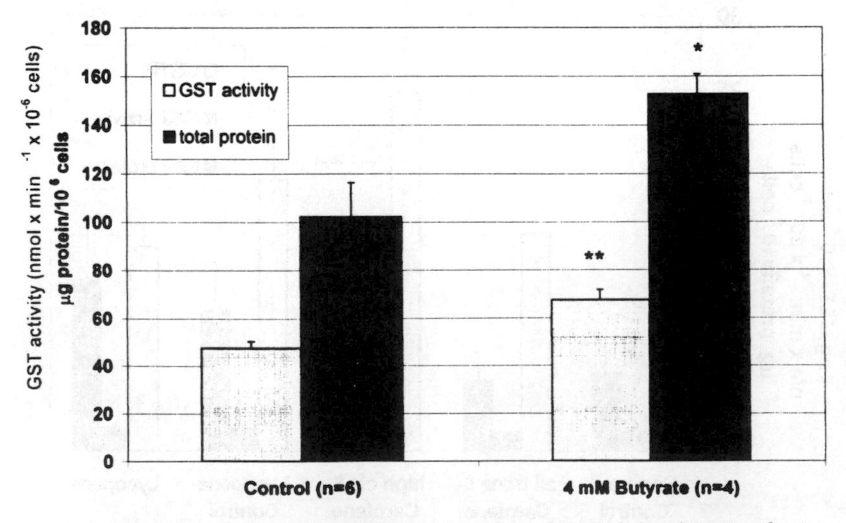

Figure 3 *GST and total protein in HT29 clone 19A after 72 h* in vitro-*incubation with butyrate (4 mM); mean ± SEM; unpaired, two-tailed t-test; * p < 0.05; ** p > 0.01*

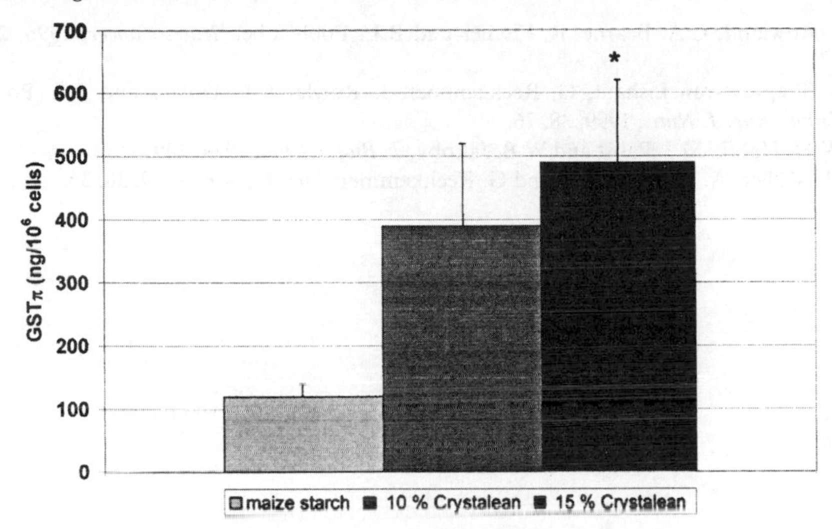

Figure 4 *Effects of resistant starch on GSTπ levels in rat colon cells (*in vivo*); mean ± SEM; n = 3; unpaired, one-sided t-test; * $p < 0.05$*

In rats, GST of the colon was significantly increased by feeding of amylase resistant starch, a prebiotic, expected to increase butyrate in the gut (Figure 4).

4 Discussion and Conclusion

It is not clear, which carotenoid in the human intervention study is responsible for the increased GSTπ levels in the 'responders'. *In vitro*, *high-cis β* carotene showed a slight induction and, therefore, could contribute to the increased levels in the intervention study.

The GSTπ induction in rat colon cells by Crystalean may be explained by the fact that complex carbohydrates are fermented by the colon microflora to yield increased levels of short chain fatty acids such as butyrate. *In vitro*, butyrate increased the GST activity in HT29 clone 19A cells 1.4 fold.

This type of mechanism-directed *in vitro*/*in vivo* approach should help increase our knowledge on the nature of protective properties by complex plant foods.

Acknowledgements

This work was supported by ECAIR-2-CT94-0933 and ERB-IC-15-CT-961012.

5 References

1 B.L. Pool-Zobel, A. Bub, U.M. Liegibel, S. Treptow-van Lishaut and G. Rechkemmer, *Cancer Epidemiol. Biomarkers Prev.*, 1998, **7**, 891.

2 I. Rowland, C.A. Bearne, R. Fischer and B.L. Pool-Zobel, *Nutr. Cancer*, 1996, **26**, 37.
3 S. Treptow-van Lishaut, G. Rechkemmer, I. Rowland, P. Dolara and B.L. Pool-Zobel, *Eur. J. Nutr.*, 1999, **38**, 76.
4 W.H. Habig, M.J. Pabst and W.B. Jakoby, *J. Biol. Chem.*, 1974, **249**, 7130.
5 H. Müller, A. Bub, B. Watzl and G. Rechkemmer, *Eur. J. Nutr.*, 1999, **38**, 35.

4.4

Potential Chemopreventive Mechanisms of Chalcones

C. Gerhäuser, E. Heiss, C. Herhaus and K. Klimo

GERMAN CANCER RESEARCH CENTER, DIVISION OF
TOXICOLOGY AND CANCER RISK FACTORS (C0200),
IM NEUENHEIMER FELD 280, 69120 HEIDELBERG, GERMANY

1 Introduction

Chalcones are ring-open biosynthetic precursors of flavonoids and contain an α,β-unsaturated carbonyl moiety flanked by two phenyl groups as a common structural feature. The chemical structure of the unsubstituted chalcone is shown in Figure 1. Chalcones have been reported to inhibit skin, oral, pulmonary and mammary carcinogenesis in rodent models. These effects have been attributed to a variety of anti-carcinogenic properties including anti-inflammatory,[1,2] anti-proliferative,[3] anti-oxidant[4] and radical-scavenging[5] mechanisms as well as induction of detoxification enzymes.[6]

As an approach to further determine chemopreventive mechanisms, we have performed a detailed analysis of modulatory effects of selected chalcones on xenobiotic metabolism. Induction of NAD(P)H:quinone oxidoreductase (QR) as a model phase 2 detoxification enzyme was tested in Hepa 1c1c7 murine hepatoma cells.[7,8] The mode of induction was determined using northern blotting and transient transfection techniques.[8] In addition, we have investigated the influence of chalcones on induction of Cyp1A as a phase 1 cytochrome P450 enzyme and tested the potential of chalcones to inhibit Cyp1A activity.[9]

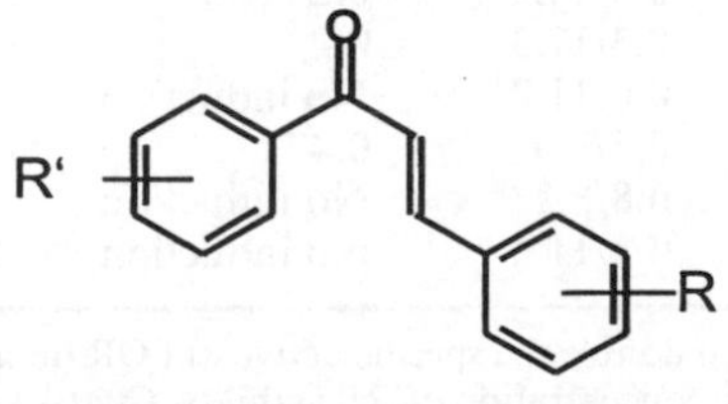

Figure 1 *Schematic representation of the chalcone structure*

Excessive production of nitric oxide (NO) and prostaglandins in inflammation is thought to be a causative factor of cellular injury and cancer. Elevated levels of the inducible forms of nitric oxide synthase (iNOS) and cyclooxygenase (Cox-2) have been implicated in the pathogenesis of many disease processes including acute and chronic neuro-degenerative diseases, rheumatoid arthritis and carcinogenesis.[10–12] Thus, we have determined the potential of chalcones to suppress lipopolysaccharide (LPS)-mediated NO formation in murine Raw macrophages.[13] Western blot analyses were performed to examine the influence on iNOS and Cox-2 protein levels.

2 Results and Discussion

We have tested a series of about twenty synthetic chalcones with hydroxy-, methoxy-, methyl- and bromo-substitution for their potential to induce QR in cultured Hepa1c1c7 cells (Herhaus *et al.*, in preparation). These compounds were identified as potent inducers of QR activity and displayed CD values (concentration required to double the specific activity of QR) below 1 μM. Results obtained with a selection of six chalcones Ch1 to Ch6 are summarised in Table 1.

The QR-inducing potential of two 2′-hydroxy-substituted chalcones Ch1 and Ch2, which were additionally found to potently elevate GSH levels in Hepa 1c1c7 cell culture, was analysed in more detail. Northern blotting experiments revealed transcriptional regulation of QR induction in a time- (0 to 15 h) and dose-dependent (0.2 to 25 μM) manner (data not shown). To determine the mode of induction, namely monofunctional or bifunctional mechanisms, transient transfection experiments with HepG2 cells were performed, using a construct containing all of the known regulatory elements of the rat QR gene (pDTD-1097-CAT) and a XRE (*x*enobiotic *r*esponsive *e*lement)-CAT construct, respectively. After treatment with 10 μM of Ch1 and Ch2 for 48h, chloramphenicol acetyltransferase (CAT) levels (determined by CAT-ELISA,

Table 1 *Summary of potential chemopreventive activities of selected chalcones*

Chalcone		*QR induction*	*Cyp1A induction*	*Cyp1A inhibition*	*NO inhibition*
Name	*Substitution*	*CD[a]/IC$_{50}$[b]*	*CD*	*IC$_{50}$*	*IC$_{50}$*
Ch 1	2′-OH, 2,6-OMe	0.4/14.4	0.2	0.3	3.6
Ch 2	2′-OH, 2-OMe	0.3/17.3	0.2	0.2	8.9
Ch 3	2′-OMe, 2-OMe	0.6/11.2	No induction	0.4	2.1
Ch 4	2′-OMe, 2,6-OMe	0.5/9.0	0.47	0.2	1.1
Ch 5	2′,6′-OMe	0.8/8.2	No induction	>5	0.7
Ch 6	2′,6′-OMe, 2-OMe	0.2/11.4	No induction	0.4	2.1

[a] CD: Concentration required to double the specific activity of QR (in μM).
[b] IC$_{50}$: Half-maximal inhibitory concentration of cell viability, Cyp1A activity or NO production (in μM).

Boehringer Mannheim) were significantly induced about 10-fold ($P > 0.0001$) in comparison with a solvent control in cultures transfected with either construct. This indicated an A*h* (aryl hydrocarbon) receptor-mediated and thus bifunctional mode of induction, *i.e.* concomitant induction of A*h*-receptor-dependent phase 1 and 2 enzymes. Consequently, weak induction of Cyp1A activity by Ch1, Ch2, and Ch4 was also observed in intact Hepa 1c1c7 cells by fluorimetric determination of the time-dependent dealkylation of 3-cyano-7-ethoxycoumarin (CEC) to 3-cyano-7-hydroxycoumarin,[9] although the compounds have no structural similarity to known ligands of the A*h*-receptor (*e.g.* polychlorinated biphenyls, polyaromatic arylhydrocarbons). Interestingly, Cyp1A-inducing potential was abolished by replacement of the 2′-OH-group by a methoxy-group (Ch2 vs. Ch3). In addition to Cyp1A induction, however, these compounds were also identified as potent inhibitors of Cyp1A enzyme activity using lysates of β-naphthoflavone-induced rat hepatoma (H4IIE) cells as an enzyme source. The results are summarised in Table 1.

To further investigate chalcone-mediated chemopreventive mechanisms, effects on NO production[13] and on the expression of iNOS and Cox-2 were examined. Chalcones were found to potently inhibit LPS-mediated NO production in Raw 264.7 macrophages with IC_{50} values ranging from 0.7 to 2 μM (Table 1). Western blot analyses revealed that Ch1 time- and dose-dependently (0 to 12 h incubation; dose range 0.4 to 25 μM) suppresses LPS-mediated up-regulation of iNOS and Cox-2 protein levels in Raw macrophages (data not shown). This effect was measurable when the compound was added up to 4 h post LPS treatment, indicating a rather early target (activation of NF-κB?) in the signal transduction pathway from LPS stimulation to protein expression. These novel results suggest that the known anti-inflammatory activity of chalcones may not only be mediated by inhibition of relevant enzyme activities like Cox-1,[1] but also by inhibition of the expression of enzymes involved in inflammatory reactions. Taken together, our investigations provide additional data to demonstrate potential chemopreventive activity of chalcones by multiple anti-initiating and -promoting effects.

3 References

1 C.N. Lin, T.H. Lee, M.F. Hsu, J.P. Wang, F.N. Ko and C.M. Teng, *J. Pharm. Pharmacol.*, 1997, **49**, 530–538.

2 F. Herencia, M.L. Ferrandiz, A. Ubeda, J.N. Dominguez, J.E. Charris, G.M. Lobo and M.J. Alcaraz, *Bioorg. Med. Chem. Lett.*, 1998, **8**, 1169–1174.

3 S. Ducki, R. Forrest, J.A. Hadfield, A. Kendall, N.J. Lawrence, A.T. McGown and D. Rennison, *Bioorg. Med. Chem. Lett.*, 1998, **8**, 1051–1056.

4 R.J. Anto, K. Sukumaran, G. Kuttan, M.N. Rao, Subbaraju and R. Kuttan, *Cancer Lett.*, 1995, **97**, 33–37.

5 A.T. Dinkova-Kostova, C. Abeygunawardana and P. Talalay, *J. Med. Chem.*, 1998, **41**, 5287- 5296.

6 L.L. Song, J.W. Kosmeder II, S.K. Lee, C. Gerhäuser, D. Lantvit, R.C. Moon, R.M. Moriarty and J.M. Pezzuto, *Cancer Res.*, 1999, **59**, 578–585.

7 H.J. Prochaska and A.B. Santamaria, *Anal. Biochem.*, 1988, **169**, 328–336.

8 C. Gerhäuser, M. You, J. Liu, R.M. Moriarty, M. Hawthorne, R.G. Mehta, R.C. Moon, and J.M. Pezzuto, *Cancer Res.*, 1997, **57**, 272–278.
9 C.L. Crespi, V.P. Miller and B.W. Penman, *Anal. Biochem.*, 1997, **248**, 188–190.
10 T.P. Misko, J.L. Trotter and A.H. Cross, *J. Neuroimmunol.*, 1995, **61**, 195–204.
11 N.A. Simonian and J.T. Coyle, *Annu. Rev. Pharmacol. Toxicol.*, 1996, **36**, 83–106.
12 M. Takahashi, K. Fukuda, T. Ohata, T. Sugimura and K. Wakabayashi, *Cancer Res.*, 1997, **57**, 1233–1237.
13 A.H. Ding, C.F. Nathan and D.J. Stuehr, *J. Immunol.*, 1988, **14**, 2407–2412.

4.5

Induction of Glutathione-S-Transferases in Humans by Vegetable Diets

Hans Steinkellner,[1] Gerhard Hietsch,[1] Lakshmaiah Sreerama,[2] Gerald Haidinger,[1] Andrea Gsur,[1] Michael Kundi[3] and Siegfried Knasmüller[1]

[1] INSTITUTE FOR TUMOUR BIOLOGY-CANCER RESEARCH, UNIVERSITY OF VIENNA, VIENNA, AUSTRIA
[2] DEPARTMENT OF PHARMACOLOGY, UNIVERSITY OF MINNESOTA MEDICAL SCHOOL, MINNEAPOLIS, USA
[3] INSTITUTE FOR ENVIRONMENTAL HYGIENE, UNIVERSITY OF VIENNA, VIENNA, AUSTRIA

1 Introduction

Glutathione-S-transferases (GSTs) play a major role in the detoxification of a large group of both natural and xenobiotic DNA-reactive compounds.[1] Conjugation of electrophilic xenobiotics with glutathione leads to increased water solubility which decreases their half life in the body and reactivity towards cellular proteins and DNA. It is known that GST-activity can be induced by foreign compounds. More than 100 different compounds have so far been identified as GST inducers in laboratory rodents. An important group are phytochemicals that are part of the human diet.[1] Very potent GST inducers in both mice and rats are isothiocyanates, breakdown products of glucosinolates which are occuring in cruciferous vegetables such as broccoli, cabbage and Brussels sprouts.[2–5] Wattenberg and Loub[6] demonstrated that feeding of glucosinolate metabolites (indole-3-acetonitrile, 3,3-diindolylmethane and indole-3-carbinol) inhibited the formation of benz[*a*]pyrene induced forestomach tumors in rats. In further experiments with mice, strong induction of GST through benzyl isothiocyanate was associated with a strongly reduced incidence of benz[*a*]pyrene induced neoplasia in the forestomach.[7]

Several authors emphasised that uptake of glucosinolates in man through vegetables in Western diet could have a beneficial effect in regard to chemoprevention of cancer in man;[8-10] nevertheless evidence for GST induction in humans consuming cruciferous vegetables is scarce (Table 1).

Table 1 *Current data base on human intervention studies on GST-induction by cruciferous vegetables*

Vegetable diet (type of study)	*No. of subjects*[a]	*Treatment*[b]	*Methods*[c]	*Results*	*Reference*
Brussels sprouts (intervention study)	5 (3 m/2f) diet *versus* 5 (2m/3f) control	300 g/d; 7d	ELISA	increase of α-GST 1.5 fold in plasma of m; no increase of π-GST in plasma and urine	11
Broccoli pills (intervention study)	11 diet *versus* 17 control	6 tablets (30 g/d); 14 d	spectrophotometry	no increase in lymphocytes	12
Broccoli (intervention study)	1 f	300 g/d; 12 d;	spectrophotometry	3-fold increase in saliva	13
Diet high in cruciferous vegetables (matched control study)	12 f vegans *versus* 12 f control	60 g/d; *versus* 21 g/d	radioimmunoassay	no significant increase of α-GST in plasma	14
Brussels sprouts (intervention study)	5 (3 m/2f) diet *versus* 5 (3m/2f) control	300 g/d; 7 d	spectrophotometry; ELISA	no increase in lymphocytes, rectal and duodenum cells; slight increase of α- and π-GST in rectal cells	14
Brussels sprouts (intervention study)	5 m diet *versus* 5 m control	300 g/d; 21 d	radioimmunoassay	1.4-increased of α-GST in plasma in sprouts group	16

[a] m = male, f = female; [b] g = grams, d = day; [c] spectrophotometry = spectropotometric determination of overall GST activity according to Habig *et al.*,[17] ELISA = enzyme-linked immunoabsorbent assay according to Nijhoff *et al.*,[11] radioimmunoassay = radioimmunoassay with delayed tracer addition according to Bogaards *et al.*[16]

The aim of our study was to broaden the current knowledge about GST induction in humans and furthermore to clarify the putative influence of gender[11] and GST-μ genotype on induction of glutathione-S-transferases in man.

2 Design of the Studies

Three intervention studies were carried out. For each, 10 healthy adult male and female non-smoking volunteers aged from 23 to 54 years were recruited. From a total of 21 people participating in the three studies 8 were male, 13 female. Some subjects repeatedly participated in the studies. All volunteers were in good health and not under the influence of permanent medication. Vegetarians were not included in the studies. Informed consent was obtained from all participants. Two weeks before the begin of and during each trial volunteers had to abstain from eating brassica vegetables and drinking coffee. All vegetables (Brussels sprouts 'Cyrus', red cabbage 'Roxy', broccoli 'Montop'), were obtained from Novartis Seeds, The Netherlands. Vegetables were steamed (330 ml water was added per kg vegetables) for 10 min. in a steel pot and 300 g meals were given as a lunch to each participant for 5 consecutive days. On days one and five of each intervention trial saliva and blood were collected from each participant.

3 Materials and Methods

Blood (20 ml) was aspired by venipuncture, collected in heparinized syringes and centrifuged at $2000 \times g$ for 15 min. Plasma samples were stored in liquid nitrogen. 5 ml of saliva were collected from each participant. Dithiotreitol was added to a final concentration of 5 mM and the samples were centrifuged at $9000 \times g$ for 15 min. The supernatants were stored in liquid nitrogen. GST – activities were determined in triplicates in plasma and saliva following the protocol of Habig *et al.*[17] using 1-chloro-2,4-dinitrobenzene and glutathione as substrate. The glutathione conjugate (2,4-dinitrophenyl-glutathione) formed in the presence of GST was quantified spectrophotometrically at a wavelength of 340 nm with a Beckman 640 spectrophotometer. All chemicals were obtained from Sigma-Aldrich. Content of α-isozyme in plasma samples was determined in duplicates with a GST-α specific immunoassay (High sensitivity Hepkit-α, Biotrin Co., Ireland) in plasma. For the determination of GST-μ genotypes with polymerase chain reactions (PCR), DNA was isolated from saliva using the Qiagen Blood Kit (Qiagen, Germany). PCR was then carried out according to Bell *et al.*[18] The reactions were performed in a Perkin Elmer 2400 thermocycler. Amplification products were resolved on a 2% ethidium bromide-stained gel.

4 Statistics

To analyse the overall induction of GST-activity in plasma and saliva and the increase of α-GST content in plasma of the probands in the three trials,

Wilcoxon tests were performed. To evaluate correlations between gender and GST-μ genotype and overall induction of GST-activity in plasma and saliva and increase of α-GST content in plasma of the probands in the three trials, Mann-Whitney-U tests were performed. P values < 0.05 were considered statistically significant.

5 Results and Discussion

The results obtained in the three intervention studies are summarised in Table 2. The basic values (before intervention) for plasma were substantially lower (about two orders of magnitude) than those measured under similar experimental conditions in saliva. These findings are in line with an earlier report.[13] It is also notable that the individual variations of plasma GST activities were distinctly lower (up to 30%) than those seen in saliva (up to 50%). Note that differences in the baseline values (values before intervention) reflect individual differences between the participants involved in the studies). With one vegetable, namely red cabbage, a pronounced (*ca.* 1.8 fold) increase of enzyme activity was measured which was paralleled by an increase of enzyme activity in saliva (*ca.* 1.4 fold). With Brussels sprouts only a weak (*ca.* 1.2 fold) but highly significant increase of GST activity could be measured. No effect could be observed under identical conditions in saliva. Broccoli was ineffective in both body fluids. The correlation between gender and GST induction (in plasma and saliva) and α-GST induction was analysed in Mann-Whitney-U tests and consistently P-values were below statistical significance levels. Likewise no correlations were

Table 2 *Statistical analysis of the intervention studies with vegetables*

Vegetable	*No of subjects*[a]	*Parameter*[b]	*Before trial*[c] M ± *S.D.*	*After trial*[c] M ± *S.D.*	*P-value*[d]
Brussels sprouts	9 (6f/3m)	GST activity in plasma	0.05 ± 0.01	0.06 ± 0.02	0.015*
	10 (7f/3m)	GST activity in saliva	30.3 ± 19.0	16.0 ± 10.0	0.462
	9 (6f/3m)	α-GST content in plasma	1.7 ± 1.6	3.0 ± 2.9	0.165
Red cabbage	9 (5f/4m)	GST activity in plasma	0.055 ± 0.01	0.095 ± 0.03	0.005*
	10 (6f/4m)	GST activity in saliva	24.2 ± 10	35.0 ± 16.3	0.024*
	9 (5f/4m)	α-GST content in plasma	0.29 ± 0.11	0.26 ± 0.12	0.405
Broccoli	9 (4f/5m)	GST activity in plasma	0.04 ± 0.01	0.04 ± 0.01	0.314
	10 (5f/5m)	GST activity in saliva	17.7 ± 24.4	9.3 ± 7.0	0.757
	9 (4f/5m)	α-GST content in plasma	3.0 ± 1.0	2.0 ± 1.0	0.260

[a] f = female, m = male; [b] GST activity in plasma and saliva was determined spectrophotometrically according to Habig *et al.*[17] using 1-chloro-2,4-dinitrobenzene and glutathione as a substrates; the amount (in μmol) of the product (2,4-dinitrophenyl-glutathione) formed per minute per mg protein is the unit for enzyme acitivity. [c] α-GST content in plasma was determined using the Hepkit, Alpha, Biotrin International; the unit is nanomoles per ml plasma (nm ml^{-1}). [d] Wilcoxon tests were performed to evaluate statistically significant differences between human samples before and after a vegetable diet in a group of 10 probands.
* Indicates statistical significance (P-value < 0.05).

seen between GST-μ genotype and increase of either overall GST activity or α-GST levels.

According to our knowledge, this is the first comprehensive comparative study on GST induction by cruciferous vegetables in humans. In contrast to most earlier studies (see Table 1) a higher number of individuals was included. We compared GST induction in the individual participants before and after intervention whereas in other studies enzyme induction was compared between groups that received different diets. Since considerable differences of GST levels do exist in humans, we think that intra-individual comparisons are more suitable to detect induction effects. As described above we measured the effects of three different cruciferous vegetables on overall GST activity and on the levels α-GST in two body fluids (plasma, saliva) and furthermore analysed correllations between enzyme induction and sex or GST-μ genotype. The results of the experiments show that consumption of Brussels sprouts and red cabbage induced GST, whereas broccoli yielded no effects. In this context it is notable that in animal studies GST was induced in several organs by glucosinolates and isothiocyanates, compounds contained in cruciferous vegetables such as broccoli and Brussels sprouts.[2–5] Although in two previous studies[12,13] induction of GST in humans after consumption of broccoli was reported we could not find any effects of broccoli in our study. Induction of the α-GST isozyme in human plasma by Brussels sprouts was in two Dutch publications.[11,16] Using a similar study design to that of Nijhoff *et al.*[11] and using the same commercially available test kit, we failed to find any increase of the α-GST isozyme in all of our experiments. In the same publication[11] the authors postulated that the α-GST isozyme level is better inducible in males than in females whereas in all our experiments no link between gender and GST induction could be observed. With red cabbage also in saliva an increase in GST activity could be measured. However, the inter-individual differences seen in saliva were high and therefore only very pronounced effects can be detected. Therefore we think that this technique is less useful as a biomarker for GST induction than plasma. Several studies indicated that expression of GST-isozymes in humans play a role in cancer aetiology and in addition findings from animal experiments give evidence that dietary constituents increase GST levels and that this induction effect is associated with protection from chemically induced cancer (for review see Hayes and Pulford[1]).

The results of this study clearly show that consumption of certain cruciferous vegetables indeed leads to induction of this cancer-protective enzyme and supports the assumption that consumption of cruciferous vegetables is chemopreventive in man.

6 References

1 J.D. Hayes and D.J. Pulford, *Crit. Rev. Biochem. Mol. Biol.*, 1995, **30**(6), 445.

2 R.M.E. Vos, M.C. Snoek, W.J.H. van Berkel, F. Müller and P.J. van Bladeren, *Biochem. Pharmacol.*, 1987, **37**(6), 1077.

3 J.J.P. Bogaards, B. van Ommen, H.E. Falke, M.I. Willems and P.J. van Bladeren, *Food Chem. Toxic.*, 1990, **28**(2), 81.
4 R. Moser, T. Oberley, T. D. Daggett, D.A. Friedman, A. L. Johnson and F.L. Siegel, *Toxicol. Appl. Pharmacol.*, 1995, 131.
5 H.M. Wortelboer, E.C.M. van der Linden, C.A. de Kruif, J. Noordhoek, B.J. Blaauboer, P.J. van Bladeren and H.E. Falke, *Food Chem. Toxicol.*, 1992, 30.
6 L.W. Wattenberg and W.D. Loub, *Cancer Res.*, 1978, **38**, 1410.
7 V.L. Sparnins and L.W. Wattenberg, *J. Natl. Cancer Inst.*, 1981, **66**, 769.
8 R. Lange, R. Baumgrass, M. Diedrich, and M. Henschel, *Ernähr. Umsch.*, 1992, **39**, 292.
9 K. Sones, R.K. Heaney and G.R. Fenwick, *J. Sci. Food Agric.*, 1984, **35**, 712.
10 M.R.A. Morgan and G.R. Fenwick, *Lancet*, 1990, **336**, 1492.
11 W.A. Nijhoff, T.P.J. Mulder, H. Verhagen, G. van Poppel and W.H.M. Peters, *Carcinogen.*, 1995, **16**, 955.
12 M.L. Clapper, C.E. Szarka, G.R. Pfeiffer, T.A. Graham, A.M. Balshem, S. Litwin, E.B. Goosenberg, H. Frucht and P.F. Engstrom, *Clin. Cancer Res.*, 1997, **3**, 25.
13 L. Sreerama, M.W. Hedge and N.E. Sladek, *Clin. Cancer Res.*, 1995, **1**, 1153.
14 H. Verhagen, A.L. Rauma, R. Törronen, N. de Vogel, G. C. D. M. Bruijntjes-Rozier, M. A. Dreve, J. J. P. Boogards and H Mykkänen, *Human Exp. Toxicol.*, 1996, **15**, 821.
15 W.A. Nijhoff, M.J.A.L. Grubben, F.M. Nagengast, J.B.M.J. Jansen, H. Verhagen, G. van Poppel and W.H.M. Peters, *Carcinogen.*, 1995, **16**, 2125.
16 J.J.P. Bogaards, H. Verhagen, M.I. Willems, G. van Poppel and P.J. van Bladeren, *Carcinogen.*, **15**, 1073.
17 W. H. Habig, M.J. Pabst and W.B. Jakoby, *J. Biol. Chem.*, 1974, **249**, 7130.
18 D. A. Bell, C.L. Thompson, J. Taylor, C.R. Miller, F. Perera, L.L. Hsieh and G.W. Lucier, *Env. Health. Perspect.*, 1992, **98**, 113.

4.6

Dietary Administration of a Non-polar Extract of Rosemary Leaves Produces a Monofunctional Induction of Phase II Enzymes in Rat Liver

P. Debersac,[1] M F. Vernevaut,[1] J. Del Campo,[2] J.M. Heydel[3] and M.H. Siess[1]

[1] UNITÉ DE TOXICOLOGIE NUTRITIONNELLE, INRA, 17 RUE SULLY, 21034 DIJON CEDEX, FRANCE
[2] STATION DE TECHNOLOGIE DES PRODUITS VÉGÉTAUX, INRA, DOMAINE SAINT PAUL, SITE AGROPARC, 84914 AVIGNON CEDEX 9, FRANCE
[3] UNITÉ DE BIOCHIMIE, PHARMACOLOGIE, TOXICOLOGIE, FACULTÉ DE MEDECINE ET DE PHARMACIE, 7, BD JEANNE D'ARC, 21033 DIJON, FRANCE

1 Introduction

The strong antioxidant activity of rosemary can be mainly attributed to non-polar compounds, and more precisely to phenolic diterpenes such as carnosol and carnosic acid.[1] Many substances with antioxidant or reactive oxygen scavenging properties have been proved to inhibit carcinogenesis or biochemical events associated with it.[2] Few publications about the effects of rosemary extract on carcinogenesis are available, especially those dealing with the modulation of carcinogen metabolism.[3,4] The aim of this study was to give an over-view of the effect of an extract of rosemary high in phenolic diterpenes upon carcinogen metabolizing enzymes, in order to evaluate the potential protective action against carcinogenesis. In our study, we focused on the modulation of phase I enzymes (P450) and phase II enzymes (glutathion-S-transferase (GST), NAD(P)H-quinone reductase (QR), UDP-glucuronosyl-transferase (UGT)) in rat liver. Moreover, identification of specific isoforms of GST was carried out by Western blot analysis whereas the expression of mRNA specific UGT1A6 isoform was performed by RT-PCR.

2 Materials and Methods

Preparation of the rosemary extract. Leaves of *Rosmarinus officinalis*, crushed into powder, were extracted with dichloromethane (DCM), filtrated and finally evaporated. The extract was dissolved in methanol for HLPC analysis (not shown). A lyophilization of this extract permits its incorporation into the diet of rats.

Animals and treatments. Six male SPF Wistar rats, four weeks old, were fed *ad libitum* with a diet containing or not 0.5% (w/w) of DCM-extract (DCME) for 14 days.

Preparation of liver fractions. Rats were killed and the livers were immediately removed. Cytosol and microsomal fractions were prepared by differential centrifugation.

Enzyme assays. Activities of P450 1A1, 1A2, 2B1,2, and 2E1 were respectively evaluated by ethoxy, methoxy and pentoxyresorufine-O-dealkylation and p-nitrophenol hydroxylation activities. Phase II enzyme activities were determined with specific substrates: 1-chlorodinitrobenzene (CDNB) for GST, p-nitrophenol (PNP) for UGT1, 4-hydroxybiphenyl for UGT2 and menadione for QR.

Immunoblot analysis of GST. Polyclonal antibodies raised against rGSTA1/A2, A3/A5, M1, M2 and P1 were used. The quantification by image analysis can be carried out only for GST subunits present in control and/or treated rats. For the rGSTP1 subunit, the amount of protein is only qualitatively estimated.

mRNA level of UGT1A6. Liver tissue samples of six animals are pooled within each group (two pools of three rats) and total RNA are extracted from pools with commercial kit RNAxel containing RNAbind (Eurobio). An aliquot of reverse transcription products is used for PCR co-amplification and for semiquantitative analysis with specific primers for UGT1A6 isoform (507 pb fragment) and for β-actin internal standard according to the sequences determined by Iyanagi *et al.*[5]

3 Results

The data indicated that dietary administration of DCM extract for two weeks at 0.5% (w/w) did not modify any P450 activities. On the contrary, it significantly stimulated the activities of all phase II enzymes: GST (1.7 fold), UGT2 (1.4 fold) and the increase of UGT1 and QR activities being the most pronounced (3.7 fold and 4.1 fold) (Figure 1).

Immunoblot carried out on GST subunits showed that DCME enhanced all subunit levels in the liver, particularly rGSTA3/A5 ($\times$4.1) and rGSTP1 (Figure 2). Concerning the other subunits, the level was moderately increased, rGSTM1 ($\times$ 1.2), rGSTM2 ($\times$ 1.4) and rGSTA1/A2 ($\times$ 1.5).

Figure 3 shows the levels of UGT 1A6 mRNA in control and in DCME-treated rat liver using as internal standard the β-actin mRNA levels (Figure 3A) which remains unchanged whatever the treatment. By comparing the relative amount of UGT1A6 mRNA/β-actin mRNA between DCME treated and

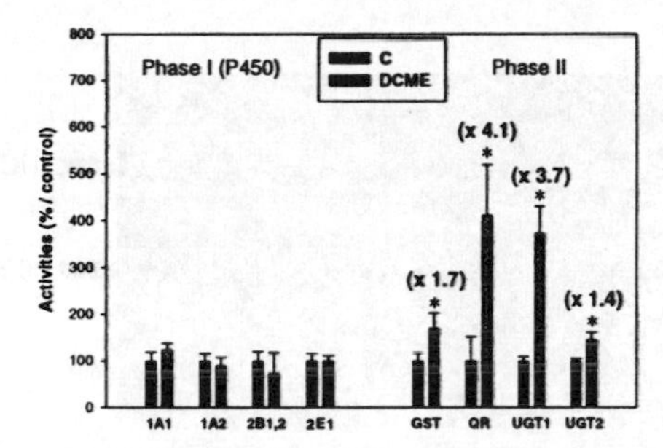

Figure 1 *Effect of rosemary DCME on drug metabolizing enzymes activities in rat liver. The results are expressed in % of control (means ± SEM of six rats), * significantly different from the control (Dunnett's test, p 0.05)*

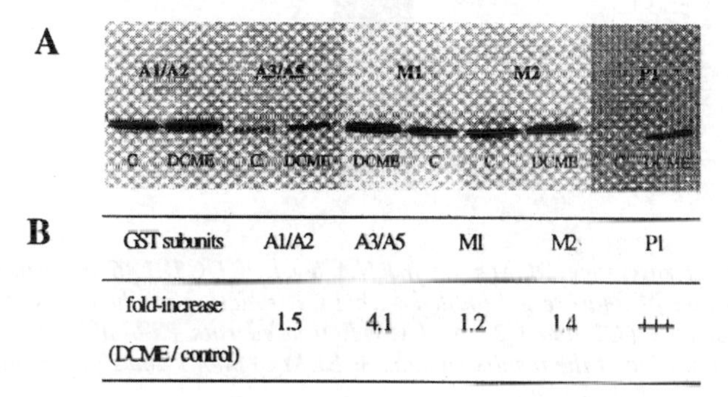

GST subunits	A1/A2	A3/A5	M1	M2	P1
fold-increase (DCME / control)	1.5	4.1	1.2	1.4	+++

Figure 2 *Estimation of the amount of liver glutathion-S-transferase subunits from rats treated with the rosemary DCME. The quantification is carried out by comparing blot density* (A) *between DCME-treated rats and controls rats (fold-increase) excepted rGSTP1 which amount is only appreciated by qualitative criteria* (B). *The fold-increase in the subunit levels in DCME-treated rats is calculated from means of six rats*

control rats, we provide evidence that DCME is a strong inducer of the UGT 1A6 (11.1 fold) (Figure 3B) and it seems to be as powerful as the typical inducer 3-methylcholanthrene (3-MC).[5]

4 Conclusion

In this study, we demonstrated that dietary exposure to dichloromethane extract of rosemary produced a monofunctional induction of phase II enzymes in rat liver. The increase of GST CDNB-activity was accompanied by an elevation of alpha, pi and mu GST levels, rGST A3/A5 and P1 being the most induced. Finally, we underlined the induction of the UGT1A6 isoform which seems to be as strong as the typical inducer 3-MC. To our knowledge, this is the first study to describe the effect *in vivo* of such rosemary extract on phase I and phase II enzymes. It must be emphasized that this extract of rosemary strongly

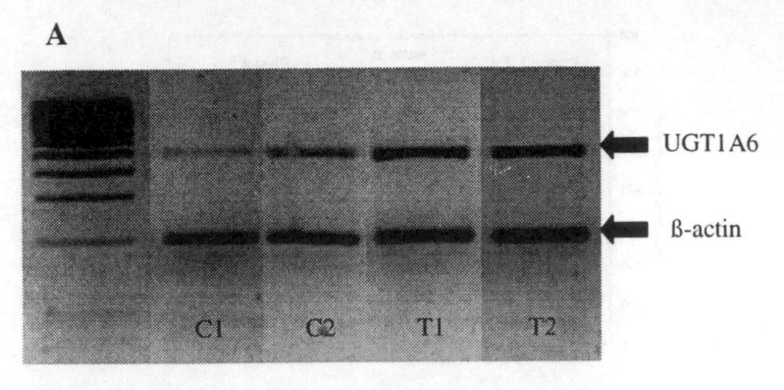

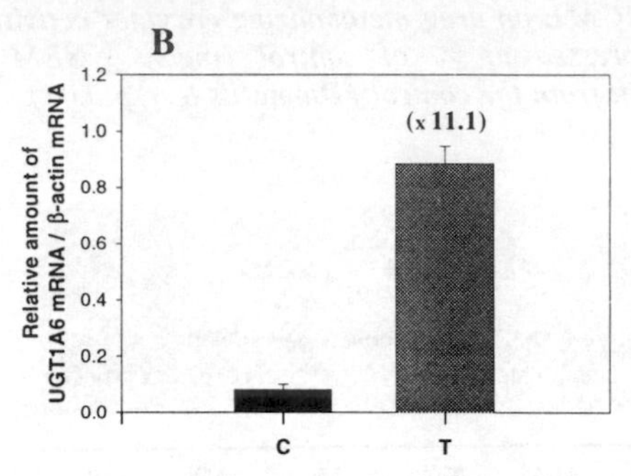

Figure 3 *Effect of rosemary DCME on mRNA level of UGT 1A6 isoform in rat liver.* A, *picture of agarose gel loaded with PCR reaction products (pool 1 and 2 for control rats (C1 and C2) or DCME treated rats (T1 and T2)).* B, *graphic representation of the results (means ± SEM of pool 1 and 2) for control (C) and DCME treated (T) rats*

induces enzymes involved in the detoxication of carcinogens. Therefore, it would be of great interest to explore whether this kind of induction can protect against carcinogenesis.

5 References

1 O.I. Aruoma, B. Halliwell, R. Aeschbach and J. Loligers, *Xenobiotica*, 1992, **22**, 257.
2 T.J. Slaga, *Crit. Rev. Food. Sci. Nutr.*, 1995, **35**, 51.
3 E.A. Offord, K. Mace, C. Ruffieux, A. Malnoe and A.M. Pfeifer, *Carcinogenesis*, 1995, **16**, 2057.
4 K.W. Singletary, *Cancer Lett.*, 1996, **100**, 139.
5 T. Iyanagi, M. Haniu, K. Sogawa, Y. Fujii-Kuriyama, S. Watanabe, J.E. Shively and K.F. Anan, *J. Biol. Chem.*, 1986, **261**, 15607.

4.7

Experimental Studies on Cancer Chemoprevention by Tea Pigments

Chi Han and Yunyun Gong

INSTITUTE OF NUTRITION AND FOOD HYGIENE, CAPM, BEIJING 100050, CHINA

1 Introduction

The anticarcinogenic properties of tea have received much attention recently. Many animal studies have shown that tea has a strong protective effect on tumor formation in many organs.[1,2] Tea polyphenols are the major ingredients in green tea, and it has been suggested that their anti-carcinogenic effect is a result of their antioxidant activities, modulation of immunity, and activation of detoxification enzymes. Recent studies demonstrated that theaflavins and thearubigins, oxidised products of polyphenols which form the major constituents of black tea pigments, are also strong antioxidants. Based on the multi-stage-theory of carcinogenesis,[3,4] we developed a batch of short-term tests to examine and compare the effects of tea pigments and tea polyphenols on the initiation, promotion and progression stages during carcinogenesis. We also examined the inhibitory effect of tea pigments on liver tumor formation in rats. In addition, we investigated the effects of tea polyphenols and tea pigments on phase II detoxification enzymes to further elucidate their anti-tumor mechanism.

2 Materials

Green tea polyphenols (purity 40%) and tea pigments, which are the oxidized products of 40% green tea polyphenols, were provided by the Institute of Tea Science and Research, Chinese Academy of Agricultural Sciences. TPA, diethylnitrosamine (DEN), cytomycin B, bisacrylamide, and 6-thioguanine (6-TG) were purchased from Sigma company. Acrylamide is a product of Fluka Co. AP-labeled secondary IgG antibody was purchased from Bio-Rad Co. GST 1-1, 1-2 and 3-3 polyclonal antibody were gifts from Hirosaki University School of Medicine, Japan.

3 Methods

3.1 Short-term Screening Assays

Tests on the induction of forward gene mutation by mitomycin C in V79 cells was based on the methods reported by Cajelli *et al.*[5] Micronuclei formation in V79 cells induced by mitomycin C (MMC) was measured by the cytokinesis-block method as described previously (Krishna *et al.*).[6] Metabolic cooperation in V79 cells induced by TPA was based on the method reported by Yitti *et al.*[7] Tests on TPA induced inflammatory epidermal edema in Kunming mouse ear was carried out according to the method described by Gschwendt.[8] Tests on the viability of Hela cells and growth ability of Hela cells in soft agar test were based on the methods reported by Wu.[9] Tests on the growth of S180 solid tumor in Kunming mice were conducted following Yamaguchi *et al.*[10]

3.2 Determination of Detoxifying Enzymes

3.2.1 Cells culture and treatment. Hep G2 cells were cultured in 96 well plate at a density of 1×10^4 cells/well in 200 ml of DMEM supplemented with 10% fetal calf serum, 100 U ml^{-1} of penicillin G, 100 mg ml^{-1} of streptomycin, and 0.1% DMSO. The cells were grown for 24 hours in a humidified incubator in 5% CO_2 at 37 °C. The medium was decanted, and each well was re-fed with 100 ml of medium and tea products (2, 10 mg ml^{-1} designed for each sample). Control cells were fed only with media. The cells were exposed to test compounds for 24 hours for quinone reductase activity assay.[11,12] Hep G2 cells were grown in DMEM supplemented with 10% fetal calf serum. Tea polyphenols and tea pigments at concentrations of 0, 20, 50, 100 mg ml^{-1} were added separately when the cells were in logarithmic growth rate. The cells were collected after 24 and 48 hours of treatment. After homogenizing with PBS, the cell suspension was centrifuged twice and the supernatant was store at −80 °C for glutathione-S-transfrase (GST) assay.

3.2.2 Animal studying and treatment.[13] Male Wistar rats, 80–100 g b.w., were purchased from the Animal Center of Chinese Academy of Medical Sciences, Beijing. Rats were fed with regular chow, and divided into 4 groups randomly. The positive-control group was treated as follows: given DEN (10 mg kg^{-1} b.w.) i.p. once a day from day 7 to 13, 20 ml CCl_4 (0.5 ml/rat) i.p. once a day at day 40 and 41, and then subjected to 2/3 hepatectomy under ether anesthesia at the end of the sixth week. The rats were killed at day 56. Besides the same treatment as the positive control group, tea polyphenols-DEN group and tea pigments-DEN group were provided with 0.1% tea polyphenols and 0.1% tea pigments respectively as drinking fluid during the whole procedure. The rats were killed and the livers were collected and homogenized. Cytosol fraction was prepared for analyzing GST and isoenzymes.

3.2.3 Enzyme assays. Quinone reductase activity assay: After 24 hours treatment, the cells were lysed by incubation at 37 °C for 10 min with 50 ml in each well of solution containing 0.8% digitonin. The plates were then shaken for an additional 10 min at 25 °C. Then 200 ml of the complete reaction mixture was added to each well, the reaction was arrested after 5 min by adding 50 ml of a solution containing 0.3 mM dicoumarol. The plates were then scanned at 610 nm. The number of cells or the amount of protein in each microtiter was determined by the method of Delong.[12]

GST assay: Cytosolic GST activity was analyzed based on the rate of GS-CDNB conjugate formation using CDNB as substrate as described by Habig.[14] 0.30 mg of each cytosolic protein was subjected to SDS-PAGE according to the method of Laemmli.[15] The separated protein was then transferred onto nitrocellulose membranes and detected with antibodies against rat GST 1-1 or 1-2 or 3-3. Densitometric analysis of immunoblots are performed to quantify the relative level of each band.

4 Results

4.1 Effects on Initiation Stage of Carcinogenesis

As shown in Table 1, in the test of forward gene mutation induced by mitomycin in V79 cells, tea polyphenols and tea pigments both exhibited inhibitory effects on forward gene mutation at concentrations of 10–100 mg ml^{-1} and the inhibitory effect has significant dose-response relationship ($p < 0.01$). The same pattern of inhibition was observed on the micronuclei formation in V79 cells induced by mitomycin C. Table 2 shows that inhibition was significant at concentrations of 10–100 mg ml^{-1} of both tea polyphenols and tea pigments. Dose-response relationships were also observed as the concentration increased. These results showed that tea polyphenols and tea pigments both have a significant antimutagenic effect and blocking effect on the initiation phase of carcinogenesis.

Table 1 *The effects of tea polyphenols and tea pigments on the induction of forward gene mutation induced by MMC in V79 cells (average of triplicate assay)*

Treatment	*Concentration (mg ml^{-1})*	*Mutation rate (per 10^6 cells)*	*Inhibition rate (%)*
MMC	–	188.3	
Tea polyphenols	10	132.7*	29.5
	50	105.3*	45.3
	100	76.3*	59.5
MMC	–	219.0	
Tea pigments	10	79*	43.9
	50	73*	66.7
	100	51**	76.7

* Compare with MMC group, x^2-test, $p < 0.05$.
** Compare with MMC group, x^2-test, $p < 0.01$.

Table 2 *The effects of tea polyphenols and tea pigments on the micronuclei formation in V79 cells induced by MMC (average of triplicate assay)*

Treatment	*Concentration (mg ml^{-1})*	*Micronuclei (per 1000)*	*Inhibition rate (%)*
MMC	–	73.2	0
Tea polyphenols	10	52.9*	27.5
	50	41.7*	42.2
	100	27.4*	62.0
MMC	–	70.0	0
Tea pigments	10	48.0*	33.3
	50	42.0*	43.1
	100	37.0*	50.8

* Compare with the MMC group, x^2-test, $p < 0.05$.

4.2 Effects on the Promotion Stage of Carcinogenesis

The effects of tea polyphenols and tea pigments on the metabolic cooperation test are summarized in Table 3 and 4. When tea polyphenols or tea pigments were added at concentrations ranging from 10 to 20 mg ml^{-1} with TPA, the survival rate of M cells was reduced. A dose-response relationship was observed as the concentration increased. In the test of inflammatory epidermal edema induced by TPA, tea pigments alleviated the degree of edema. The effect of tea pigments showed significant dose-response relationship. These results showed that tea polyphenols and tea pigments both have a significant inhibitory effect on the promotion stage of carcinogenesis.

Table 3 *The effects of tea polyphenols and tea pigments against the inhibition of metabolic cooperation in V79 cells induced by TPA*

Treatment	*Concentration (mg ml^{-1})*	*Survival cells (%)*	*Blocking rate (%)*
NC		20.8 ± 10.5	
TPA	–	79.5 ± 9.9	
Tea polyphenols	10	81.8 ± 2.1	0
	20	68.5 ± 2.0*	18.7
	50	62.5 ± 6.1*	29.0
NC		22.2 ± 7.9	
TPA	–	83.6 ± 8.4	
Tea pigments	4	83.4 ± 18.8	0
	10	75.0 ± 19.2	10.3
	20	66.6 ± 12.5*	20.3

* Compare with the TPA group, x^2-test, $p < 0.05$.

Table 4 *The effects of tea pigments on the inhibition rate of TPA induced mouse ear edema*

Sex	*Treatment group*		*Right-Left (mg)*	*Inhibition rate (%)*
Male	Tea pigments	Control	34.7 ± 4.0	0
		12.5 mg	30.7 ± 4.0*	11.5
		50.0 mg	27.4 ± 6.0*	21.0
Female	Tea pigments	Control	31.6 ± 4.0	0
		12.5 mg	27.6 ± 5.0	12.7
		50.0 mg	24.5 ± 5.0*	22.4

* Compare with thc TPA group, x^2-test, $p < 0.05$.

4.3 Effects on the Proliferation of Cancer Cells

As shown in Tables 5 and 6, tea polyphenols and tea pigments both have significant inhibitory effects on the survival rate and grow ability of Hela cells in soft agar medium. The results are consistent with the results shown in

Table 5 *The effects of tea polyphenols and tea pigments on the survival of hela cells*

Treatment	*Concentration ($mg\ ml^{-1}$)*	*No. survival colonies*	*Inhibition rate (%)*
Control		476.3 ± 19.5	0
Tea polyphenols	20	456.7 ± 11.0	4.1
	50	426.0 + 7.9*	10.6
Control		226.0 ± 5.5	0
Tea pigments	10	221.0 ± 5.6	2.2
	50	176.0 ± 16.3*	22.1

* Compare with the TPA group, x^2-test, $p < 0.05$.

Table 6 *The effects of tea polyphenols and tea pigments on the growth ability of hela cells in soft agar*

Treatment	*Concentration ($mg\ ml^{-1}$)*	*No. of colonies*	*Inhibition rate (%)*
Control		284.0 ± 9.7	0
Tea polyphenols	20	269.1 ± 21.2	5.0
	50	250.3 ± 16.8	11.9
	100	180.7 ± 9.7*	36.4
Control		226.0 ± 5.5	0
Tea pigments	20	221.0 ± 5.6	2.2
	50	176.0 ± 16.3	22.1
	100	149.0 ± 14.4*	34.1

* Compare with the control group, x^2-test, $p < 0.05$.

Table 7 *The effects of tea pigments on the growth of S180 solid tumor*

Sex	*No. of animals*	*Concentration (mg ml^{-1})*	*Tumor weight (g, mean ± SD)*	*Inhibition rate (%)*
Male	30	Control	2.682 ± 0.964	0
	28	1.25	2.073 ± 0.724*	22.7
	28	2.50	2.065 ± 0.389*	23.0
	29	5.00	1.859 ± 0.597**	30.7
Female	28	Control	2.424 ± 0.706	0
	28	1.25	2.197 ± 0.621	9.3
	25	2.50	1.900 ± 0.753*	21.6
	26	5.00	1.680 ± 0.945*	30.7

* Compare with the control group, t-test, $p < 0.05$.
** Compare with the control group, t-test, $p < 0.01$.

Table 7, that tea pigments have an inhibitory effect on S180 solid tumor formation in mice.

4.4 Effects on Antioxidant and Detoxication Enzymes

As shown in Table 8, tea pigments, theaflavins and thearubigins significantly induced QR activity in Hep G2 cells.The results in Table 9 showed that tea polyphenols and tea pigments induced GST activity in Hep G2 cells. A significant effect was observed after 24 hours, and GST activity continued to increase to 48 hours. The effects exhibited both dose-response and time-response relationships. As shown in Table 10, oral administration of 0.1% tea polyphenols and 0.1% tea pigments as drinking water induced cytosolic GST activity by 25% and 18% in rat liver with precancerous lesions induced by DEN as compared with the positive control group. At the same time, GST 1-1, 1-2, and 3-3 protein expression level in liver cytosol of these two groups were also increased (Tables 11 and 12). Tea pigments showed stronger induction on GST

Table 8 *Induction of QR activity in Hep G2 cell*

Groups	*Concentrations (mg ml^{-1})*	*QR activity (mmol g^{-1})*	*Induction rate*
Control	–	248.6 ± 62.6	0
Tea polyphenols	2	336.8 ± 43.8	35.8
	10	430.4 ± 71.8*	73.0
Tea pigments	2	394.8 ± 97.4**	58.8
	10	399.0 ± 57.6**	60.4
Thearubigens	2	298.2 ± 43.6	19.9
	10	324.2 ± 29.0**	30.3
Theaflavins	2	244.4 ± 47.0	0
	10	294.8 ± 51.8	18.5

* Compare with control, t-test, $p < 0.05$.
** Compare with control, t-test, $p < 0.01$.

Table 9 *The effects of tea polyphenols and tea pigments on the activity of cytosol glutathione S-transferase in Hep G2 cells*

Concentration ($mg\ ml^{-1}$)	GST activity ($U\ min\ g^{-1}$ prot.) Tea polyphenols 24 hour	Tea polyphenols 48 hour	Tea pigments 24 hour	Tea pigments 48 hour
0	47.11 ± 2.18	45.65 ± 1.51	47.11 ± 2.18	45.65 ± 1.51
20	46.41 ± 2.07	46.24 ± 1.99	50.78 ± 1.07	48.61 ± 1.20
50	52.20 ± 0.61	61.65 ± 0.48**	56.62 ± 2.67	67.78 ± 1.79**
100	57.78 ± 1.21*	75.14 ± 2.02**	63.64 ± 2.09*	84.15 ± 2.65**

* Compare with control, t-test, $p < 0.05$.
** Compare with control, t-test, $p < 0.01$.

Table 10 *The effects of tea polyphenols and tea pigments on the activity of cytosol glutathione S-transferase in rats with precancerous liver lesions*

Group	No. of animals	GST activity ($U\ min\ g^{-1}$ prot.) Mean ± SD	Induction rate (%)
DEN	12	459.38 ± 84.68	0
DEN + tea polyphenols	12	573.10 ± 95.58*	25
DEN + tea pigments	12	542.93 ± 80.81*	18
Negative control	12	514.82 + 77.88*	–

* Compare with the DEN group, t-test, $p < 0.05$.

Table 11 *The effects of tea polyphenols and tea pigments on GST 1-1 and 1-2 in rats with precancerous liver lesions (Western blot assay)*

Group	No. of animals	GST 1-1 Relative level	GST 1-1 Induction rate (%)	GST 1-2 Relative level	GST 1-2 Induction rate (%)
DEN	6	1.095 ± 0.163	0	0.742 ± 0.255	0
DEN + Tea polyphenols	6	1.510 ± 0.288**	37.2	1.280 ± 0.102**	72.5
DEN + Tea pigments	6	1.454 ± 0.245**	32.7	1.148 ± 0.089**	54.7
NC	6	1.003 ± 0.114	–	1.055 ± 0.123	–

** Compare with the DEN group, t-test, $p < 0.01$.

Table 12 *The effects of tea polyphenols and tea pigments on GST 3-3 in rats with precancerous liver lesions (Western blot assay)*

		GST 3-3	
Group	*No. of animals*	*Relative level*	*Induction rate (%)*
DEN	6	0.816 ± 0.131	0
DEN + Tea polyphenols	6	1.146 ± 0.165**	40.4
DEN + Tea pigments	6	1.337 ± 0.374*	63.8
NC	6	0.986 ± 0.031	–

* Compare with the DEN group, t-test, $p < 0.05$.
** Compare with the DEN group, t-test, $p < 0.01$.

3–3 than tea polyphenols, and tea polyphenols greatly increased the level of GST 1-2. Tea pigments and tea polyphenols had approximately the same effect on the induction of GST 1-1.

5 Discussion

In short-term tests, tea pigments and tea polyphenols have similar inhibitory effects on the four short term screening assays employed – forward gene mutation, micronuclei formation in V79 cell induced by mitomycin C, V79 cells metabolic cooperation and survival and growth ability of Hela cells in soft agar media. Tea pigments also showed a strong inhibitory effect on S180 solid tumor formation in mice. These results are consistent with our earlier results of studies on black tea.[16,17]

The data from the Hep G2 cell system showed tea pigments could induce the activity of antioxidant enzymes and phase II enzymes, QR and GST, thus enhancing carcinogen detoxification *in vivo*. The results also showed that antioxidant activity of tea pigments is comparable to that of tea polyphenols. In the QR activity assay, the effects of theaflavins or thearubigins were weaker than that of tea pigments; their sum still being less than that of the latter.

In the rat liver precancerous model initiated by DEN and promoted with CCl_4 as well as partial hepatectomy, the activity of cytosolic GST was less than that of untreated negative controls. We observed a significant increase of GST activity in the tea polyphenols and tea pigments groups. The Western blot study showed tea pigments and polyphenols increasing levels of GST 1-1, 1-2 and 3-3 isoenzymes. GST, especially GSTm, is important for the detoxification of active metabolites produced during bio-transformation of DEN. GSTa also inhibits and detoxifies the activity of certain electrophilic substances by directly conjugating with these substances. GSTm catalyzes the conjugation of GSH with epoxy-metabolites of DMBA, B[a]P and 1-nitropyrene.[18] It is important for metabolism of nitrosamines such as DMN and the result of the Western Blot assay showed that tea pigments and polyphenols could increase the level of DEN.[19] Whilst the level of GSTa induced by tea polyphenols is higher than that of tea pigments, the reverse situation was found for the induction of GSTm. This suggests that tea polyphenols may be greater inhibitors of lipid peroxide formation, with tea pigments having the stronger effect on detoxification of DEN and its metabolites. The inductive effect of tea polyphenols and tea pigments on the GST activity and its isoenzymes a and m may be the major mechanism in its cancer chemoprevention.

The amount of tea pigments in black tea is similar to that of tea polyphenols in green tea, but the former are more stable chemically and may be ideal chemopreventive agents. As with tea polyphenols, tea pigments have several components – theaflavins, thearubigins and theabrownin. It will be necessary to compare the effects of individual components in further studies of tea pigments. Since the results showed tea pigments to have promising effects on prevention of cardiovascular diseases,[20] it is thought worthwhile to further study and exploit this newly-recognized component.

Acknowledgements

Supported by the Chinese National Natural Science Foundation research grant No 39330189. The author would like to thank Prof. Qikun Chen of the Institute of Tea Science, Chinese Academy of Agricultural Science for supplying tea polyphenolstea pigments and tea ingredients; Dr. Kimihiko Satoh of Hirosaki University School of Medicine, Japan for providing GST 1-1, 1-2 and 3-3 polyclonal antibodies.

6 References

1 Han C and Xu Y (1990) The effect of Chinese tea on the occurrence of esophageal tumor induced by N-nitrosomethylbenzylamine in rats. *Biomed. Environ. Sci.* **3**, 35.
2 Yang CS and Wang ZY (1993) Tea and cancer. *J. Natl. Cancer Inst.* **85**, 1038.
3 Bertram J, Kolone LN, Meyskens FL (1987) Rationale and strategies for chemoprevention of cancer in humans. *Cancer Res.* **47b, 328.**
4 Lutz WK and Majer P (1988) Genotoxic and epigenetic chemical carcinogenesis: one progress, different mechanisms. *Trends pharmacol. Sci.* **9**, 322.
5 Cajelli E, Canonero R and Martelli A (1987) Methylglycoxal induced mutation to 6-thioguanine resistance in V79 cells. *Mutat. Res.* **190b, 47.**
6 Krishnaa G, Kropko ML and Theiss JC (1989) Use of the cytokinesis-block method for the analysis of micronuclei in V79 Chinese hamster lung cells, results with mitomycin C and cyclophosphamid. *Mut. Res.* **222**, 63.
7 Yitti LP, Chang CC and Trosko JE (1979) Elimination of metabolic cooperation in Chinese hamster cells by a tumor promoter. *Science.* **206**, 1089.
8 Puignero V, Turull A and Queralt (1998) Arachidonic acid (AA) and tetradecanoylphorbol acetate (TPA) exert systemic effects when applied topically in the mouse. *J. Inflammation.* **22**, 307.
9 Wu DF (1986) The inhibition effects of cordyceps on the growth of cervical cancer cell *in vitro* (in Chinese). *Cancer.* **5**, 337.
10 Yamaguchi T, Ikezaki K, Kishiye T *et al.* (1984) Potentiation of anticancer agents by

new synthetic isoprenoids. II. Inhibition of the growth of transplantable murine tumors. *J. Natl. Cancer Inst*. **73**, 903.

11 Prochaska HJ, Delong MJ and Talalay P (1985) On the mechanisms of induction of cancer-protective enzymes: a unifying proposal. *Proc. Natl. Acad. Sci*. **82**, 8232.

12 Delong MJ, Prochaska HJ and Talalay P (1986) Induction of NAD(P)H-quinone reductase in murine hepatoma cells by phenolic antioxidants, azo dyes, and other chemoprotectors: a model system for the study of anticarcinogens. *Proc. Natl. Acad. Sci*. **83**, 787.

13 Mao R, Ding L and Chen JY (1993) The inhibitory effects of epicatechin complex on diethylnitrosamine induced initiation of hepatocarcinogenesis in rats. *Chung-Hua-Yu-Fang-I-Hsueh-Tsa-Chih (Chinese J. Prev. Med.)*. **27**, 201.

14 Habig WH, Pabst MJ and Jokoby WB (1974) Glutathione S-transferase. The first enzymatic step in merapturic acid formation. *J. Biol. Chem*. **249**, 7130.

15 Laemmli UK (1970) Cleavage of structural proteins during the assembly of the head of bacteriophage T4. *Nature*. **227**, 680.

16 Yang GY, Liu ZJ, Darren N. Seril *et al*. (1997) Black tea constituents, theaflavins, inhibit 4-(methynitrosamino)-1-(3-pyridyl)-1-butanone (NNK)-induced lung tumorigenesis in A/J mice. *Carcinogenesis*. **18**, 2361.

17 Lu YP, Lou YR, Xie JG *et al*. (1997) Inhibitory effect of black tea on the growth of established skin tumors in mice: Effects on tumor size, apoptosis, mitosis and bromodeoxyuridine incorporation into DNA. *Carcinogenesis*. **18**, 2163.

18 Zangar RC, Springer DL, McCrary JA *et al*. (1992) Changes in adult metabolism of aflatoxin B1 in rats neonatlly exposed to diethylstilbestrol. Alterations in a-class glutathione S-transferases. *Carcinogenesis*. **13**, 2375.

19 Smith MT, Evans CG, Doane-Setzer P *et al*. (1989) Denitrosation of 1,3-bis(2-chloroethyl)-1-nitrosourea by class GST m and its role in cellular resistance in rat brain tumor cells. *Cancer Res*. **49**, 2621.

20 Lou FQ, Yang ZC, Yuan WL (1983) Study on the effect of tea pigments on experimental atherosclerosis in rabbits and treating patients of hyper-fibrinogenemia. *Chin. Med. J*. **63**, 632.

Section 5

Defence Systems: Antioxidant Mechanisms

5.1

Dietary Antioxidants in Cancer Prevention: Future Directions for Research

D.G. Lindsay

INSTITUTE OF FOOD RESEARCH, NORWICH RESEARCH PARK, COLNEY, NORWICH NR4 7UA, UK

1 Introduction

The hypothesis that reactive oxidative species (ROS) are a major contributory factor in the initiation and progression of a number of chronic diseases including cancer is a highly attractive one. ROS are generated in all living species as a result of normal oxidative metabolism. In addition both plants and animals use them as a defence mechanism against attack by pathogens. In humans phagocytic cells such as neutrophils, monocytes and macrophages combat infection through the induced synthesis of large amounts of nitric oxide (NO) and the superoxide radical ($O_2^{\bullet-}$) as well as species derived from these molecules, resulting in localised cell death. Some of the species are the same as those that are generated through radiation exposure. If these highly toxic radicals are not de-activated then they are capable of directly damaging membrane lipids, DNA, and proteins in the cell.

It is inconceivable that life as we know it could have developed unless a highly sophisticated set of control mechanisms existed to de-activate ROS. There are a wide range of antioxidant mechanisms, both constitutive and induced, that limit such damage.

Enzymatic mechanisms include catalase, glutathione reductase, superoxide dismutase, and haem oxygenase. Non-enzymatic mechanisms involve glutathione, ascorbic acid and α-tocopherol. The role of many more potential antioxidants remains to be fully elucidated. However, in spite of the many mechanisms that exist to neutralise the toxicity of ROS, the process of regulation is not 100% efficient and oxidised bases, lipids and proteins are formed in increasing amounts as humans age.[1,2] It is a possibility that these systems become less well controlled under periods of persistent oxidative stress which can occur following infection and inflammation. Once a threshold is passed there is an increasing probability of further oxidative damage as the

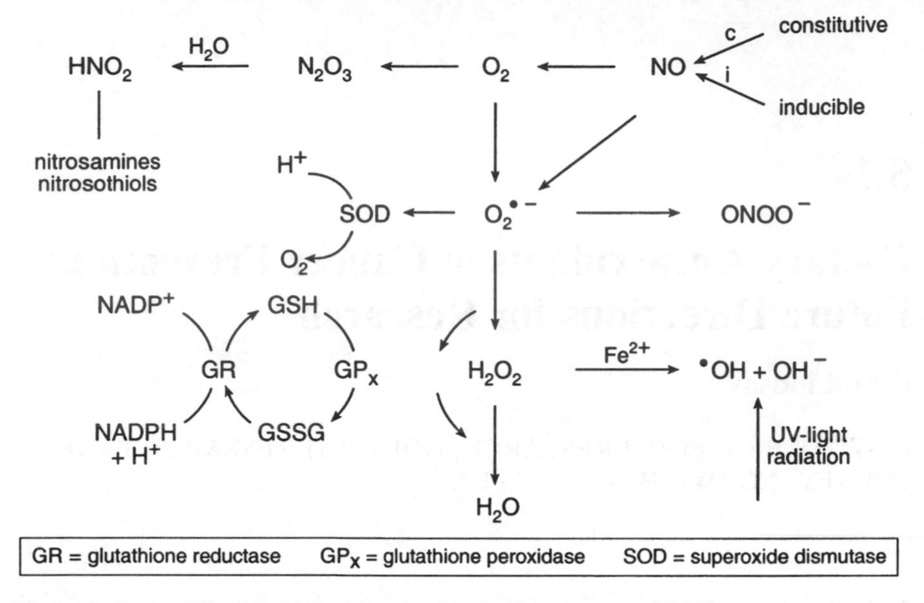

Figure 1 *Antioxidant mechanisms involved in the deactivation of ROS*

mechanisms of control become less efficient through mutation or damage. Nonetheless in animals it is possible to promote oxidative stress solely by dietary means.

In the specific case of DNA damage there are enzymes that are capable of repairing the damage caused by ROS but the efficacy and fidelity of these processes decline with age.[3] The mutagenic potential of ROS increases with age as these repair processes themselves are damaged. This is probably an important explanation for the exponential increase of cancer with age.

2 Free Radical Theory of Ageing

The free radical theory of ageing was first proposed in 1956 by Harman,[4] who argued that damage by ROS is critical in determining life span. Numerous experiments since that time reinforce the view that oxygen and other radicals play a role in degenerative diseases and senescence.[5] The theory would argue that any increase in the rate of energy metabolism would shorten life since the opportunity to generate oxidative damage is increased. The evidence does support the view that the variable life span of mammals is due to their differing metabolic rates (Figure 2). Species with longer lifespans generate superoxide and hydrogen peroxide at lower rates, accrue less oxidative damage, and resist induced oxidative stress.[6]

The ROS theory of ageing and degenerative disease initiation provides an intellectual framework around which to develop a comprehensive strategy of disease prevention and improvements in public health. One of the most

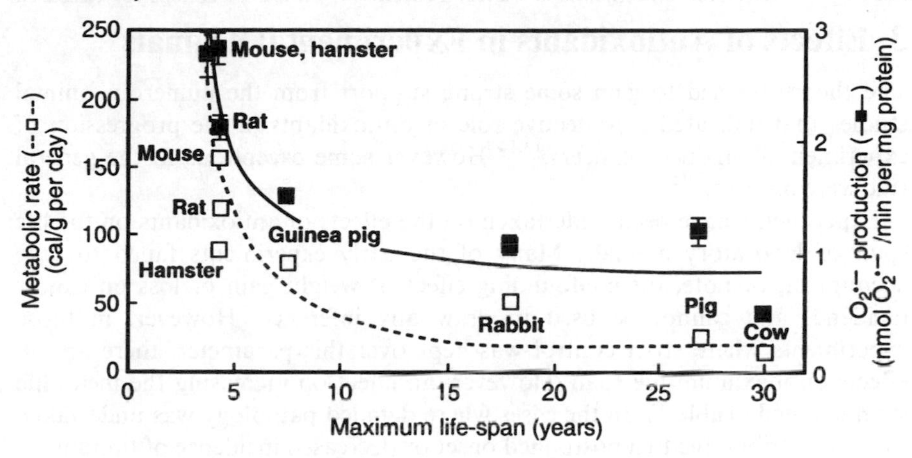

Figure 2 *The relationship of life span to metabolic rate in mammals*

attractive and practical avenues to pursue is to adequately define the role that components of the diet might play in neutralising the damaging effects of ROS generation.

As a means of testing the hypothesis there are two consistent observations that are relevant. The first of these is the observation that, in all living animal species that have been tested to date, there is increased longevity and protection from disease if they are restricted in food intake rather than allowed to feed *ad lib.*[7] The second observation is the known protective effect of increased fruit and vegetable consumption, particularly against cancers of the gastrointestinal system and the lung.[8]

The increased longevity observed in irradiated animals following dietary restriction has been shown to arise, in part, through a reduction in the number of tumours.[9–11] It has also been shown that it results in an increase in DNA repair activity.[12] This observation is also found in non-irradiated animals. The evidence suggests that this is not due to a decrease in the intake of exogenous dietary mutagens but is due to some underlying endogenous process.[7] It indicates the important role that diet and ROS production plays in the initiation and/or progression of tumours in experimental animals. However, to date there has been no unequivocal demonstration that caloric restriction results in a decreased rate of ROS production.

The ROS theory has also been invoked in suggesting a mechanism for this observation that diets rich in fruit and vegetables are protective. Phytochemicals and antioxidant vitamins in these foods have been implicated as agents in the reduction of ROS damage. This could occur either through a direct antioxidant effect or indirectly through the induction of antioxidant enzyme systems.

The theory has been examined by the administration of dietary antioxidants to experimental animals and humans on the assumption that they should be effective chemopreventants.

3 Effects of Antioxidants in Experimental Animals

The theory seemed to gain some strong support from the numerous animal studies that indicated a protective role of antioxidants in the progression of experimentally-induced cancers.[13,14] However some exceptions to the general rule were observed.[15]

Experiments have been undertaken on the effects of antioxidants on the life span of laboratory animals. Many of the early experiments failed to take account of, or note, the confounding effect of weight gain or loss on cancer incidence and cannot be used to draw any inference. However, in those experiments where strict control was kept over this parameter, there are no effects on maximum life span. However, an effect on increasing the mean life span is noted (Table 1). In the cases where detailed pathology was undertaken this was attributable to a postponed onset or decreased incidence of tumours.

As far as human studies are concerned, again the results are contradictory. Many of the early studies were based on a comparison of populations with high and low antioxidant vitamin plasma status and indicated a potential benefit.[19] But such studies are unacceptable in relating cause to effect for cancer risk. The use of random, placebo-controlled, intervention trials is regarded as essential. Preferably these trials should be based on an end point of cancer occurrence or mortality. Sadly they are very expensive to undertake and intermediate markers, such as pre-cancerous lesions, have been widely used. However, there is often no clear demonstration that these lesions truly reflect the certainty that cancer will result. Again no consistent pattern has emerged.

When intervention trials have been undertaken that are accepted as well conceived, they too show inconsistent results. A five year study in China with β-carotene, vitamin E and selenium, showed a significant reduction in mortality for stomach cancers.[20] A Finnish study on heavy smokers showed that β-carotene alone or in combination with α-tocopherol resulted in a significant increase in lung cancer.[21] A study undertaken concurrently with this on workers

Table 1 *Effect of antioxidant supplemented diets on mortality*

Species	*Antioxidant*	*Effect*
Rat	ubiquinone	None *Mech. of Ageing & Dev. 1980,* **13***:1*
Mouse	mixture	+ *Gerontology 1981,* **27***:133*
Rat	*dl*-α-tocopherol	Mean survival + Max. survival – *Mech. of Ageing & Dev. 1980,* **13***:1*
Mouse*	2-mercaptoethanol	Mean survival + Max. survival – *Mech. of Ageing & Dev. 1984,* **13***:341*

* Postponed onset and decreased incidence of tumours.

at high risk of lung cancer (asbestos exposure and smokers) showed that there was an increased risk of lung cancer in the group given β-carotene and vitamin E although the risks for former smokers were significantly decreased. In a physician's health study β-carotene was shown to have no effect on cancer incidence.[22] An intervention study with vitamin C, E and β-carotene supplements showed a significant decrease in oxidative DNA damage in lymphocytes but this cannot be related to cancer risk.[23]

From the experimental data the administration of single antioxidants in amounts that bear little or no relation to the current dietary intakes is not a satisfactory approach to deciding whether dietary antioxidants are of use in protecting against cancer risks, or to define optimal intakes.

The effects of antioxidants cannot be considered solely in terms of cancer risk or prevention. There is increasing evidence that antioxidants do have an effect on the protection from cardiovascular disease, from UV-induced skin cancer, and in the prevention of cataract.[24,25] They also have an important effect on the stimulation of the immune response.[26] Indeed for many sectors of the population the intakes of vitamins C, E and β-carotene are significantly less than would be the case if an ideal diet based on the recommended intakes of fruit and vegetables were consumed (Figure 3).[27]

There is also growing evidence for their preventive effects in other diseases of ageing. It is also clear that the toxicity of many of the dietary antioxidants is low

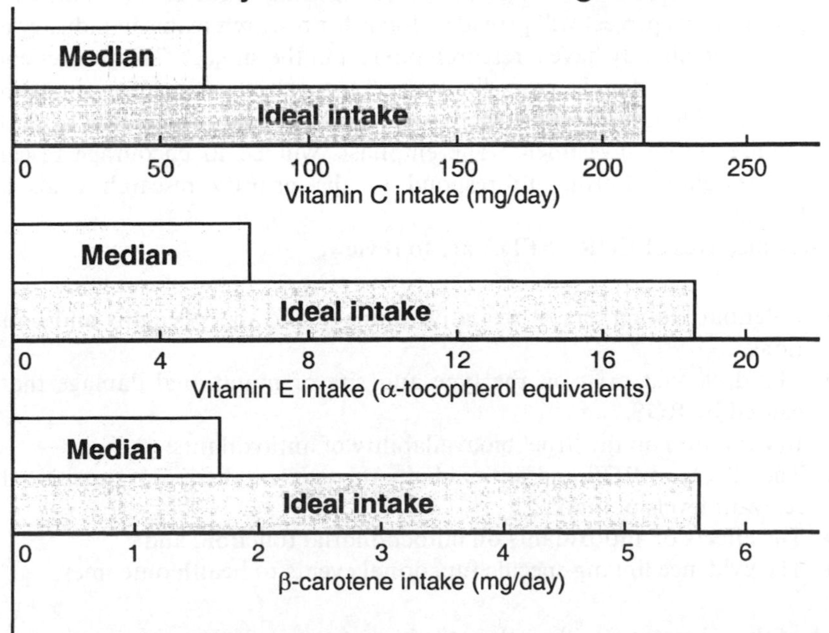

Figure 3 *Average population plasma levels of antioxidant vitamins*

and intakes could be increased without any adverse effects. It will be essential to consider an optimal balance of intakes to assure maximal protection. However, the science necessary to define these is still urgently needed. The biochemistry of dietary and cellular antioxidants needs to be better understood. It is only then that there will be maximal benefit in undertaking dietary intervention studies. These studies will need to take into account not only end point morbidity measurements but also include studies on cellular effects.

For some years the UK Ministry of Agriculture, Fisheries and Food have been supporting a multi-disciplinary strategic research programme on antioxidants. A major focus of the programme has been to develop suitable functional indices of oxidative damage. Evaluation of the programme's outputs, and analysis of the challenges of developing the research into an instrument of public policy, led them to conclude that what had been developed at the national government level should be promoted at the European level. The European Union's Fifth Framework Programme, with the emphasis this has given to promoting the key theme of 'The Quality of Life and the Management of Living Resources' and on improving competitiveness and enhancing the quality of life of the EU citizen, has provided the opportunity to pursue this objective.

4 European Research on the Functional Effects of Antioxidants 'EUROFEDA'

At the present time a concerted action project is being negotiated with the EU. It is hoped that this project will provide a focus for research teams in industry and academia who already have a research interest in the subject. The project could provide a platform for the co-ordination of research, the exchange of information, and the identification of where important gaps in information exist that will require future investment. The emphasis will be to encourage research teams throughout Europe to respond to the priority research goals and collaborate.

The objectives of EUROFEDA are to review:

- Potential biomarkers of oxidative damage and the strengths and limitations;
- The data that exist on the type and site of mutational damage that is caused by ROS;
- Information on the 'true' bioavailability of antioxidants;
- The effects of ROS and antioxidants on gene expression at nutritionally-relevant levels of intake;
- The effects of antioxidants on mitochondrial function; and
- The evidence linking specific functional events to health outcomes.

Each of these key tasks will result in the identification of the priorities for future research.

It is hoped is to organise the project through a series of Task Groups that will cover the specific areas of:

- Biomarkers of oxidative damage
- Bioavailability of antioxidants; and
- Gene expression and mitochondrial function

5 Biomarkers of Oxidative Damage

As the first stage in the definition of the functional effects of antioxidants there is an urgent need to develop reliable and reproducible methods for the measurement of oxidative damage. The fact that ROS can damage DNA, lipids and proteins provides a basis for the quantification of this damage and their application in dose-response studies.

The measurement of products of DNA oxidation, particularly of 8-oxo-7,8-dihydro-2′-deoxyguanine (8-oxodG), is the most widely developed of all of the methods that are applied. It was recognised by experts in the field that there were errors in the results obtained by various laboratories. Efforts by a group of 20 laboratories in Europe to identify methodological weaknesses and standardise methods has begun through the European Standards Committee for Oxidative DNA Damage (ESCODD). It is hoped to work with this group to encourage the dissemination of information on the methods that show greatest promise.

In the context of understanding the specific functional effects of antioxidants at the cellular level, it is likely to be more valuable to study the effects of DNA oxidative damage at the level of the genome. The feasibility of doing this for specific regulatory genes in target cells will need to be evaluated by the group.

Other techniques that are increasingly being considered as biomarkers of oxidative damage include protein and lipid degradation products. However, there has been no systematic European effort to review the methods that currently exist, or draw conclusions about the relative advantages in terms of relevance, reliability and reproducibility, in comparison with the methods of estimating DNA damage. It is hoped to encourage such an evaluation.

6 Bioavailability of Antioxidants

In order to assess the relative importance of various dietary antioxidants in terms of their ability to decrease oxidative damage, much more information is needed about their bioavailability. The term covers the processes of ingestion, absorption, distribution, utilisation and degradation. In terms of assessing the efficacy of individual antioxidants it must be known what function they are fulfilling at the target tissue(s).

Most of the limited work so far undertaken has been confined to the common antioxidants, such as vitamins C and E. Even in these cases the majority of the data that are available is confined to a measurement of the percentage of an oral dose that can be detected in plasma over a period of time. Although this provides information about the absorption and possible degradation, it says little about the functionality at certain target sites.

In order to measure the 'true' bioavailability of an antioxidant methods are required that will enable its fate to be determined. This requires methods that will:

- Be applicable to accessible tissues
- Enable a dose-response relationship to be established in a specific and predictable way
- Ensure the response measured relates directly to a health/disease outcome.

The Task Group on the bioavailability of antioxidants will draw together all the information that exists at the present time, especially in terms of the most suitable methodologies. It will highlight the most important research needs and hopefully will stimulate further research to be undertaken across Europe.

7 Gene Expression and Mitochondrial Function

It is now apparent that anti-oxidant defences are interactive and co-ordinately regulated. Up or down regulation of a single antioxidant may disrupt the redox balance of a cell and cause an increase in oxidative stress. Rather like the function of NO as a secondary messenger, it has been established that ROS act as secondary messengers controlling gene transcription. ROS serve as messengers when present in low concentrations in cells. They are induced in many cell types in response to invading pathogens or stress at levels which are insufficient to cause cell death.

ROS, amongst other factors, are able to induce a response from three types of transcription factors NF-κB, AP-1 and the heat shock factor (HSF). Hydrogen peroxide production can lead to the activation of NF-κB and antioxidants have been shown to inhibit this activation.[28] AP-1 is a ubiquitous transcription factor capable of inducing a large number of genes involved in proliferation and differentiation. Its activation is also redox sensitive and these actions can be blocked by low levels of hydrophilic antioxidants. Unlike NF-κB, however, high concentrations of antioxidants can stimulate transcription. Heat shock proteins, capable of associating with and stabilising proteins that have been affected by cellular stress, are induced by HSF. HSF is induced by a wide range of stress factors, including ROS, or through changes in the overall redox state of the cell.

Interestingly the inflammatory reactions and altered immune responses that occur in certain pathologies, and ageing, and are associated with NF-κB-dependent protein synthesis. Dietary or pharmacological strategies that can modulate this stress or associated cytokine induction may offer new approaches for disease treatment or prevention. Alterations in NF-κB expression have been noted in the ageing process. Chronic oxidative stress producing low levels of ROS has been hypothesised to contribute to the ageing process.

In addition mitochondrial oxidative damage is thought to play an important role in the ageing process, particularly in bioenergetic cells, such as those of the cardiac, skeletal. and central nervous system. Damage to mitochondrial DNA by ROS is significantly higher than that which occurs to nuclear DNA with the likelihood of a loss of renewal of mitochondrial populations. Mitochondrial-encoded genes, such as manganese-dependent superoxide dismutase, are known to be induced by compounds produced in response to oxidative stress. ROS production can be regulated by mitochondrial proton conductance.[28,29] Increased proton linkage lowers the reduced state of ubiquinone, which is the major source of electrons for superoxide production. Alterations in proton conductance might be possible by dietary means and offers a potential to slow the ageing process. Nothing is known as yet about the effects of dietary antioxidants on the process.

Much of the work to date on studying the relationship of dietary intake of the dietary antioxidants, their effects on the redox state, and the consequent effects on the expression of genes activated by the transcription factors has been studied *in vitro*. It remains to be determined whether they are capable of mediating such changes at the cellular level in humans.

The difficulty in making major advances in understanding the functional effects of antioxidants has been associated with the lack of any ability to follow any effects on gene regulation at the tissue level. *In vitro* techniques are severely limited in their predictive value particularly if care is not exercised to ensure that the redox state of the cell reflects that which might occur *in vivo*.

It is anticipated that within the next few years the human genome will have been sequenced. An attempt will have been made to understand the functional role of specific genes in development and the factors that control expression, or repression, of specific genes in the tissues and organs of the adult. This will have a profound impact on the development of the nutritional sciences. The intimate connection between energy expenditure, metabolic turnover and the effects of specific antioxidants on the functioning of cells and homeostasis will be able to be determined at the molecular level using emerging genomic and proteomic techniques. Transgenic reporter models will provide *in vivo* information about the site and mode of action of dietary antioxidants. In order to place this work in context, in terms of human nutrition, such work will need to be complemented with information on exposure, bioavailability and metabolic studies.

It is clear that there will continue to be a major focus on research into the effects of antioxidants on health and disease in Europe in the years ahead. This paper has highlighted both the beneficial and adverse effects that result from ROS production in cells. It is clear that much more information is required about the mechanisms that may be disregulated in disease and whether sustained conditions of oxidative stress can be ameliorated by antioxidants. There is an urgent need for research into the functional effects of antioxidants in order to understand how the benefits that have been shown in experiments with antioxidants in animals might be translated into human health benefits. Given the pivotal role that ROS play in important metabolic processes the potential health benefits could be considerable.

8 References

1 S. Papa, S. Scacco, M. Schliebs, J. Trappe and P. Siebel, *Mol. Aspects Med.*, 1996, **17**, 513.

2 D. Mecocci, G. Fano S. Fulle, U. MacGarvey, *et al.*, *Free Rad. Biol. Med.*, 1996, **26**, 303.

3 M. Barnett and C. M. King, *Mutat. Res.*, 1995, **338**, 115.

4 D. Harman, *J. Gerontol.*, 1956, **2**, 298.

5 K. Beckman and B. N. Ames, *Physiological Rev.*, 1998, **78**, 547.

6 R. Weindruch and R. S. Sohal, *New Eng. J. Med.*, 1997, **337**, 986.

7 R. Weindruch and R. L. Walford, 'The Retardation of Ageing and Disease by Dietary Restriction', C. C. Thomas, Springfield, Ill., 1988.

8 G. Block, B. Patterson and A. Subar, *Nutr. Cancer*, 1992, **18**, 1.

9 L. Gross and Y. Dreyfuss, *Proc. Natl. Acad. Sci. USA*, 1984, **81**, 7596.

10 K. Yoshida, K. Inoue, K. Nojima, Y. Hirabayashi and T. Sado, *Proc. Natl. Acad. Sci. USA*, 1997, **94**, 2615.

11 K. Yoshida, K. Inoue, Y. Hirabayashi, K. Nojima and T. Sado, *J. Nutr. Health Aging*, 1999, **3**, 121.

12 Z. V. Haley and A. Richardson, *Mutn. Res.*, 1993, **295**, 237.

13 G. A. Block, *Nutr. Rev.*, 1992, **50**, 207.

14 US National Academy of Sciences. 'Diet, Nutrition and Cancer', Washington, DC, 1982.

15 N. I. Krinsky, 'Antioxidants in Human Health and Disease', ed. B. Frei, Academic Press, p. 239.

16 E. A. Porta, N. S. Joun and R. T. Nitta, *Mech. Ageing Dev.*, 1980, **13**, 1.

17 A. D. Blackett and D. A. Hall, *Gerontology*, 1981, **27**, 133.

18 M. L. Heidrick, L. C. Hendricks and D. E. Coo, *Mech. Ageing Dev.*, 1984, **27**, 341.

19 E. W. Flagg, R. J. Coates and R. S. Greenburg, *J. Am. Coll. Nutr.*, 1995, **14**, 419.

20 W. J. Blot, J. Y. Li, R. R. Taylor, *et al.*, *J. Natl. Cancer Inst.*, 1993, **89**, 1483.

21 ATBC: Alpha Tocopherol, Beta Carotene Cancer Prevention Study Group. *N. Engl. J. Med.*, 1994, **330**, 1029.

22 S. Taylor Mayne, *FASEB J.*, 1996, **10**, 690.

23 S. J. Duthie, A. Ma, M. A. Ross and A. R. Collins, *Cancer Res.*, 1996, **56**, 1291.

24 A. T. Diplock, J.-L. Charleux, G. Crozier-Willi, *et al.*, *Brit. J. Nutr.*, 1998, **80**, 577.

25 B. N. Ames, *Tox. Lett.*, 1998, **102**, 5.

26 S. N. Meydani, M. Meydani, J. B. Blumberg, *et al.*, *J. Am. Med. Ass.*, 1997, **277**, 1380.

27 G. Block and L. Langseth, *Food Technol.*, 1994, **48**, 80.

28 K. Schulze-Osthoff, M. K. A. Bauer, M. Vogt and S. Wesselborg, *Int. J. Vit. Res.*, 1997, **67**, 336.

29 A. J. Kowaltowski and A. B. Vercesi, *Free Rad. Biol. Med.*, 1998, **26**, 463.

30 D. F. S. Rolfe, J. M. B. Newman, J. A. Buckingham, *et al.*, *Am. J. Physiol.*, 1999, **45**, C692.

5.2

Application of Oxidative DNA Damage Measurements to Study Dietary Antioxidant Factors

Okezie I. Aruoma

DEPARTMENT OF BIOLOGY, IMPERIAL COLLEGE OF SCIENCE, TECHNOLOGY AND MEDICINE, SIR ALEXANDER FLEMING BUILDING, IMPERIAL COLLEGE ROAD, LONDON SW7 2AZ, UK

1 Introduction

Plant extracts and their antioxidant components such as flavonoids and other phenolic acids, vitamin E, vitamin C, and carotenoids are increasingly proposed as important dietary antioxidant factors (DAFs). The potential antioxidant actions of DAF can be assessed *in vitro* by examining the profile of products of DNA oxidation as illustrated by data on hydroxytyrosol and ergothioneine.[1–3]

Excessive production of reactive oxygen species (ROS), beyond the antioxidant defence capacity of the body cause oxidative stress.[4] The increased activity of certain enzymes (*e.g.* the free radical generating xanthine oxidase) and/or increased levels of their substrates (*e.g.* hypoxanthine), disruption of electron transport chains, increased electron leakage from superoxide radical ($O_2^{\bullet -}$), release of 'free' metal ions from sequestered sites, activation of cyclooxygenase and lipoxygenase, and release of heme proteins (hemoglobin, myoglobin) are known to precipitate oxidative stress *in vivo*. Antioxidants, when present at low concentrations compared with those of an oxidizable substrate such as fats, proteins, carbohydrates or DNA, significantly delay or prevent oxidation of a susceptible substrate. Acidic compounds (including phenols) used in foods which can readily donate an electron or a hydrogen atom to a peroxyl or alkoxy radical to terminate a lipid peroxidation chain reaction or to regenerate a phenolic compound, or which can effectively chelate a pro-oxidant transition metal, also exhibit antioxidant actions.[5]

Several plant extracts and bioactive components of the same, *e.g* the phenolic antioxidants such as carnosic acid or carnosol from rosemary or hydroxytyrosol from olives, are advocated as good protectors against oxidative stress. The view for a concerted development of experimental tools for characterizing the potential prooxidant and/or antioxidant activities of DAFs is that this will facilitate delineating their *in vivo* contribution to the modulation of the pathological consequences of oxidative stress. There is also the prospect of the methods being applied in evaluating potential uses of DAFs in food processing. The *in vitro* data may also aid decisions on more rigorous molecular biology research into antioxidant efficacy.

2 Oxidative DNA Damage and Measurement of Antoxidant Actions

Hydroxyl radical ($OH^•$) can be generated in the test tube using a reaction mixture containing ascorbate, hydrogen peroxide (H_2O_2) and Fe^{3+}-EDTA at pH 7.4.[6,7] The addition of ascorbic acid greatly increases the rate of $OH^•$ generation by reducing Fe^{3+} to, and maintaining a supply of, Fe^{2+}. The hydroxyl radicals that escape being scavenged by the chelator molecule (EDTA in this example) will become available in free solution. Compounds that are able to mimic the actions of ascorbate will increase free radical generation in this test system. The extent of inhibition by antioxidants, however, will be dependent on the effective concentrations of the molecule compared with the target molecule and on its rate constant for reaction with $OH^•$. When ascorbate is omitted from the reaction mixture, the ability of added compounds to reduce the Fe^{3+}-chelator complex can be tested. This idea led to the proposal to use assays involving DNA damage to specifically test for the abilities of dietary antioxidants to exert prooxidant actions, different from their intended abilities to minimise lipid oxidation.[5]

Mechanisms involving Fenton chemistry, ionising radiation and nuclease activation are known to account for much of the DNA damage that occurs in biological systems.[8–10] Generation of $OH^•$ by Fenton chemistry involving reaction of H_2O_2 with the transition metal ions already bound onto the DNA can lead to strand breakage, base modification and deoxysugar fragmentation. Nuclease activation mechanisms lead to the inactivation of Ca^{2+}-binding by endoplasmic reticulum, inhibition of plasma membrane Ca^{2+}-extrusion systems, and the release of Ca^{2+} from mitochondria resulting in increases in the levels of intracellular free calcium ions. Here the nuclease dependent oxidative stress leads to DNA fragmentation without the base modification observed in the Fenton mechanism.

Hydroxyl radicals induce extensive damage to all the four bases in DNA to yield a variety of products. The ability of antioxidants to induce $OH^•$-dependent base modification may therefore be used as a tool for assessing prooxidant potentials. It also follows that antioxidants (protecting lipids against oxidation) may be assessed for their ability to affect DNA base modifications *in vitro*. The results of GC-MS analysis of modified bases in DNA can be expressed as nanomoles of modified bases per milligram of DNA (equivalent to pmol μg^{-1} DNA) and these can be converted into the actual number of bases modified.

This approach has been applied to investigate the abilities of hydoxytyrosol and ergothioneine to protect DNA against oxidation by peroxynitrite.[1,2]

3 The Bleomycin-Iron Dependent Oxidation of DNA Damage

Bleomycin, an anti-tumor antibiotic, binds to DNA through its bithiazole and terminal amine residues, and complexes with metals (such as iron) using the β-aminoalanine-pyrimidine-β-hydroxy histidine portion of the molecule. Bleomycin binds iron ions and the bleomycin-iron complex will degrade DNA in the presence of O_2 and a reducing agent such as ascorbic acid. The reaction occurs by attack of a ferric bleomycin peroxide (BLM-Fe(III)-O_2H^-) on the DNA. The ferric peroxide can be formed by direct reaction of ferric-bleomycin with hydrogen peroxide, or from a BLM-Fe(III)-O_2^- complex. It is possible that under certain conditions the BLM-Fe(III)-O_2^- might decompose to yield $O_2^{\bullet -}$, and BLM-Fe(III)-O_2H^- to release $OH^{\bullet}$.[11,12] Hydroxyl radical is not necessarily the major DNA damaging species in the bleomycin system. The bleomycin-iron (III) complex by itself is inactive in inducing damage in DNA. Oxygen and a reducing agent or hydrogen peroxide are required for the damage to DNA to occur.

DNA cleavage by bleomycin releases some free bases and base propenals in amounts that are stoichiometric with strand cleavage.[13] When heated with thiobarbituric acid (TBA) at low pH, base propenals rapidly decompose to give malondialdehyde (MDA), which combines with TBA to form a pink $(TBA)_2MDA$ adduct. A positive test is obtained when the compound is able to reduce bleomycin-Fe^{3+}-DNA complex to the more active bleomycin-Fe^{2+}-DNA complex (in the presence of oxygen) in the absence of added ascorbate in the reaction mixture resulting in DNA damage.

4 The Copper-Phenanthroline Dependent Oxidation of DNA

The copper-1,10-phenanthroline complex has nuclease activity and has been used for structural studies upon DNA[14] as it can induce strand breakage. Hydrogen peroxide is implicated in the mechanism of the DNA damage by the copper-phenanthroline system. Hydroxyl radicals are involved in the damage to DNA caused by the copper-phenanthroline system.[15] Unlike the bleomycin-iron mediated damage to DNA, damage in the copper-phenanthroline system is confined mainly to the DNA bases. When a reducing agent is omitted from the reaction mixture, no damage to DNA occurs in this system. Increasing the concentrations of the reducing agents such as ascorbate and/or mercaptoethanol leads to increased DNA damage.

The assays involving DNA damage to assess prooxidant actions have unique features. The positive prooxidant actions rely on the ability of the compounds to promote reduction of Fe^{3+} to Fe^{2+} chelates and hence $OH^{\bullet}$ formation in the

Table 1 *Effect of Herbor 025 and Spice Cocktail Provençal on copper-1,10-phenanthroline-dependent DNA damage*

Addition of reaction mixture (%, v/v)	*Extent of DNA damage (A532 m)*
None	0.000
Ascorbate (240 mM)	0.201
Herbor 025	
0.05	0.090
0.10	0.159
0.20	0.244
Spice cocktail provençal	
0.05	0.035
0.10	0.075
0.20	0.122

Values are the means from duplicate experiments that varied no more than 5%. Ascorbate at a final concentration of 0.24 μM was used as a positive control. Herbor 025 and the spice cocktail at the concentrations tested, weakly stimulated DNA damage.[16] Although this may be a feature of the constituent compounds, the plant extracts are potent inhibitors of lipid peroxidation. *In vivo*, the low levels of the absorbed components would suggest that the prooxidant effects are unlikely to present physiological problems.

presence of H_2O_2. They also rely on the ability of reductants to reduce either the iron-bleomycin-DNA or copper-1,10-phenanthroline-DNA complex. In one comprehensive study,[1] it became apparent that the ability of antioxidants to modulate prooxidant actions in this system varies. For example, the concentration of hydroxytyrosol needed to induce the same level of DNA damage caused by ascorbic acid *in vitro* is not physiological, *i.e.* it is unlikely that high concentrations of hydroxytyrosol can be absorbed from average daily consumption of olive fruits and extra virgin olive oil. Table 1 shows the application of the colorimetric version of the copper-1,10-phenanthroline-DNA assay to assess the antioxidant action of rosemary extracts. Circumventing potential prooxidant action could contribute to increased protective ability of dietary antioxidants towards susceptible substrates. For example, proteins protect DNA against the prooxidant actions of some flavonoids and polyphenolic compounds *in vitro*.[17] Thus bioavailability considerations are important in understanding the nutritional pharmacology of DAFs.

5 Future Perspective

The dietary antioxidant factors (DAFs) may act directly by scavenging reactive oxygen species *in vivo*. For the proposed antioxidant to have a physiologically meaningful effect *in vivo* it must become absorbed and presented to the site of intended action at a concentration that actually exerts an antioxidant effect. The effect of antioxidants on cell signalling is critical if there are prophylactic intentions. The major risk factors for tumorigenesis include defects in genes that control DNA damage (*e.g.* the p53 gene), chromosome stability and proliferative capacity of normal cells.

Oxidative stress is mutagenic to mammalian cells and mechanisms involving reactive oxygen species have been suggested to be promoters of carcinogenesis from the standpoint of their ability to damage DNA, producing alterations in gene expression. However, malignant transformations involves mutations, deletions, translocation and duplication of different genes.

Intracellular communications are critical in determining cell and whole organ responses to environmental injury. Reactive oxygen species are intracellular second messengers that modulate signal transduction pathways. The redox activation of various signal transduction factors (*e.g.* NF-κB, AP-1, SAPK/JNK, p53, p38 and c-myc) are important in controlling such responses as growth factor gene expression, cell cycle, programmed cell death (apoptosis) and the activation of cytokines. Indeed the efficacy of DAFs or their role in functional food science would impinge on how they affect these processes. Can the complex mechanisms of redox activated signal transduction cascade affect levels of DNA adducts and lipid peroxidation marker, isoprostanes, in response to environmental chemicals including plant foods? The products of oxidative DNA damage continue to be advocated as potential markers of cancer risks. Rigorous evaluation will determine a consensus.

6 References

1 O.I. Aruoma, M. Deiana, A. Jenner, B. Halliwell, H. Kaur, S. Banni, F.P. Corongiu, A.M. Dessi and R. Aeschbach, *J. Agric. Food Chem.*, 1998, **46**, 5181.
2 M. Deiana, O.I. Aruoma, M.L.P. Bianchi, J.P.E. Spencer, H. Kaur, B. Halliwell, R. Aeschbach, S. Banni, A.M. Dessi and F.P. Corongiu, *Free Radical Biol. Med.*, 1999, **26**, 762.
3 O.I. Aruoma, J.P.E. Spencer and N. Mahmood, *Food Chem. Toxicol.*, 1999, **37**, 1043.
4 H. Sies, *Exp. Physiol.*, 1997, **82**, 291.
5 O.I. Aruoma, *Food Chem. Toxicol.*, 1996, **73**, 1617.
6 B. Halliwell, J.M.C. Gutteridge and O.I. Aruoma, *Anal. Biochem.*, 1987, **165**, 215.
7 O.I. Aruoma, *Meth. Enzymol.*, 1994b, **233**, 57.
8 A.P. Breen and J.A. Murphy, *Free Radical Biol. Med.*, 1995, **18**, 1033.
9 L.H. Breimer, *Mol. Carcinogen.*, 1990, **3**, 188.
10 B. Halliwell and O.I. Aruoma, *FEBS Lett.*, 1991, **281**, 9–19.
11 Y.T. Sigiura, Suzuki, J. Kuwahara and H. Tanaka, *Biochem. Biophys. Res. Commun.*, 1982, **105**, 1511.
12 D.H. Petering, R.W. Byrnes and W.E. Antholine, *Chem. Biol. Interact.*, 1990, **73**, 133.
13 R.M. Burger, J. Peisach and S.B. Horwitz, *Life Sci.*, 1981, **28**, 715.
14 D.S. Sigman, *Acc. Chem. Res.*, 1986, **19**, 180–186.
15 M. Dizdaroglu, O.I. Aruoma and B. Halliwell, *Biochemistry*, 1990, **29**, 8447.
16 O.I. Aruoma, J.P.E. Spencer, P. Rossi, R. Aeschbach, A. Khan, N. Mahmood, A. Munoz, A. Murcia, J. Butler and B. Halliwell, *Food Chem. Toxicol.*, 1996, **34**, 449.
17 O.I. Aruoma, *Food Chem. Toxicol.*, 1994, **32**, 671.

5.3

Chemoprotective Effects of Resveratrol Against Hydrogen Peroxide-induced Oxidative Damage

Jung-Hee Jang, Seong-Su Han and Young-Joon Surh*

COLLEGE OF PHARMACY, SEOUL NATIONAL UNIVERSITY, SEOUL 151-742, SOUTH KOREA

1 Introduction

The health benefits of moderate consumption of red wine (French Paradox) have been attributed to the antioxidant activity of polyphenols present in grapes. Resveratrol (3,4′,5-trihydroxystilbene), one of the major antioxidative constituents found in the skin of grapes, has been considered to be responsible in part for the protective effects of red wine against coronary heart diseases. In line with this notion, resveratrol inhibits platelet aggregation and eicosanoid synthesis.[1,2]

Oxidative stress induced by reactive oxygen intermediates (ROIs) including hydrogen peroxide, superoxide anion, and hydroxyl radical has been considered as a major cause of cellular injuries in a variety of clinical abnormalities, such as neurodegenerative disorders and cancer. Recent studies have revealed that ROIs can cause cell death via induction of apoptosis.[3,4] One of the plausible ways to prevent ROI-mediated cellular injury is to augment the endogenous oxidative defense capacity through dietary or pharmaceutical intake of antioxidants. In the present work, we have examined the effects of resveratrol on oxidative DNA damage and cell death induced by hydrogen peroxide.

2 Materials and Methods

Measurement of DNA strand scission. øX174 RF1 supercoiled DNA (0.3 μg) was incubated for 1 h at room temperature with hydrogen peroxide (0.25 mM) and varying amounts of $FeSO_4$ in 10 mM Tris-HCl, pH 7.4 in the absence or

* Corresponding author: surh@plaza.snu.ac.kr

presence of resveratrol. After incubation, 3 μl of loading buffer (100 mM EDTA, 0.1% bromophenol blue, and 50% glycerol) was added and the resulting mixture was subjected to 0.8% agarose gel electrophoresis. The gels were stained with ethidium bromide and photographed under a UV transilluminator.

Determination of cell viability (MTT assay). The viability of PC12 cells were determined by the MTT dye reduction assay as described previously.[5]

Measurement of intracellular ROS. PC12 cells were treated with 0.25 mM hydrogen peroxide in the absence or presence of resveratrol (50 μM). After incubation at 37 °C for 0.5 h, cells were rinsed with Kreb's ringer solution and exposed to 10 μM 2,7-dichlorofluorescein diacetate (DCF-DA) for 15 min at 37 °C. After centrifugation, the cell pellets were subjected to lysis with DMSO and the fluorescence intensity of the cell lysates was measured using Sequoia-Turner spectrofluorometer (Model 450) at an excitation wavelength of 485 nm with emission at 530 nm.

Electrophoretic mobility shift assay (EMSA) for measuring NF-κB DNA binding activity. Nuclear extracts from resveratrol- or vehicle-treated PC12 cells were incubated with 5′-end ^{32}P-labeled double-strand oligonucleotide containing the consensus binding sequence for NF-κB (5′-AGT TGA GGG GAC TTT CCC AGG C-3′). The incubation mixtures were separated on 6% non-denaturing polyacrylamide gel electrophoresis followed by autoradiography.

3 Results and Discussion

Incubation of hydrogen peroxide with øX174 RF1 DNA in the presence of ferrous ion caused DNA-strand scission as revealed by decreased proportion of the supercolied DNA with a concomitant increase in the production of the open circular form (Figure 1). Addition of resveratrol to the reaction mixture

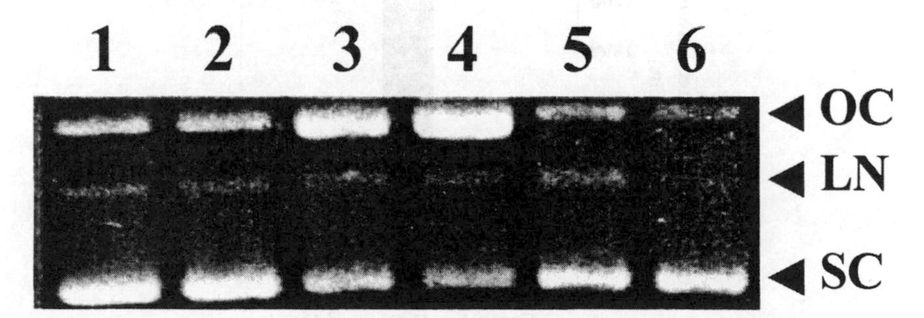

Figure 1 *Agarose gel electrophoretic pattern of ethidium bromide-stained øX174 supercoiled DNA incubated with H_2O_2 plus Fe^{2+} in the absence or presence of resveratrol. Lane 1: no treatment; lane 2: 0.25 mM H_2O_2 alone; lane 3: 0.25 mM H_2O_2 + 20 μM Fe^{2+}; lane 4: 0.25 mM H_2O_2 + 40 μM Fe^{2+}; lane 5: 0.25 mM H_2O_2 + 40 μM Fe^{2+} + 10 μM resveratrol; lane 6: 0.25 mM H_2O_2 + 40 μM Fe^{2+} + 25 μM resveratrol. Abbreviations: SC, supercoiled; LN, linear; OC, open circular*

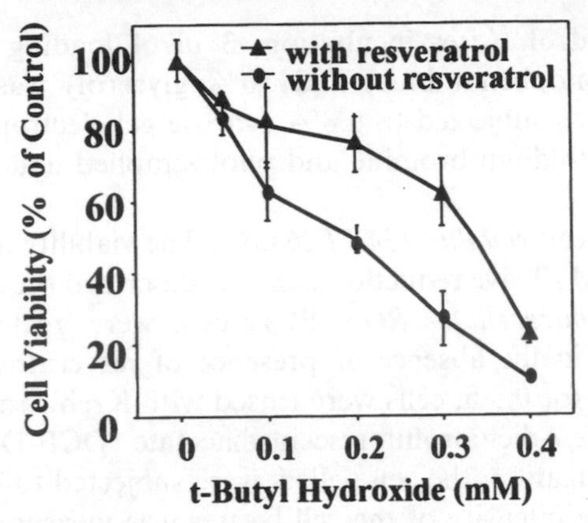

Figure 2 *Protective effect of resveratrol on* tert-*butyl hydroperoxide-induced toxicity in PC12 cells. Cells were incubated with indicated concentrations of* tert-*butyl hydroperoxide in the absence or presence of 25 μM resveratrol for 24 h*

abolished the hydrogen peroxide-induced DNA strand cleavage. PC12 cells treated with *tert*-butyl hydroperoxide exhibited reduced viability, which was attenuated by resveratrol pretreatment (Figure 2). Likewise, resveratrol protected against hydrogen peroxide-induced cytotoxicity in the same cells (data not shown). PC12 cells treated with hydrogen peroxide displayed intense fluorescence inside the cells after staining with DCF dye that detects ROIs of the peroxide type. Intracellular ROI accumulation resulting from H_2O_2 treatment was significantly reduced when resveratrol was added to the media (Figure 3). There has been accumulated evidence supporting that the eukaryotic

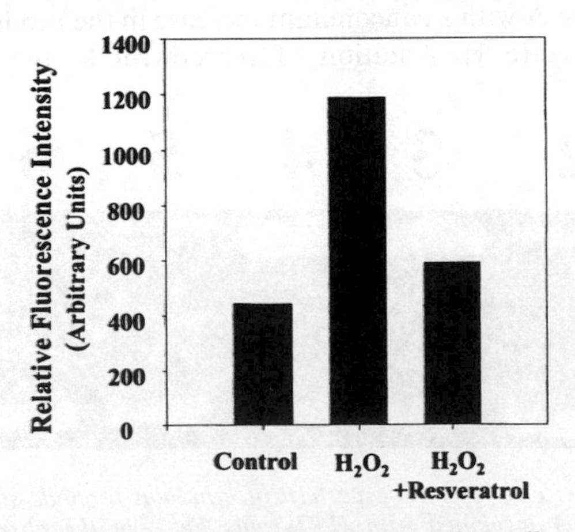

Figure 3 *Effects of resveratrol on ROI accumulation in PC12 cells treated with hydrogen peroxide. PC12 cells were treated with 0.25 mM H_2O_2 for 30 min at 37 °C in the absence or presence of resveratrol (50 μM). The intracellular accumulation of peroxides was determined using the fluorescent dye DCF as described in Materials and Methods*

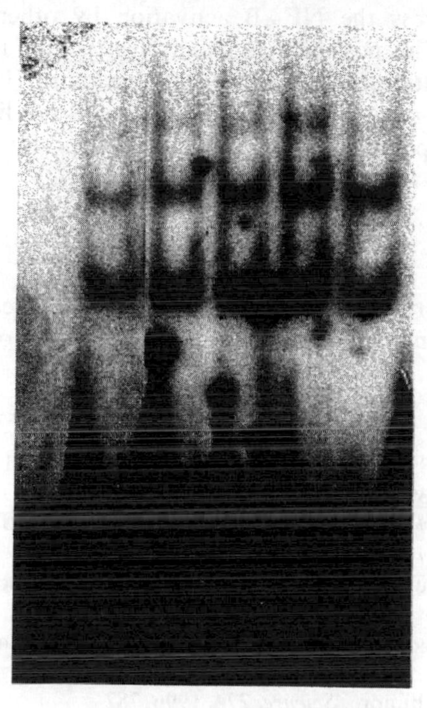

Figure 4 *Effects of resveratrol on constitutive NF-κB DNA binding activity in PC12 cells. Confluent PC12 cells were treated with 0 μM (lane 1), 1 μM (lane 2), 5 μM (lane 3), 25 μM (lane 4) or 100 μM (lane 5) of resveratrol at 37 °C for 1h. After incubation, the nuclear extracts were subjected to EMSA as described in Materials and Methods.* **P**: *probe only*

transcription factor NF-κB plays a crucial role in the cell survival.[6,7] Thus, overexpression of NF-κB/Rel promotes cell survival by inhibiting induction of apoptosis. Conversely, NF-κB inhibitors (*e.g.*, pyrrolidine dithiocarbamate) have been found to increase the cell death by stimulating apoptosis.[8] In the present study, we found that resveratrol can influence the constitutive NF-κB activation in PC12 cells. Thus, resveratrol treatment led to enhancement of NF-κB DNA binding activity, which peaked at 25 μM (Figure 4). Resveratrol at 100 μM did not cause additional enhancement of the NF-κB DNA binding activity in PC12 cells, but rather decreased the NF-κB activation, presumably due to its cytotoxicity at this concentration. The activation of constitutive NF-κB in PC12 cells by relatively non-toxic concentrations of resveratrol appears to be associated with its prooxidant activity. Low levels of oxidative stress have been shown to cause reversible alterations of intracellular signal transduction and subsequent gene expression.[9,10] Several lines of evidence suggest that the generation of ROIs is an important event in the activation of NF-κB[11] and that ROIs serve as common second messengers in many NF-κB-activating conditions.[12] Although resveratrol alone may generate ROIs, and thereby trigger NF-κB activation which is redox sensitive,[13,14] the same compound is anticipated to suppress the NF-κB activation by other oxidants, such as hydrogen peroxide. This possibility is under investigation in our laboratory. In this context, it is interesting to note that resveratrol is protective against cytotoxicity and NF-κB activation induced by oxidized lipoproteins in PC12 cells.[15] Additional experiments will be required to elucidate the molecular basis of modulation of NF-κB activation by resveratrol.

Acknowledgement

This work was supported by the Genetic Engineering Research Grant from the Ministry of Education (1998–019-F00073), Republic of Korea.

4 References

1 C.R. Pace-Asciak, S. Hahn, E.P. Diamandis, G. Soleas and D.M. Goldberg, *Clin. Chim. Acta*, 1995, **235**, 207.
2 E.N. Frenkel, A.L. Waterhouse and J.E. Kinsella, *Lancet*, 1993, **341** (April 24), 1103.
3 M.D. Jacobson, *Trends Biochem. Sci.*, 1996, **83**, 86.
4 S. Bhaumik, R. Anjum, N. Ranarj, B.V.V. Pardhasaradhi and A. Khar, *FEBS Lett.*, 1999, **456**, 311.
5 H.-J. Kim, H.-R. Yoon, S. Washington, I.I. Chang, Y.J. Oh and Y.-J. Surh, *Neurosci. Lett.*, 1997, **238**, 95.
6 A.A. Beg and D. Baltimore, *Science*, **274**, 1996, 782.
7 C.Y. Wang, M.W. Mayo and A.S. Baldwin, Jr., *Science*, 1996, **274**, 784.
8 K. Ozaki, H. Takeda, H. Iwahashi, S. Kitano and S. Hanazawa, *FEBS Lett.*, 1997, **410**, 297.
9 A. Barchowsky, S.R. Munro, S.J. Morana, M.P. Vincenti and M. Treadwell, *Am. J. Physiol.*, 1995, **269**, L829.
10 X. Wang, J.L. Martindale, Y. Liu and N.J. Holbrook, *Biochem. J.*, 1998, **333**, 291.
11 T. Henkel, T. Machleidt, I. Alkalay, M. Kronke, Y. Ben Neriah and P. Bauerle, *Nature*, 1993, **365**, 182.
12 H.L. Pahl and P.A. Baeuerle, *FEBS Lett.*, 1996, **392**, 129.
13 M.E. Ginn-Pease and R.L. Whisler, *Free Radic. Biol. Med.*, 1998, **25**, 346.
14 C.K. Sen and L. Packer, *FASEB J.*, 1996, **10**, 709.
15 B. Draczynska-Lusiak, Y.M. Chen and A.Y. Sun, *NeuroReport*, 1998, **9**, 527.

5.4

Antioxidant Activity of Polyphenol Containing Foods as Affected by Processing and Storage Conditions

Lara Manzocco,[1] Sonia Calligaris,[1] Dino Mastrocola[1] and Maria Cristina Nicoli[2]

[1] DIPARTIMENTO DI SCIENZE DEGLI ALIMENTI, UNIVERSITÀ DEGLI STUDI DI UDINE, VIA MARANGONI 97, 33100 UDINE, ITALY
[2] DIPARTIMENTO DI SCIENZE AMBIENTALI AGRARIE E BIOTECNOLOGIE AGRO-ALIMENTARI, UNIVERSITÀ DI SASSARI, VIALE ITALIA 39, 07100 SASSARI, ITALY

Recent studies have highlighted the antioxidant properties of polyphenols contained in fruit and vegetables and their significant role in the maintenance of health and in protecting from coronary diseases and certain cancers.[1–4] Polyphenols can easily undergo oxidation reactions having enzymatic and/or chemical origin. Both the reactions can induce a progressive polymerisation of phenols leading, in the advanced stages, to the browning of the product.[5–7] In most cases, polyphenol oxidation is greatly enhanced as a consequence of processing. For this reason processed fruit and vegetables are believed to have lower health protecting capacity than the fresh ones. However, *in vitro* trials showed that the antioxidant properties of enzymatically oxidised phenols increase during the formation of uncoloured intermediate products.[8,9]

The aim of this work was to study the changes in antioxidant activity as a consequence of processing and storage in polyphenol containing foods. In particular, the chain-breaking activity was measured following the methodology reported by Brand-Williams *et al.*[10] as modified by Manzocco *et al.*[11] on: (a) pasteurised green and black tea aqueous extracts stored at 25 °C for up to 30 days; (b) apple puree blanched or stirred in open air for up to 30 min; (c) pasteurised apple nectar stored at 25 °C for up to 4 months; (d) Montepulciano d'Abruzzo wines from 1973, 1995 and 1996 vintages, produced without the addition of sulphur dioxide. In order to simulate the oxidation of polyphenols, a catechin aqueous solution, stirred at 25 °C for increasing lengths of time in the presence (enzymatic oxidation) or in the absence (chemical oxidation) of polyphenoloxidase, was also considered.

Results allowed us to schematically describe the changes in antioxidant activity and colour of phenol containing foods as a consequence of enzymatic and chemical oxidation (Figures 1 and 2). It can be observed that, in both cases, phenols having an intermediate oxidation state exhibited higher antioxidant activity than the original phenols and the brown polymers. After reaching

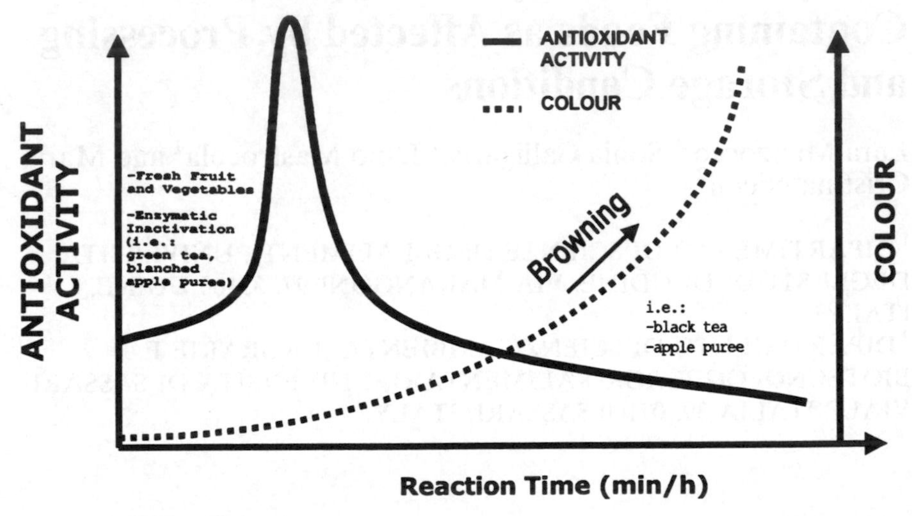

Figure 1 *Changes in antioxidant activity and colour during enzymatic oxidation of polyphenol containing foods*

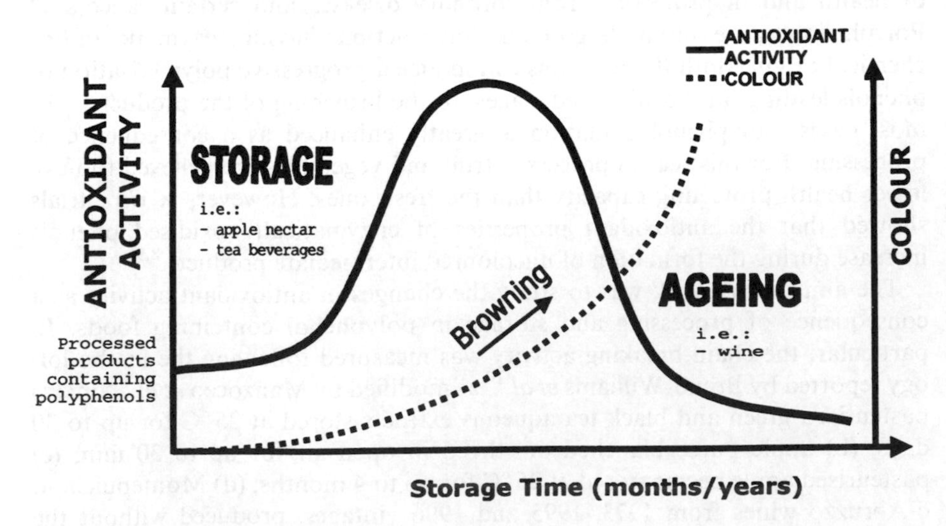

Figure 2 *Changes in antioxidant activity and colour during chemical oxidation of polyphenol containing foods*

maximum, the decrease in antioxidant activity was found to be associated with the increase in absorbance at 420 nm. Although enzymatic oxidation, due to the catalytic activity of the enzyme, proceeds to its latest phases in a shorter time as compared with chemical oxidation, they both seem to follow a common chemical pathway leading to similar changes in antioxidant activity. The higher antioxidant activity of partially oxidised phenols could be attributed to their increased ability to donate a hydrogen atom and/or to support the unpaired electron throughout delocalisation. Depending on their technological history, polyphenol containing foods can exhibit dramatic differences in their antioxidant properties.

The experimental results regarding the antioxidant activity of foods and beverages considered are shown in Figures 1 and 2 in relation to their phenol's oxidation state.

In conclusion, enzymatic and chemical browning, as caused by processing operations such as peeling, cutting, slicing, storage, *etc.*, can lead to the depletion of phenol's antioxidant properties in different time ranges. Thus, any technological treatment leading to enzymatic inactivation can inhibit polyphenol browning and allow the original antioxidant activity to be retained. However, a partial oxidation of phenols (*i.e.* prior to the formation of brown melanoidins) is also responsible for a significant increase in the antioxidant properties of the product. This information must be considered in order to find those technological conditions able to preserve or improve polyphenol original activity.

In addition, a lack of a clear correlation between *in vivo* and *in vitro* antioxidant activity of polyphenols, as reported by several authors,[12] could be attributed to the observed increase and/or decrease in polyphenol antioxidant activity induced by processing.

References

1 S. Renaud and M. de Longeril, *Lancet*, 1992, **339**, 1523.
2 J.E. Kinsella, B. Frankel, B. German and J. Kanner, *Food Technol.*, 1993, **4**, 85.
3 E.N. Frankel, A.L. Waterhause and P.L. Teissedre, *J. Agric. Food Chem.*, 1995, **43**, 890.
4 P.L. Teissedre, E.N. Frankel, A.L. Waterhause, H. Peleg and J.B. German, *J. Sci. Food Agric.*, 1996, **70**, 55.
5 V. Cheynier, C. Osse and J. Rigaud, *Food Sci.*, 1988, **53**, 1729.
6 J.J.L. Cilliers and V.L. Singleton, *J. Agric. Food Chem.*, 1989, **37**, 890.
7 J. Oszmianski and C.Y. Lee, *J. Agric. Food Chem.*, 1990, **38**, 1202.
8 H.S. Cheigh, S.H. Um and C.J. Lee, *ACS Sym. Series*, 600, 1995, 200.
9 N. Saint-Cricq de Gaulejac, C. Provost and N.J. Vivas, *Agric. Food Chem.*, 1999, **47**, 425.
10 W. Brand-Williams, M.E. Cuvelier and C. Berset, *Lebensm. Wiss. Technol.*, 1995, **28**, 25.
11 L. Manzocco, M. Anese and M.C. Nicoli, *Lebensm. Wiss. Technol.*, 1998, **31**, 694.
12 J.F. Young, S.E. Nielsen, J. Haraldsdottir, B. Daneshvar, S.T. Lauridsen, P. Knuthsen, L.O. Dragsted and B. Sandstrom, FAIR-CT 95–0158.

5.5

Antioxidant Properties of Wine Phenolic Fractions

C. Gómez-Cordovés,[1] B. Bartolomé,[1] W. Vieira[2] and V.M. Virador[2]

[1] INSTITUTO DE FERMENTACIONES INDUSTRIALES, CSIC, JUAN DE LA CIERVA 3, 28006 MADRID, SPAIN
[2] NATIONAL INSTITUTES OF HEALTH, BETHESDA, MD 20892, USA

1 Introduction

Part of the beneficial effects of wine in human health has been attributed to its phenolic components. *In vitro* antioxidant activity of wine phenolics against free radicals and human low-density lipoproteins (LDL) has been reported.[1,2] In addition, supplementation of caffeic acid, one of the main hydroxycinnamic acids in wines, has been shown to increase the α-tocopherol concentration in both plasma and lipoprotein in rats.[3]

Melanin concentration has been reported to correlate inversely with skin malignancies and UV photodamage.[4] In this paper we have studied the effect of wine phenolics in melanogenesis and melanocyte proliferation in cultured cells. For that, wine phenolics were extracted and fractionated.[2] A sorghum preparation rich in highly-polymerised condensed tannins[5] was also tested for its melanogenic effects.

2 Characterization of the Phenolic Fractions

The wine phenolic fractions were found to contain mainly anthocyanins and condensed tannins (anthocyanin fraction); alcohols, flavan-3-ols, procyanidins and flavonols (flavonoid fraction) and benzoic and cinnamic acids (acid fraction). Main individual phenolic compounds identified in each fraction are reported in Table 1. This diversity in composition among the fractions explains the differences found in their radical scavenging capacity against DPPH$^{\bullet}$ (Table 1).[6]

Table 1 *Characterisation and antioxidant activity of the phenolic fractions*

Phenolic fraction	*Main phenolic compounds present (% of total phenolic composition)*	*Radical scavenging capacity (C_{50})[a]*
Wine anthocyanin fraction	87% of the anthocyanins corresponding to single glucosilated forms: D-3-G (11.4%), Cy-3-G (0.3%), Pt-3-G (11.9%), Pn-3-G (2.4%), M-3-G (51.3%)	109.5
Wine flavonoid fraction	Alcohols [hydroxytyrosol (4.8%), tyrosol (7.4%), tryptophol (11.7%)]; flavan-3-ols [(+)-catechin (11.2%), (−)-epicatechin (5.1%)]; procyanidins [B2 (0.6%), B5 (1.7%), others (1.2%); flavonol glucosides [Q-G (4.5%)]; stilbenes [resveratrol (8.1%)]	79.8
Wine acid fraction	Benzoic acids [gallic acid (15.6%), protocatechuic (2.8%), *p*-hydroxybenzoic (1.4%), vanillic (2.9%), syringic (9.9%)]; cinnamic acids [caffeic (22.3%), *p*-coumaric (23.7%), ferulic (0.6%)]; cinnamic derivatives [*p*-coumaric esters (5.3%)]	77.7
Sorghum tannins	Polymers greater than procyanidin pentamers	63.0

[a] Expressed as the phenolic concentration (m equivalent of gallic acid l^{-1}) necessary to decrease the initial DPPH$^{\bullet}$ concentration by 50%. Total phenolic concentration was determined by the Folin-Ciocalteu method.[7]

3 Activity on Melonogenesis and Melanocyte Proliferation

The effect of the phenolic fractions on melanogenic activitiy (tyrosinase activity), total melanin and melanocyte proliferation was evaluated by the STOPR protocol[8] (Figure 1). Melanogenic activity of wine fractions at concentration that minimally affect proliferation (100 mg l^{-1} for anthocyanin fraction, 40 mg l^{-1} for flavonoid fraction and 200 mg l^{-1} for acid fraction) seems to be related to their antioxidant activity (see Table 1). This result may be explained by direct competitive inhibition of melanocyte tyrosinase by some of the phenolic compounds contained in these fractions and/or by uncertain antioxidant mechanisms affecting tyrosinase expression. Surprisingly, sorghum tannins at 20 mg l^{-1} increased melanogenic activity although no increase was found in total melanin. Since tannins show the highest antioxidant activity, melanin production may be inhibited inside the cells but radioactive melanin can be enzymatically produced from solubilized tyrosinase in the melanogenic assay.

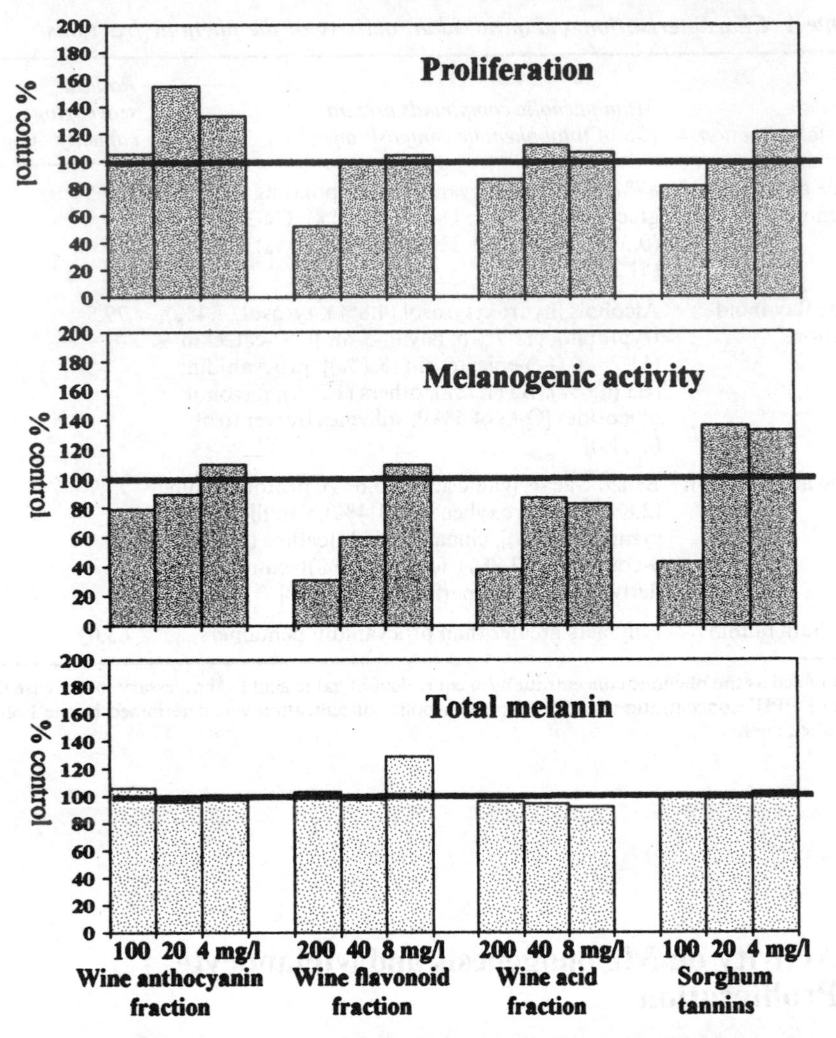

Figure 1 *Effect of phenolic fractions on melanogenesis and melanocyte proliferation*

Acknowledgements

The authors are grateful to Miss Isabel Izquierdo and Mr. Luis Piñal for technical assistance and to the Spanish Comisión Interministerial de Ciencia y Tecnología (CICYT) for financial support (Project ALI97–0590).

4 References

1 E. N. Frankel, J. Kanner, J. B. German, E. Parks and J. E. Kinsella, *Lancet*, 1993, **341**, 454.

2 A. Ghiselli, M. Nardini, A. Baldi and C. Scaccini, *J. Agric. Food Chem.*, 1998, **46**, 361.

3 M. Nardini, F. Natella, V. Gentili, M. Di Felice and C. Scaccine, *Arch. Biochem. Biophys.*, 1997, **342**, 157.
4 W. C. Quevedo Jr, T. Holstein, J. Dyckman and E. L. Isaacson, 'Melanin: Its Role in Human Photoprotection', Valdenmar, Overland Park, 1995, p. 221.
5 L. M. Jimenez-Ramsey, J. C. Rogler, T. L. Housley, L. G. Butler and R. G. Elkin, *J. Agric. Food Chem.*, 1994, **42**, 963.
6 W. Brand-Williams, M. E. Cuvelier and C. Berset, *Lebensm. Wiss. Technol.*, 1995, **28**, 25.
7 V. Singleton and J. Rossi, *J. Sci. Food Agric.*, 1965, **10**, 144.
8 V. M. Virador, N. Kobayashi, J. Matsunaga and V. J. Hearing, *Anal. Biochem.*, 1999, **270**, 207.

5.6

In vitro Superoxide Anion Radical Scavenging Ability of Honeybush Tea (*Cyclopia*)

M.E. Hubbe and E. Joubert

ARC INFRUITEC-NIETVOORBIJ, PRIVATE BAG X5013, STELLENBOSCH 7599, SOUTH AFRICA

1 Abstract

The superoxide anion radical ($O_2^{\bullet -}$) scavenging ability of aqueous extracts from honeybush tea, as affected by species and fermentation, and phenolic compounds from fermented *C. intermedia*, was evaluated. Unfermented *C. sessiliflora* showed the strongest $O_2^{\bullet -}$ scavenging ability. Fermentation significantly ($P \leqslant 0.05$) decreased the $O_2^{\bullet -}$ scavenging ability of the different species. A strong structure-activity relationship was illustrated by the relative $O_2^{\bullet -}$ scavenging efficacy of the phenolic compounds. The flavone luteolin and the flavanone eriodictyol showed excellent $O_2^{\bullet -}$ scavenging activity compared to the flavanones, hesperidin and hesperetin. The isoflavone formononetin and coumestan medicagol were not effective scavengers of this radical.

2 Introduction

Honeybush, represented by *ca.* 24 species of genus *Cyclopia*, is endemic to the Cape fynbos region in South Africa.[1] Processing ('fermentation') of the plant material is essential to develop the desired sensory properties of this traditional herbal beverage.[2] Fermented *C. intermedia* contains flavanones, flavones, isoflavones, coumestans and xanthones.[3]

3 Materials and Methods

3.1 Processing of Plant Material and Preparation of Test Material

Plant material was processed ('fermented') according to a standardised processing procedure.[4] Unfermented material was tunnel-dried (40 °C/12 h) directly after cutting of the leaves and stems into small pieces. Steeping of material in freshly boiled, deionised water (*ca.* 20 g/500 ml) for exactly 5 min afforded the aqueous extracts. The soluble solid contents of the cooled, filtered extracts were determined gravimetrically. The total polyphenol contents were determined according to the Folin Ciocalteu method.[5] Freeze-dried methanol extracts were prepared from pulverized material of *C. intermedia*, pre-washed with chloroform and acetone.[3] Phenolic compounds were either kindly supplied by Prof. D. Ferreira (Department of Chemistry, University of the Orange Free State) or obtained commercially. All phenolic compounds were dissolved in dimethyl sulphoxide (DMSO) for testing.

3.2 Superoxide Anion Radical ($O_2^{\bullet -}$) Scavenging Ability of Aqueous Extracts

$O_2^{\bullet -}$, non-enzymatically generated in a β-NADH/PMS system, was measured at 560 nm as the reduction product of nitroblue tetrazolium.[6] Results were expressed as IC_{50} values. Low IC_{50} values indicate high $O_2^{\bullet -}$ scavenging activity, and *vice versa*.

4 Results and Discussion

All the *Cyclopia* extracts scavenged $O_2^{\bullet -}$ but with varying efficacies (Table 1). The poorer $O_2^{\bullet -}$ scavenging activity of fermented *Cyclopia* species (high IC_{50} values) than their unfermented counterparts is attributed to the effect of oxidation on their phenolic compounds during processing. For aqueous extracts, unfermented *C. sessiliflora* with soluble solids showing the highest total polyphenol content was the most effective $O_2^{\bullet -}$ scavenger (lowest IC_{50} value) ($P \leqslant 0.05$). Fermented *C. intermedia* and *C. maculata* with soluble solids containing the lowest total polyphenol contents, showed the weakest $O_2^{\bullet -}$ scavenging activity (highest IC_{50} values) ($P \leqslant 0.05$). Selective extraction of *C. intermedia* with methanol decreased its $O_2^{\bullet -}$ scavenging ability. Luteolin and eriodictyol, which were comparable to the antioxidant enzyme, superoxide dismutase (SOD), were the most effective $O_2^{\bullet -}$ scavengers (lowest IC_{50} values), except for the reference compound, the flavonol quercetin. The effectivity of luteolin and eriodictyol to scavenge $O_2^{\bullet -}$ is attributed to the *o*-dihydroxy structure in the B-ring in combination with the 5-OH and 4-carbonyl groups in the A- and C-rings.[7] Hesperidin, a flavanone glycoside, showed the weakest activity.

5 Conclusion

Aqueous extracts, the methanol fraction of unfermented *C. intermedia*, and several phenolic compounds from honeybush tea (*Cyclopia*) were effective scavengers of non-enzymatically generated $O_2^{\bullet -}$ radicals. Further testing is necessary to determine the *in vivo* antiradical and antioxidant activity of this commodity.

Table 1 *$O_2^{\bullet -}$ scavenging activity of Cyclopia aqueous extracts, crude methanol fractions from* C. intermedia, *and phenolic metabolites from fermented* C. intermedia

Test material	*IC_{50}*[c]	
Aqueous extracts	*Unfermented*	*Fermented*
C. sessiliflora	60.79 c[d]	97.01 d[d]
C. genistoides	109.89 d	132.42 e
C. subternata	96.68 d	151.81 f
C. maculata	158.82 f	249.26 g
C. intermedia	170.50 f	238.21 g
methanol fraction (*C. intermedia*)	226.05 g	>2000 m
luteolin[a]	39.31 b[d]	
eriodictyol	39.53 b	
mangiferin	331.97 h	
naringenin	515.35 i	
hesperetin	880.88 j	
hesperidin	1209.87 k	
formononetin	not effective	
medicagol	not effective	
SOD (antioxidant enzyme)[b]	35.25 b	
quercetin	16.77	

[a] Phenolic metabolites from fermented *C. intermedia*.
[b] Reference compounds.
[c] μg antioxidant per ml reaction mixture necessary to inhibit NBT reduction by $O_2^{\bullet -}$ with 50%.
[d] Each value represents the means of three replicates (n = 3). Means within the table followed by the same letter are not significantly different ($P \leqslant 0.05$) as determined by pairwise Student's t-tests.

6 References

1 B.-E. Van Wyk, B. Van Oudtshoorn and N. Gericke, 'Medicinal Plants of South Africa', Briza Publications, South Africa, 1997, p. 100.
2 J. Du Toit, E. Joubert and T.J. Britz, *J. Sust. Agric.*, 1998, **12**, 67.
3 D. Ferreira, I.B. Kamara, V.E. Brandt and E. Joubert, *J. Agric. Food Chem.*, 1998, **46**, 3406.
4 J. Du Toit and E. Joubert, *J. Food Qual.*, 1999, **22**, 241.
5 V.L. Singleton and J.A. Rossi, *Am. J. Enol. Vitic.*, 1965, **16**, 144.
6 J. Robak and R.J. Gryglewski, *Biochem. Pharmacol.*, 1988, **37**, 837.
7 W. Bors, W. Heller, C. Michel and M. Saran, *Methods Enzymol.*, 1990, **186**, 343.

5.7

Selenium-dependent Phospholipid Hydroperoxide Glutathione Peroxidase Protects Against Lipid, Protein and DNA Damage

Yongping Bao and Gary Williamson

CELLULAR METABOLISM SECTION, DIET, HEALTH AND CONSUMER SCIENCES DIVISION, INSTITUTE OF FOOD RESEARCH, COLNEY, NORWICH NR4 7UA, UK

1 Introduction

Reactive oxygen species (ROS) can oxidize all types of biomolecules such as lipid, protein and DNA.[1] Lipid hydroperoxides are toxic primary products of lipid peroxidation, and their breakdown products such as malondiadehyde and 4-hydroxynonenal can modify DNA and protein.[2,3] ROS can directly target proteins in cells and generate protein hydroperoxide, which can further produce new free radicals, inactive enzymes and cross link DNA.[4] The antioxidant defences which protect against oxidative damage in the cell include the non-enzymatic molecules such as tocopherol, ascorbate, flavonoids and enzymes such as superoxide dismutase (SOD), catalase, glutathione transferases (GSTs) and glutathione peroxidases (GPXs).[5] The selenium-dependent enzyme phospholipid hydroperoxide glutathione peroxidase (Se-PHGPx, GPX4) is distinct from classical cGPx (GPX1) because Se-PHGPx can not only act on H_2O_2, but also reduce phospholipid hydroperoxides, cholesterol hydroperoxides, oxidized low-density lipoproteins and thymine hydroperoxide[6–8] which are all poor substrates for cGPx. We have recently found that Se-PHGPx can reduce albumin hydroperoxide generated by free radicals *in vitro*.

2 Materials and Methods

Materials. 1-Palmitoyl-2-linoleoyl-L-3-phosphatidylcholine (PLPC), soybean lipoxidase, thymine, human albumin, NADPH, glutathione reductase, glutathione (GSH), H_2O_2 were purchased from Sigma. 2,2′-azobis (2-amidinopropane) dihydrochloride (AAPH) was from Wako Pure Chemical Industries Japan. Ultracarb 5 ODS column (250 × 4.6 mm) was from Phenomenex, UK. Ultrafree-0.5 centrifugal filter was from Millipore, MA.

Methods. Phospholipid hydroperoxide (PLPC-OOH) was prepared from PLPC using soybean lipoxidase according to ref. 9. Human liver Se-PHGPx was purified as described by Chambers *et al.*[10] The Se-PHGPx activity towards phospholipid hydroperoxides was measured by a novel HPLC assay based on the direct separation of phospholipid hydroperoxide and hydroxide.[11] Thymine hydroperoxide was prepared using H_2O_2 according to ref. 12. The Se-PHGPx activity on thymine hydroperoxide was measured by a coupled assay based on the oxidation of NADPH by glutathione reductase.[8,9] Albumin hydroperoxide was prepared by using AAPH (50 mM) in 10 mM phosphate buffer (pH 7.4) containing 10 mg ml^{-1} albumin reacted at 50 °C for 1 h.[13] After desalting using an Ultrafree-0.5 centrifugal filter, the hydroperoxy group of albumin was measured by an iodometric assay at 353 nm.[14] The reaction of Se-PHGPx on albumin hydroperoxide was carried out in 0.1 M Tris-HCl, pH 7.4, 2 mM EDTA, 1 mM GSH at 37 °C. The reduction of albumin hydroperoxy group by Se-PHGPx was measured using an iodometric assay after desalting by an Ultrafree-0.5 centrifugal filter.

3 Results and Discussion

In DNA, thymine residues have the highest electron affinity and therefore are the most likely sites of free radical damage, giving rise to thymine hydroperoxides. Se-PHGPx shows an activity of 460 μmol min^{-1} per mg protein on thymine hydroperoxide which is about four orders of magnitude higher than that of glutathione transferases.[8] Se-PHGPx exhibits activities of 336μmol min^{-1} per mg protein on phospholipid hydroperoxide (Figure 1) and 40 μmol/min per mg protein on albumin hydroperoxide (Figure 2). Protein hydroperoxides, lipid hydroperoxides and breakdown products such as aldehydes formed during oxidative stress can react with DNA to cause strand breakage and adduct formation. Se-PHGPx therefore diminishes the formation of alkoxyl and peroxyl radicals through the reduction of lipid hydroperoxides and protein hydroperoxides to the corresponding less reactive corresponding hydroxides.

4 Conclusion

In conclusion, Se-PHGPx can protect lipid, protein and DNA damage (i) by detoxification of lipid hydroperoxides and protein hydroperoxides to prevent the formation of toxic aldehydes (ii) by directly reducing thymine hydroperoxide to thymine hydroxide and repairing oxidatively damaged DNA. Se-PHGPx is present in almost all human organs and tissues, but testes contain the highest levels. In rat testis nuclei, Se-PHGPx is bound to chromatin.[15] The location and large amount of Se-PHGPx in testes suggest that Se-PHGPx may

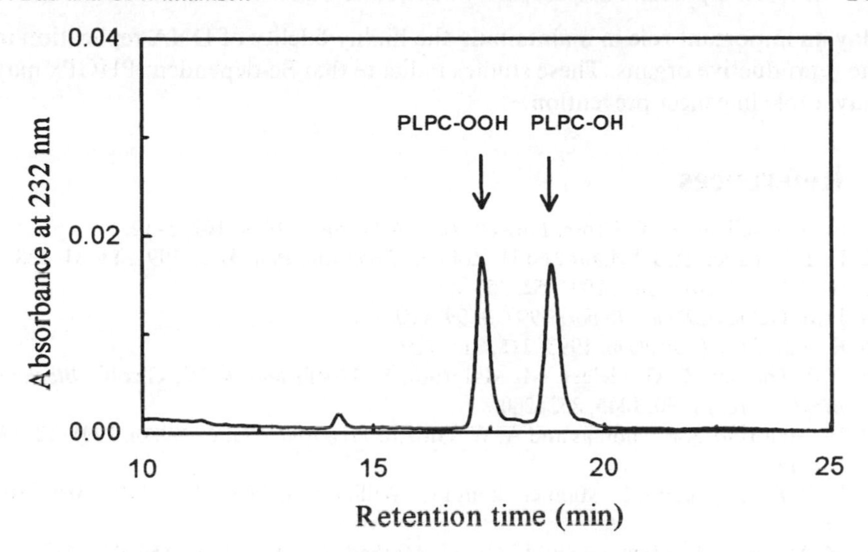

Figure 1 *Se-PHGPx activity on phospholipid hydroperoxide. The assay mixture contained 0.1 M Tris-HCl, pH 7.4, 2 mM EDTA, 1 mM NaN_3, 3 mM GSH, 0.12% Triton X-100 and Se-PHGPx (10 μl, 30 ng protein) in a final volume of 500 μl. The mixture was incubated at 37 °C for about 3 min and the reaction was started by addition PLPC-OOH to 25 μM. Separation of PLPC-OOH and PLPC-OH was carried out using an Ultracarb 5 ODS column. The mobile phase was acetonitrile-methanol-water (50:49.5:0.5, v/v/v) containing 10 mM choline chloride and the flow rate was 0.5 ml/min*

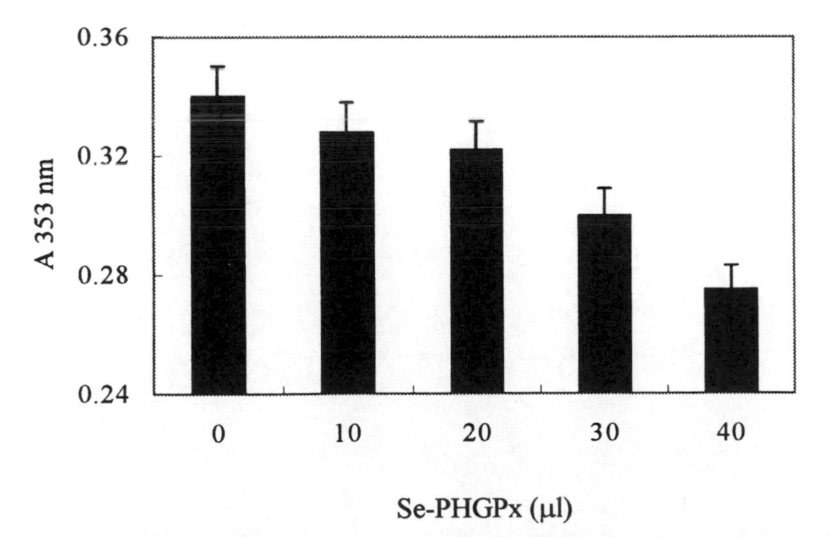

Figure 2 *Se-PHGPx activity on protein hydroperoxide. The reaction mixture contained 0.1 M Tris-HCl, pH 7.4, 2 mM EDTA, 1 mM GSH, albumin hydroperoxide (1 mg) and various amount of Se-PHGPx was carried at 37 °C. After desalting, the albumin hydroperoxy group was measured using an iodometric assay. Data are expressed as the mean ± SD from 4 replicates*

play an important role in maintaining the highly fidelity of DNA replication in the reproductive organs. These studies indicate that Se-dependent PHGPx may play a role in cancer prevention.

5 References

1 B. Halliwell and C. E. Cross, *Environ. Health Perspect.* 1994, **102**, 5–12.
2 H. Esterbauer, R. J. Schaur and H. Zollner, *Free Rad. Biol. Med.* 1991, **11**, 81–128.
3 B. Halliwell, *Nutr. Rev.* 1994, **52**, 253–265.
4 J. M. Gebicki, *Redox Report* 1997, **3**, 99–110.
5 H. Sies, *Eur. J. Biochem.* 1993, **215**, 213–219.
6 J. P. Thomas, P. G. Geiger, M. Maiorino, F. Ursini and A. W. Girotti, *Biochim. Biophys. Acta* 1990, **1045**, 252–260.
7 M. Maiorino, J. P. Thomas and A. W. Girotti, *Free Rad. Res. Commun.* 1991, **12–13**, 131–135.
8 Y. P. Bao, P. Jemth, B. Mannervik and G. Williamson, *FEBS Lett.* 1997, **410**, 210–212.
9 M. Maiorino, C. Gregolin and F. Ursini, *Methods Enzymol.* 1990, **186**, 448–457.
10 S. J. Chambers, N. Lambert and G. Williamson, *Int. J. Biochem.* 1994, **26**, 1279–1286.
11 Y. P. Bao, S. J. Chambers and G. Williamson, *Anal. Biochem.* 1995, **224**, 395–399.
12 B. S. Hahn and S. Y. Wang, *J. Org. Chem.* 1976, **41**, 567–568.
13 S. Gebicki and J. M. Gebicki, *Biochem. J.* 1993, **289**, 743–749.
14 J. A. Buege and S. D. Aust, *Methods Enzymol.* 1978, **52**, 302–310.
15 C. Godeas, F. Tramer, F. Micali, A. Roveri, M. Maiorino, C. Nisii, G. Sandri and E. Panfili, *Biochem. Mol. Med.* 1996, **59**, 118–124.

5.8

High Performance Liquid Chromatography Studies on Free Radical Oxidation of Flavonols

Dimitris P. Makris and John T. Rossiter

DEPARTMENT OF BIOLOGICAL SCIENCES, WYE COLLEGE, UNIVERSITY OF LONDON, WYE, ASHFORD, KENT TN25 5AH, UK

Flavonoids constitute a group of naturally occurring antioxidants, which over the past years have gained tremendous interest, because of their possible beneficial biological properties.[1] Among the most common and important flavonoids are flavonols, a particular class of compounds that exhibit a remarkable antioxidant activity. Numerous studies have clearly demonstrated the ability of flavonols to chelate with metals and scavenge free radicals *in vitro*. However, little is known about free radical-flavonol interactions, and how these interactions could affect the flavonol molecule. The present project was undertaken in order to investigate some aspects of free radical-mediated flavonol degradation, and the implications of such reactions in human nutrition and health.

Quercetin, morin, and rutin (Figure 1) were chosen as representative flavonols with key differences in structure, and reactions were performed in acetonitrile or aqueous acetonitrile due to poor solubility of these flavonols in aqueous media. Flavonols were treated with free radicals generated through a Cu^{2+}/H_2O_2 system, for various time periods, and then analysed by means of reversed-phase, high-performance liquid chromatography, coupled with UV-Vis detection. Spectrophotometric studies in each case were also undertaken. Treatments with oxidising agents such as sodium periodate and potassium ferricyanide, as well as with the enzymes tyrosinase and peroxidase, were also carried out for comparison reasons.

The results obtained showed that free radical oxidation of quercetin resulted in more than seven major products (Figure 2), while rutin was not acted upon. Morin gave a different HPLC trace than that of quercetin, and thus the B-ring appears as a crucial factor with respect to degradation mechanism. Also, it was found that quercetin can be oxidised with Cu^{2+} in the absence of H_2O_2, but no

Morin

Quercetin

Rutin (quercetin 3-*O*-rhamnosylglucoside)

Figure 1 *Chemical structures of the flavonols used in this study*

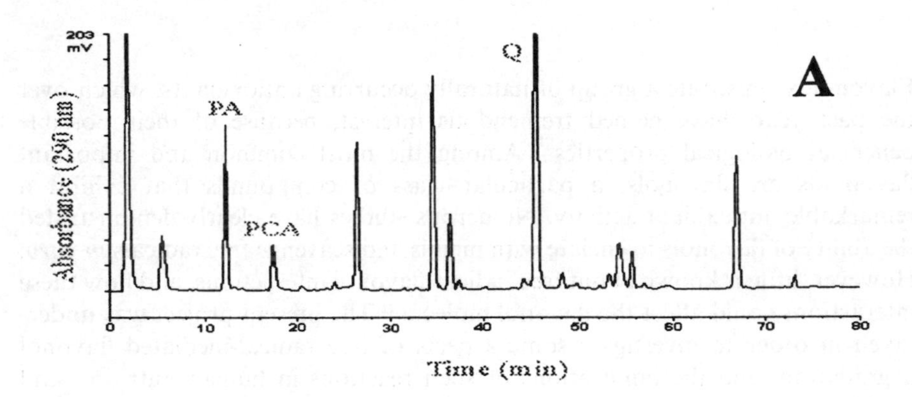

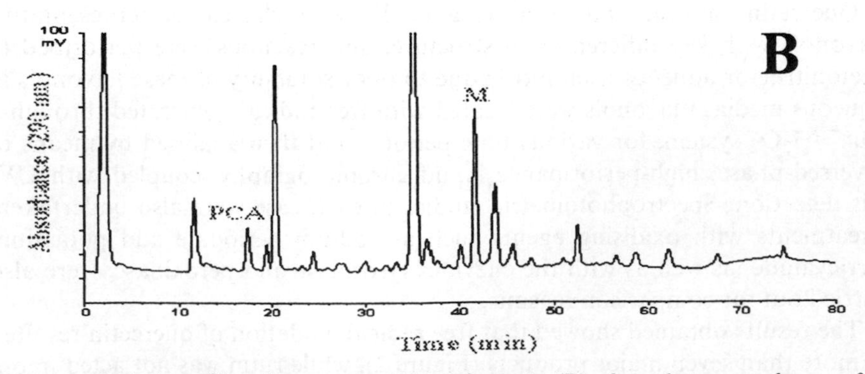

Figure 2 *HPLC profile of quercetin* (A) *and morin* (B) *degradation products, after treatment with* Cu^{2+}/H_2O_2. *Detection was accomplished at 290 nm. A water-acetonitrile gradient was employed. PA: Protocatechuic acid; PCA: Phloroglucinol carboxylic acid; Q: Quercetin; M: Morin*

reaction was observed in metal-free solutions. In contrast, morin could be oxidised with only H_2O_2.

Among the degradation products of quercetin, protocatechuic and phloroglucinol carboxylic acid could tentatively be identified. Phloroglucinol carboxylic acid was also one of morin degradation products. These compounds have been found during thermal degradation of quercetin.[2]

Oxidation of quercetin with sodium periodate showed notable similarities with free radical oxidation. On the other hand, oxidations carried out with potassium ferricyanide, tyrosinase, and peroxidase exhibited significant differences, although common features were also observed.

Conclusions

The present study demonstrated important differences in the free radical oxidation of structurally related flavonols. Particularly, it was shown that both the *o*-diphenol structure of the B-ring, as well as the availability of the 3-OH group, might substantially influence the stability and decomposition mechanism of quercetin and give rise to different set of products.

It is claimed that such approaches will further aid in elucidating the biological activities of flavonols, and could be useful in assessing flavonols as natural food additives. Furthermore, the degradation products detected may have physiological significance, as they may be generated *in vivo*. To the extent that this should hold true, the results may be directly applicable.

References

1 L. Bravo, Polyphenols: Chemistry, dietary sources, metabolism, and nutritional significance, *Nutr. Rev.*, 1998, **56**(11), 317–333.

2 D. P. Makris and J. T. Rossiter, Flavonol thermal degradation in aqueous media as a model for processed plant foods and products, in Functional Foods 99 – Claims and Evidence, conference, 14–15 April 1999, Wye College, University of London, Wye, Ashford, Kent TN25 5AH.

5.9

The Inhibition of NADPH-dependent ROS Production May Be an Important Modality for the Protective Effect of Ascorbic Acid and 6-*O*-Palmitoyl Ascorbate Against Cancer: *in situ* Tissue Culture Study

G. Rosenblat,[1] M. Graham,[2] M. Tarshis,[3] S.Y. Schubert,[1]
A. Jonas[1] and I. Neeman[1]

[1] DEPARTMENT OF FOOD ENGINEERING AND BIOTECHNOLOGY, TECHNION-ISRAEL INSTITUTE OF TECHNOLOGY, 32000 HAIFA, ISRAEL
[2] LABORATORY OF TISSUE REPAIR, DEPARTMENT OF PEDIATRICS, MEDICAL COLLEGE OF VIRGINIA CAMPUS OF VIRGINIA COMMONWEALTH UNIVERSITY, RICHMOND, VIRGINIA, 23298, USA
[3] INTERDEPARTMENTAL UNITS, HEBREW UNIVERSITY-HADASSAH MEDICAL SCHOOL, JERUSALEM, POB 12272, JERUSALEM, ISRAEL

NADPH oxidase was previously demonstrated to produce large amounts of superoxide radicals and hydrogen peroxide in human tumor cells and proto-oncogene $p21^{Ras}$ (c-Ras)-transformed NIH 3T3 fibroblasts.[1] Although the physiological role of NADPH oxidase-dependent ROS production in non-phagocytic cells is not clear, it may be involved in mitogenic signaling in cancer cells.

The protective effect of ascorbic acid (AA) and its hydrophobic derivative palmitoyl ascorbate (PA) against cancer has been intensively studied for the last decade. The precise mechanism by which AA and its derivatives inhibits cancerogenesis remains unknown.

The goal of our study was to investigate whether ascorbic acid (AA) and its acylated derivative 6-*O*-palmitoyl ascorbate (PA) would affect the intracellular NADPH-derived reactive oxygen species (ROS) production.

The experiment was performed on primary human skin or foreskin fibroblasts. The method was based on the intracellular interaction of the non-fluorescent compound 2′,7′-dichlorofluorescein diacetate (DCFH-DA) with ROS (particularly H_2O_2) to produce the fluorescence product 2′,7′-dichlorofluorescein (DCF). DCF formation was used as an index of intracellular ROS concentration. Briefly, the cells were seeded in 4-chamber tissue culture plate at densities of 10,000–20,000 cells/well and incubated in a humidified atmosphere for 72 h. Before the experiment, the medium was removed and replaced with PBS supplemented with 5 μM DCFH-DA. The experimental samples were additionally supplemented with AA or PA. *In situ* intracellular fluorescence of DCF was determined using confocal microscopy utilizing a Zeiss LSM 410 confocal laser-scanning microscope. The rate of the increase in the level of DCF was used as a quantitative measure of intracellular ROS formation in the following study.

In an initial experiment, the level of intracellular ROS production in cultured foreskin fibroblasts was evaluated. DCFH oxidation in foreskin fibroblasts was completely inhibited by 10 μM PA, thus demonstrating that in foreskin fibroblasts, superoxide radicals (or hydrogen peroxide formed under superoxide dismutation) generated by NADPH oxidase are responsible for intracellular DCFH oxidation. In the present study we demonstrated that both AA and PA are inhibitors of NADPH oxidase-derived superoxide production (Figure 1).

Palmitoyl ascorbate strongly inhibited intracellular DCFH oxidation in cultured fibroblasts probably by inhibition of NADPH-oxidase activity. A similar effect has been previously described in a study of the effect of 6-*O*-acylated AA derivatives on NADPH-oxidase activity in polymorphnonuclear neutrophils.[2,3] The PA may therefore be an important modality for mitogenic signaling and protective effects against cancer.

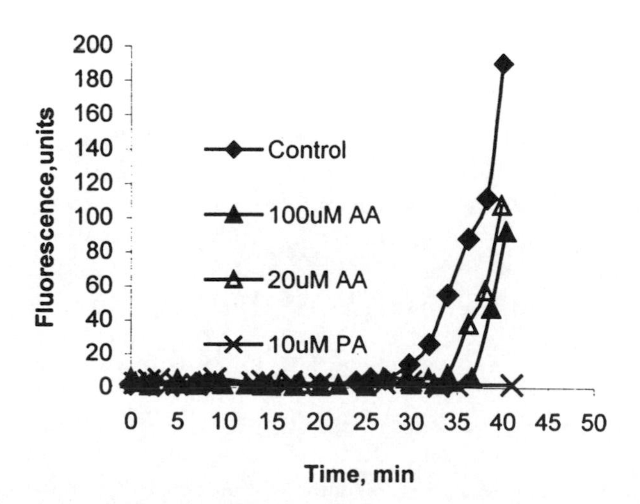

Figure 1 *The effect of ascorbic acid and palmitoyl ascorbate on intracellular DCFH-DA oxidation in foreskin fibroblasts*

References

1 K. Irani, Mitogenic signaling mediated by oxidants in Ras-transformed fibroblasts, *Science*, 1997, **275**, 1649–1652.

2 W.J. Baader, A. Hatzelmann and V. Ullrich, The suppression of granulocyte functions by lipophilic antioxidant, *Biochem. Pharm.*, 1988, **37**, 1089–1098.

3 E. Schmid, V. Figala and V. Ullrich, Inhibition of NADPH-oxidase activity in human polymorphnonuclear neutrophils by lipophilic ascorbic acid derivatives, *Mol. Pharm.*, 1994, **45**, 815–825.

5.10

Antioxidant Capacity of Polyphenolic Flavonols in Tea

S.G. Magnusdottir,[1] P.A. Burns[2] and S. Khokhar[1]

[1] DEPARTMENT OF FOOD SCIENCE, AND [2] SCHOOL OF MEDICINE, UNIVERSITY OF LEEDS, UK

1 Introduction

Recent research has suggested that dietary flavonoids may play a role in human health and disease prevention, particularly in diseases believed to involve, in part, oxidation, such as coronary heart disease, inflammation and mutagenesis leading to carcinogenesis. Flavonoids in tea are a major source of antioxidants in the human diet world wide. Tea is a rich source of the flavan-3-ol class of flavonoids, commonly known as catechins; (+)-catechin (C), (−)-epicatechin (EC), (−)-epigallocatechin EGC) (−)-epicatechin gallate (ECG) and (−)-epigallocatechin gallate (EGCG) (Figure 1). Flavonoids from tea may act by chelating metal ions or by acting as a free radical scavenger by working

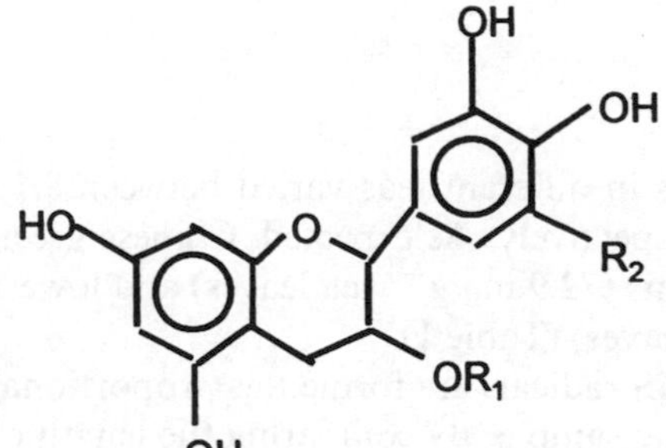

R_1	R_2	abr	name:
H	H	EC:	(epi)catechin
G	H	ECG:	epicatechingallate
H	OH	EGC:	epigallocatechin
G	OH	EGCG:	epigallocatechin-gallate

G = Gallate:

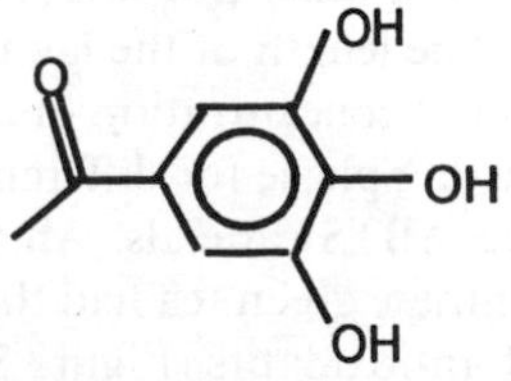

Figure 1 *Structure of catechins*

either as hydrogen or electron donors, for example by scavenging peroxyl radicals.

$$ROO^{\bullet} + AH \rightarrow ROOH + A^{\bullet}$$

DNA damage induced by a reactive oxygen species (ROS) is an important intermediate in the pathogenesis of human conditions such as cancer and ageing. ROS-induced DNA damage products are both mutagenic and cytotoxic. H_2O_2 is the most studied ROS and is produced endogenously by several physiological processes. Because H_2O_2 is freely diffusible, it is not degraded by antioxidant enzymes and can potentially reach the nucleus to interact with DNA. H_2O_2 causes strand breaks and base damage in DNA by a mechanism that requires transition metal ions, such as iron or copper, and produces an ultimate DNA-reactive species possessing the reactivity of the hydroxyl radical.[1] P53 is a tumour suppressor gene and it has the ability to suppress uncontrolled cell proliferation. It is believed that p53 gene mutations are involved in the development of about half of all human cancers.[2]

The objective of this study was to establish the antioxidant potential of polyphenolic flavonoids in tea as scavengers of free radicals ($ABTS^{\bullet}$ and $OH^{\bullet}$).

2 Materials and Methods

Total antioxidant activity and phenolic flavonoids. The total antioxidant activity was determined as $ABTS^{\bullet}$ – scavenging capacity by the method of Rice-Evans and Miller.[3] Tea catechins and total phenols were measured in different teas by using the methods of Khokhar *et al.*[4] and Singleton *et al.*,[5] respectively.

Yeast-based p53 functional assay. The pLS76 yeast expression plasmid contained a human wild type p53 cDNA and the LEU2 selectable marker. Oxidative damage was induced to the plasmid with 2.5 mM H_2O_2, 100 M $FeCl_3$ and ascorbic acid.[1] The plasmid was then expressed in the yeast (*Saccharomyces cerevisiae*, strain ylG397). Assessment of p53 status was based on colony colour mutation frequency (Figure 2).[6]

3 Results and Discussion

The levels of catechins and total phenols in different teas varied between 8.1–72.9 and 80.5–98.9 mg g^{-1} tea leaves, respectively. As expected, Chinese green tea contained highest level of total catehins (72.9 mg g^{-1} tea leaves) and lowest level of other phenols (14.1 mg g^{-1} tea leaves) (Table 1).

The length of the lag time, before ABTS radicals are formed, is proportional to the concentration of antioxidants in the sample. By comparing the length of the lag phase for different teas it is possible to estimate their ability to scavenge the ABTS radicals. All the teas studied showed different lag times (1–3 min). Chinese green tea had the longest lag phase and therefore highest concentration of antioxidants (Figure 3).

The contribution of total catechins in Chinese green tea to the antioxidant

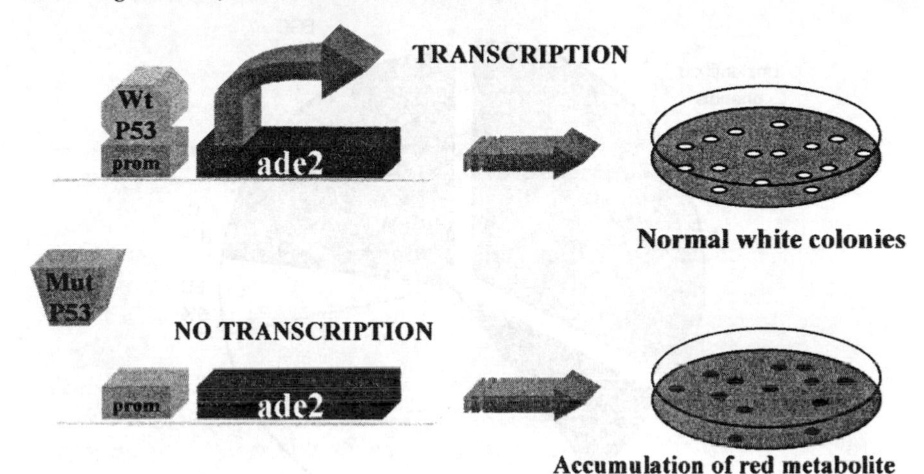

Figure 2 *Schematic diagram of the yeast based functional assay*

Table 1 *Levels of polyphenolic flavonols in green and black teas*

	Chinese green tea	*Darjeeling* *mg g^{-1} tea leaves*	*PG-tips*
EGC	21.7 ± 0.7	3.0 ± 0.5	0.2 +0.4
C	0.12 ± 0.0	0.7 ± 0.2	0.5 ± 0.2
EC	5.7 ± 0.0	2.3 ± 0.2	1.4 ± 0.2
EGCG	35.1 ± 1.1	24.9 ± 4.8	3.9 ± 0.7
ECG	8.5 ± 0.1	5.9 ± 0.7	2.1 ± 0.3
Total catechins	72.9 ± 1.9	36.9 ± 4.3	8.1 ± 1.0
Total phenols	87.0 ± 2.2	98.8 ± 2.4	80.5 ± 3.0

The values are given in mg g^{-1} tea leaves and are expressed as mean ± SD (n = 4–20).

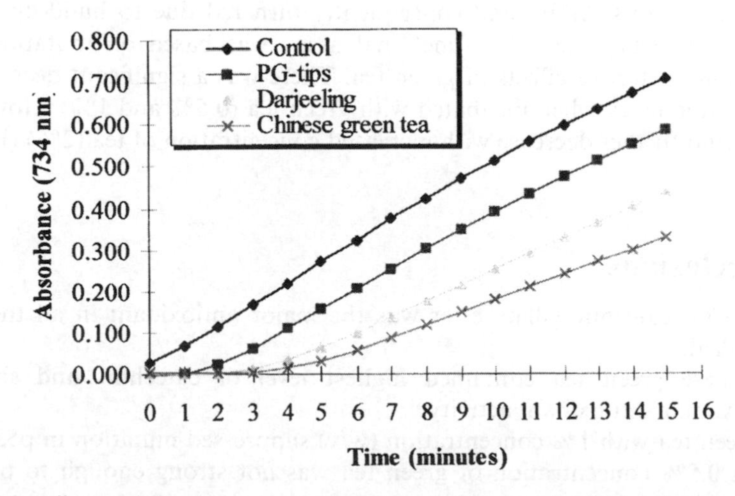

Figure 3 *Effect of different teas on the lag phase (inhibition of ABTS radical formation)*

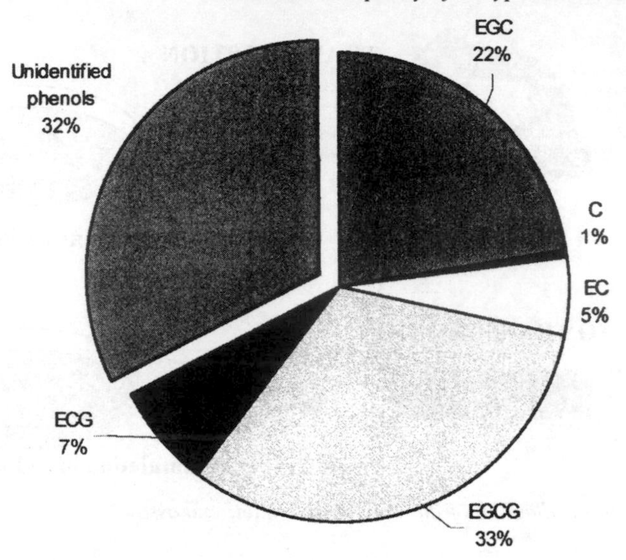

Figure 4 *The contribution of polyphenolic flavonols to the total antioxidant activity of Chinese green tea*

activity was 68% while 32% was from unidentified phenolics. The order of their contribution to the total antioxidant activity was EGCG > EGC > ECG > EC > C. EGCG and ECG were found to be the major contributors and together they constituted 55% (Figure 4).

Hydroxyl radicals ($OH^{\bullet}$) are formed when iron reacts with H_2O_2, and can attack guanine and form 8-hydroxyguanine (mutant p53). It has been reported that guanine is the most easily modified base in associated with H_2O_2-mediated DNA damaging reactions both *in vivo* and *in vitro*.[1] Cells containing mutant p53 fail to express ADE2 and consequently turn red due to build-up of an intermediate metabolite. This functional assay was based on mutation frequency and protective effects of green tea. There was a significant decrease in mutation frequency when incubated with green tea (0.5% and 1%). However, there was no further decrease with increased concentration of tea (2%) (Figure 5).

4 Conclusions

- EGCG, catechin-gallate ester was the major antioxidant in all the teas studied.
- Chinese green tea contained highest level of catechins and showed maximum antioxidant activity.
- Green tea with 1% concentration (w/v) suppressed mutation in p53 gene but 0.5% concentration of green tea was not strong enough to protect mutation.

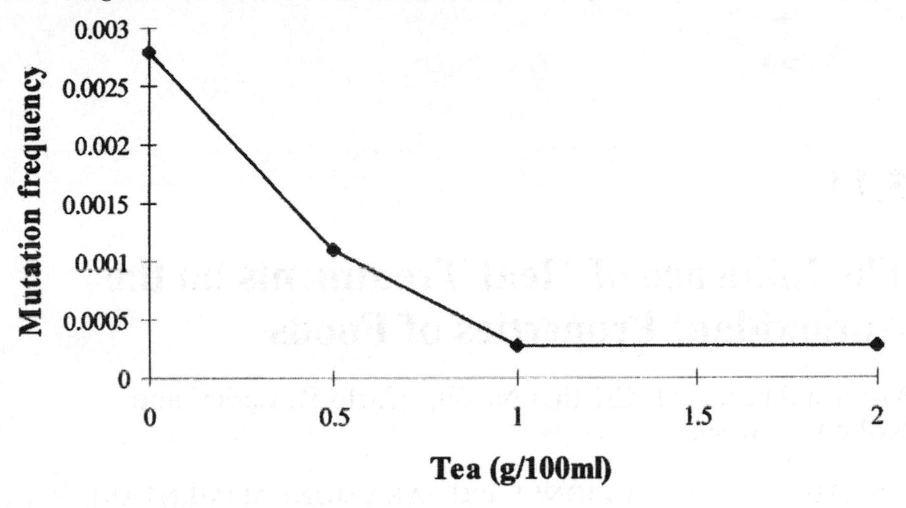

Figure 5 *Protective effect of Chinese green tea on DNA damage expressed as mutation frequency in p53 gene*

5 References

1 H. Rodriguez, G.P. Holmquist, R. D'Agostino, J. Keller and S.A. Akman, *Cancer Res.*, 1997, **37**, 2394–2404.

2 P.J. Russell, *Genetics*, Benjamin/Cummings Publishing, CA, 1998, Chapter 18, p. 585.

3 C. Rice-Evans and N.J. Miller, *Methods Enzymol.*, 1994, **234**, 279–293.

4 S. Khokhar, D.P. Venema, P.C.H. Hollman, M. Dekker and W. Jongen, *Cancer Lett.*, 1997, **114**, 171–172.

5 V.L. Singleton, R. Orthofer and R.M. Lamuela-Raventos, *Methods Enzymol.*, 1999, **299**, 152–178.

6 C. Ishioka, T. Frebourg, Y.X. Yan, M. Vidal, S.H. Friend, S. Schmidt and R. Iggo, *Nature Genetics*, 1993, **5**, 124–129.

5.11

The Influence of Heat Treatments on the Antioxidant Properties of Foods

Monica Anese,[1] M. Cristina Nicoli,[2] Carlo R. Lerici[3] and Roberto Massini[1]

[1] ISTITUTO DI PRODUZIONI E PREPARAZIONI ALIMENTARI, THE UNIVERSITY OF BARI, VIA NAPOLI 25, 71100 FOGGIA, ITALY
[2] DIPARTIMENTO DI SCIENZE AMBIENTALI, AGRARIE E BIOTECNOLOGIE AGRO-ALIMENTARI, THE UNIVERSITY OF SASSARI, VIALE ITALIA 39, 07100 SASSARI, ITALY
[3] DIPARTIMENTO DI SCIENZE DEGLI ALIMENTI, THE UNIVERSITY OF UDINE, VIA MARANGONI 97, 33100 UDINE, ITALY

1 Introduction

Despite the huge number of literature reports on the relationships between food and health, there is a lack of information on the influence and role of processing on the antioxidant properties of foods. In the majority of cases, food processing and preservation operations, carried out industrially or even during home meal preparation, may be responsible for a significant loss in natural antioxidants, being most of them scarcely stable.[1–3] However, recent findings have shown that heat-based processing may be responsible for different and in some cases opposite effects on the antioxidant properties of foods. For instance, it is well documented that moderate heating can increase the bioavailability of some natural antioxidants, such as β-carotene;[4] blanching may cause vitamin loss, but also inactivation of oxidative enzymes, thus preventing further losses during slow processing and storage;[5] heat treatments can be responsible for the formation of novel compounds with antioxidant properties (*i.e.* the Maillard reaction products, MRPs).[6–12] In the latter case, when heating is applied, the loss in natural antioxidants can be balanced or even enhanced by the formation of thermally induced MRPs.[3] Also, studies have shown that heating can be responsible for a depletion of the overall antioxidant properties of food products not only because of the degradation of naturally occurring antioxidants, but also because of the formation, in the early stages of the Maillard reaction, of uncoloured compounds with pro-oxidant properties.[13–16]

Here we present results on the influence of some heat-based treatments, including cooking, pasteurisation and dehydration, on the antioxidant properties of some foods. In particular, the overall antioxidant potential of food products was evaluated under the application of heating processes with various thermal intensity.

2 Results and Discussion

Figure 1 shows the changes in the chain breaking activity of tomato juice samples[16] heated at 70 or 95 °C for up to 6 hours. The changes in the chain breaking activity of tomato juice-olive oil samples[17] heated at 95 °C for up to 6 hours are also reported. As observed by Anese *et al.*[16] the chain breaking activity of tomato juice samples heated at 95 °C for up to 6 hours decreased within the first 3-hour heating. By prolonging heating times, a recovery and a further increase in the antioxidant properties were observed. This increase corresponded to the appearance of colour[16] and can be attributed to the formation of MRPs with antioxidant properties. When heated at 70 °C only a reduction in the chain braking activity of tomato juice was found. The reduction in the antioxidant activity in the first few hours heating both at 70 and 95 °C can

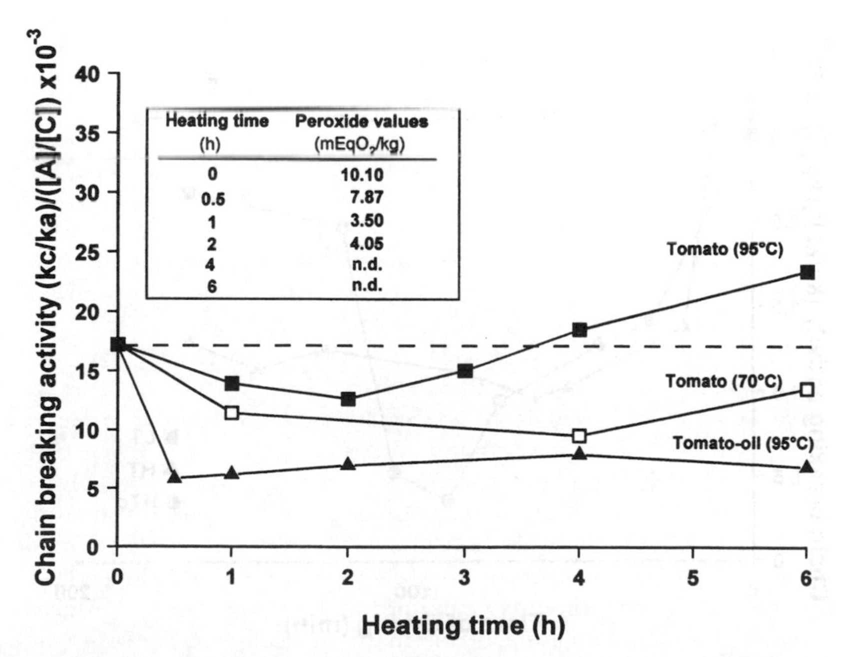

Figure 1 *Changes in the chain breaking activity of tomato juice and tomato juice-olive oil (10% w/w) samples heated at 70 or 95 °C and at 95 °C, respectively*

only partially be attributed to a loss in the naturally occurring antioxidants. As already reported by many authors,[13–16] the observed decrease in the antioxidant capacity of tomato juice can also be attributable to the formation of compounds with pro-oxidant properties during the early stages of the Maillard reaction, when no colour development is detected. Also in the case of the tomato juice added with olive oil a reduction in the chain breaking activity within the first few hours heating was detected. However, by prolonging heating times, the antioxidant properties of the tomato juice-oil samples remained low for up to 6 hours. The peroxide values referred to the tomato-oil samples, which are reported in the inset of Figure 1, unexpectedly decreased as heating time increased, up to values close to zero for the 4-hour and 6-hour heated samples. This can be explained with the fact the MRPs, which are formed during heating, can slow down the oxidative reactions acting as primary antioxidants. However, according to Arnoldi *et al.*[18] it can also be presumed that carbonyl compounds formed as a consequence of lipid oxidation are gradually 'consumed', as they can react with amino groups. Thus, an increase in the stability of the lipid fraction was achieved to the detriment of the overall antioxidant capacity as a consequence of the interactions between antioxidants (namely the MRPs) and oil.

Figure 2 shows the changes in the chain breaking activity of pasta samples[19] as a function of drying time carried out at low, high and very high temperatures. As can be observed, the chain breaking activity of the LT samples did not

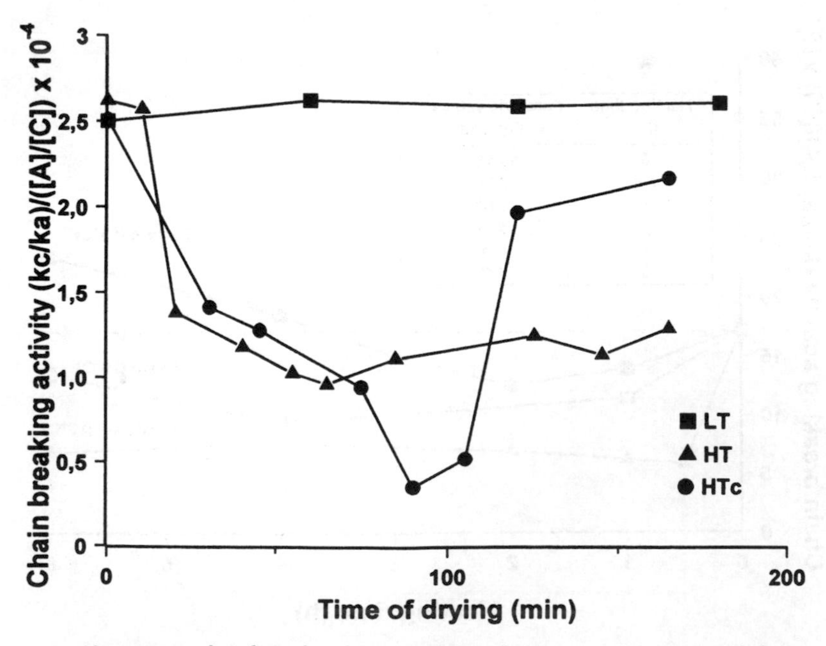

Figure 2 *Changes in the chain breaking activity of pasta as a function of drying time carried out at low, high and very high temperatures*

change. On the contrary, a strong decrease in the antiradical capacity was detected in the first stages of the HT drying process. The activity remained low for up to 70 min heating, then slightly increased in the final stages. An even stronger decrease in the chain breaking activity in the early stages of the drying process was observed in the case of pasta dried at 110 °C (HTc process), which sharply increased in the last stages, until the antioxidant capacity was almost recovered. The development of colour was detected only in correspondence of an increase in the chain breaking activity (data not shown). Thus, these results would confirm the hypothesis stating that the antioxidant properties of pasta as well as of tomato samples can be attributed to the formation of coloured MRPs, while compounds having pro-oxidant properties are formed in the early stages of the heating process.

A reduction in the original antioxidant properties through the formation of compounds with pro-oxidant properties would appear of considerable interest with regard to low temperature or short time heat treatments. The formation of pro-oxidants during the early phases of the Maillard reaction may depend upon the intensity and the duration of the heat treatment: it is likely that when low temperature heating is applied the phases which contribute to the formation of compounds with pro-oxidant properties longer last than in the case of high temperature treatments. Thus, as schematised in Figure 3, vegetable matrices subjected to heat treatment can undergo to an initial reduction in the antioxidant properties, due to a loss in the naturally occurring antioxidants and/or to the formation of compounds with pro-oxidant properties during the early stages of the Maillard reaction. By prolonging heating time this loss can be minimised by a recovery or even an enhancement in the overall antioxidant properties of the product due to the formation of MRPs with antioxidant properties.[16,18] However, only when sufficiently high temperatures are applied would a recovery in the antioxidant properties occur in a length of time useful for food processing. For instance, it must be pointed out that the drying process

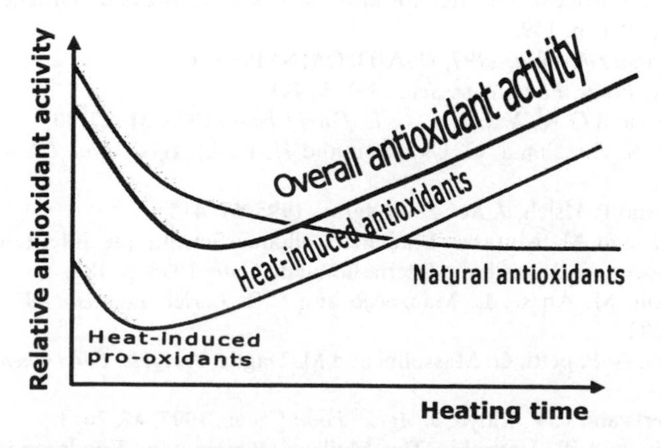

Figure 3 *Changes in the overall antioxidant activity due to loss of naturally occurring antioxidants and/or development of Maillard reaction products*

has generally been completed at the early stages of the Maillard reaction, which in our experimental conditions have been shown to correspond to the development of pro-oxidant activity.

In conclusion, results reported here clearly show that heating of foods can give rise to various effects, whose consequences are not necessarily a loss in the antioxidant properties, but more probably opposite effects on the stability of the product and on its health protecting properties. In fact, heating of foods, besides a loss in natural antioxidants, can lead to an enhancement of the antioxidant properties of food because of the formation of MRPs having antioxidant properties. But processing can also be responsible for a depletion of the overall antioxidant potential of food products because of the formation of MRPs with pro-oxidant properties or as a consequence of chemical inter-reactions (for example, those between Maillard reaction and lipid oxidation). The prevalence of one mechanism or of the other depends upon the intensity and duration of the heat treatment. These 'events' should be taken into account when considering the protective role of foods in preventing some chronic and degenerative diseases, such as cancer and cardiovascular diseases. Although more research is needed in order to better understand the processing conditions and the mechanisms responsible for the loss and/or formation of antioxidants, investigation into their effects (in particular into the effects of early and advanced MRPs as pro-oxidants and antioxidants) on human health appear to be worthwhile.

3 References

1 J.W. Erdman Jr, *Food Technol.*, 1979, 38.
2 L. Jonson, 'Nutritional and Toxicological Consequences of Food Processing', Plenum Press, New York, 1991, p. 75.
3 M.C. Nicoli, M. Anese, M.P. Parpinel, S. Franceschi and C.R. Lerici, *Cancer Lett.*, 1997, **114**, 71.
4 S. Southon, 'European Research towards Safer and Better Food', Druckerei Grasser, Karlsrhue, 1998, p. 159.
5 D.B. Rodriguez-Ayama, 1997, USAID, OMNI Project.
6 K. Eichner, *Progr. Food Nutr. Sci.*, 1981, **5**, 441.
7 H. Lingnert and G.R. Waller, *J. Agric. Food Chem.*, 1983, **31**, 27–30.
8 F. Hayase, S. Hirashima, G. Okamoto and H. Kato, *Agric. Biol. Chem.*, 1989, **53**, 3383.
9 G.C. Yen and P. Hsieh, *J. Sci. Food Agric.*, 1995, **67**, 415.
10 S. Homma and M. Murata, 'Sextieme Colloque Scientifique International sur le Café'. Association Scientifique International du Café, 1995, p. 183.
11 M.C. Nicoli, M. Anese, L. Manzocco and C.R. Lerici, *Lebensm.-Wiss. Technol.*, 1997, **30**, 292.
12 G. Gazzani, A. Papetti, G. Massolini and M. Daglia, *J. Agric. Food Chem.*, 1998, **46**, 4118.
13 R.L. Roberts and R.V. Lloyd, *J. Agric. Food Chem.*, 1997, **45**, 2413.
14 M. Namiki and T. Hayashi, 'The Maillard Reaction in Foods and Nutrition', American Chemical Society, Washington DC, 1983, p. 21.

15 M. Pischetsrieder, F. Rinaldi, U. Gross and T. Severin, *J. Agric. Food Chem.*, 1998, **46**, 2945.
16 M. Anese, M.C. Nicoli and L. Manzocco, *J. Sci. Food Agric.*, 1999, **79**, 750.
17 M.C. Nicoli, M. Anese and L. Manzocco, *Adv. Food Sci.*, 1999, **21**, 10.
18 A. Arnoldi, C. Arnoldi, O. Baldi and C. Griffini, *J. Agric. Food Chem.*, 1987, **35**, 1035.
19 M. Anese, M.C. Nicoli, R. Massini and C.R. Lerici, *Food Res. Intern.*, in press.

5.12

Melatonin in Cereal Grains as a Potential Cancer Prevention Agent

Henryk Zieliński, Halina Kozłowska and Bogdan Lewczuk[2]

[1] DIVISION OF FOOD SCIENCE, INSTITUTE OF ANIMAL REPRODUCTION AND FOOD RESEARCH OF POLISH ACADEMY OF SCIENCES, TUWIMA 10, PO BOX 55, 10-718 OLSZTYN 5, POLAND
[2] OLSZTYN UNIVERSITY OF AGRICULTURE AND TECHNOLOGY, DEPARTMENT OF HISTOLOGY, OCZAPOWSKIEGO 13, 10–719 OLSZTYN, POLAND

Melatonin, the hormonal product of the pineal gland, has recently been reported as an effective hydroxyl radical scavenger, an ability that shed some light on several of its nonreceptor-related actions. Other *in vivo* and *in vitro* studies gave evidence that the indoleamine may protect DNA from peroxidative damages by chemical carcinogens, avoid cytogenetic damage by ionizing radiations, and also protect the whole cell antioxidant defense system by increasing the activity of glutathione peroxidase and by raising the mRNA for superoxide dismutase. To date there are no studies related to foodstuffs derived from plants as a possible source of melatonin. The present paper reports the presence of melatonin in selected cereal grains as determined by radioimmunoassay (RIA). The level of melatonin was ranged from 150 to 450 pg g^{-1} of d.m. showing higher contents for oat, barley and buckweat than for wheat and rye. These findings indicate that cereal grains are more rich in melatonin in relation to vegetables and fruits. Thus, we suggest that melatonin ingested in foodstuffs of cereal grain origin may play a significant role as a cancer prevention agent; however, further research on melatonin content in processed cereal grains are required.

1 Introduction

Carcinogenesis is a multistep process consisting of three stages: initiation, promotion, and progression. Initiation takes a short period of time and is characterized by an irreversible alteration of the cellular DNA, which allows the transformation of the cell to a nonmalignant state. Promotion, which takes a longer time and may be reversible, permits the nonmalignant cell to become malignant. This involves the selection and clonal proliferation of initiated cells caused by chemical compounds or other factors. The progression stage involves the growth of malignant cells to a tumor.[1] The role of dietary factors in the prevention of cancer is under intensive investigations by many laboratories around the world. Dietary habits are regarded as possible causative factors in the development of a considerable proportion of human cancer although these observations have not been adequately explained.[7] It has not been determined whether individual vitamins, dietary fibres, antioxidants, or some other food components, or various combination of many substances, are involved in the observed protective effect of some foods in reducing cancer incidence.[8]

Melatonin (*N*-acetyl-5-methoxytryptamine), the principal hormone of the pineal gland of vertebrates, is implicated in physiological processes that are controlled by photoperiod: circadian rhythms, reproduction, and behaviour. Circulating melatonin concentrations increase in response to melatonin production during darkness. Melatonin production in the pineal gland declines progressively with age such that in elderly humans the levels of melatonin available to the organism are a fraction of that of young individuals.[9] Melatonin, which is both hydrophilic and hydrophobic,[16] is now known to be a multifaceted free radical scavenger and antioxidant. It detoxifies a variety of free radicals and reactive oxygen intermediates including the hydroxyl radical, peroxynitrite anion, singlet oxygen and nitric oxide. Additionally, it reportedly stimulates several antioxidative enzymes including glutathione peroxidase, glutathione reductase, glucose-6-phosphate dehydrogenase and superoxide dismutase; conversely, it inhibits a prooxidative enzyme, nitric oxide synthase. Melatonin has been shown to markedly protect both membrane lipids and nuclear DNA from oxidative damage. In every experimental model in which melatonin has been tested, it has been found to resist macromolecular damage and assosioated dysfunction associated with free radicals.[3]

Edible plant tissues contain melatonin and their consumption increases the circulating melatonin supply in mammals.[4] Although the significance of melatonin for plant life is unknown, it was suggested that one important function of melatonin may be the scavenging of free radicals, thereby protecting plants against oxidative stress.

The objective of this work was to determine the levels of melatonin in different cereal grains before and after extrusion cooking (process temperatures of 120–160–200 °C) being used as a model of thermal technological processes.

2 Materials and Methods

Chemicals. (O-methyl-^{3}H)-melatonin [spc. act. 87 Ci mmol^{-1}] was purchased from Du Pont NEN; antiserum G/S/704–6483 from Stockgrand Ltd, University of Surrey, Guildford, UK; sodium chloride and toluene from POCh-Poland; gelatine from Merck-Germany; *N*-acetyl-5-methoxytryptamine (cold melatonin), phosphate buffered saline (0.01 M sodium and potassium phosphate buffer, pH 7.4, containing 0.0027 M potassium chloride and 0.137 M sodium chloride) and all other reagents were from Sigma Co.

Samples. Single cereal grain samples grown in 1997 and 1998 were obtained from a local plant breeding station in North-East Poland. The samples included wheat (*Triticum aestivum L.*) cv. Almari and cv. Henika; barley (*Hordeum vulgare L.*) cv. Gregor, cv. Mobek, cv. Maresi, cv. Rodos, and cv. Rudzik; one cultivar of rye (*Secale cereale L.*) cv. Dańkowskie Złote; one cultivar of oat (*Avena sativa L.*) cv. Sławko, and one cultivar of buckwheat (*Fagopyrum esculentum* Moench) cv. Kora. Samples from two replications at each were chosen for the analysis. Whole-grain samples and extruded grain samples were ground in a laboratory mill and then they were extracted in duplicate with 0.01 M. phosphate buffered saline pH 7.4 (5 mL per 1 g of sample) during 2 hours of shaking at 37 °C. They were then centrifuged at 12,000 g (Beckman centrifuge GS-15R) and the fresh supernatant was assayed for melatonin.

Melatonin radioimmunoassay. Melatonin immunoreactivity was assayed by a slightly modified direct method of Fraser *et al.*[13] Briefly, 200 μL antiserum, diluted 1:6000 in assay buffer (tricine 0.1 mol L^{-1}, sodium chloride 9 g L^{-1}, gelatine 1g/L was added to 500 μL sample or standard (0–500 pg mL^{-1} prepared in PBS) and the mixture was incubated at room temperature for 30 min. Then 100 μL of ^{3}H-melatonin, diluted in the assay buffer to approximately 1 μCi/10 mL, was added. After overnight incubation at 4 °C, antibody-bound melatonin was separated from the free fraction by incubation with 500 μL dextran-coated charcoal (0.9 g Norit A and 60 mg dextran in 100 mL of assay buffer) for 15 min at 4 °C. After centrifugation, the radioactivity of 700 μL of supernatant was measured using a liquid scintillation method. Samples were assayed in duplicates. For each sample the non-specific binding was determined (in duplicates) by addition of tricine buffer instead of antibody solution. The assay was validated by running of the samples containing different amounts of exogenous melatonin. The melatonin concentration was calculated using four-parameter logistic curve (Immuno Fit EIA/RIA ver.3.0a firmy Beckman).

The mean standard curve of the direct assay employing antiserum G/S/704–6483 was generated during 6 assay series. The mean specific binding was 39.5% of the total radioactivity, the mean unspecific binding in PBS 1.4% of the total radioactivity, and the unspecific binding of samples varied from 4% to 12% of the total radioactivity. The sensitivity of this assay was below 5 pg mL^{-1}. The intra-assay coefficients of variation ($n = 6$) for the samples containing 28.50 pg/mL and 57.82 pg/mL of melatonin were 6.3% and 5.6%, respectively. The results of the assay of the samples before and after addition of known amounts of melatonin are shown in Table 1.

3 Results and Discussion

Currently, the presence of radioimmunoassayable melatonin in various plants[4,10] and in feverfew and other medicinal plants has been shown.[11] In this work immunologically-identifiable melatonin was found in all cereal grains studied. The data were recalculated on the matter of the investigated cereal

Table 1 *The levels of radioimmunoassayable melatonin measured in the cereal grain extracts after addition of different amounts of synthetic melatonin*

	Amounts of synthetic melatonin added (pg per tube)			
Origin of the extract	*0*	*25*	*50*	*100*
wheat cv. Almari	14.2 ± 1.3	39.8 ± 2.8	64.6 ± 4.5	107.2 ± 7.7
wheat cv. Henika	28.4 ± 2.1	53.3 ± 3.7	67.0 ± 4.4	97.1 ± 6.6
barley cv. Gregor	57.9 ± 4.0	84.3 ± 5.9	120.2 ± 8.8	157.8 ± 12.1
barley cv. Mobek	47.2 ± 3.3	75.0 ± 4.6	96.2 ± 6.3	178.1 ± 13.9
rye cv. Dańkowskie Złote	19.2 ± 1.6	45.8 ± 3.2	85.9 ± 5.9	84.8 ± 5.8
PBS	0	24.0 ± 1.8	45.5 ± 3.2	88.7 ± 6.2

grains. Results are given as the means of triplicates and the standard deviation. The level of melatonin ranged from 150 to 450 pg g^{-1} of mature grain showing higher contents for oat, barley and wheat than rye and buckweat. The amount of melatonin found in mature and extruded cereal grains was lower than previously found in medicinal plants;[11] however, it was higher than that found in edible-plant products. The amount of melatonin in extruded grains was decreased, probably due to the destruction caused via applied high temperature and pressure during used technological process (Figures 1–3). Only melatonin in

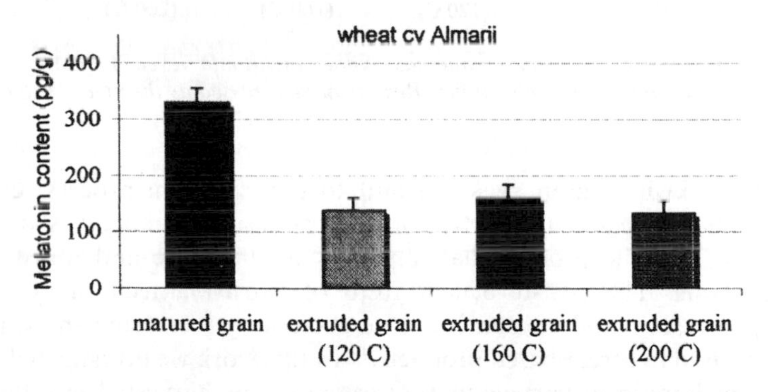

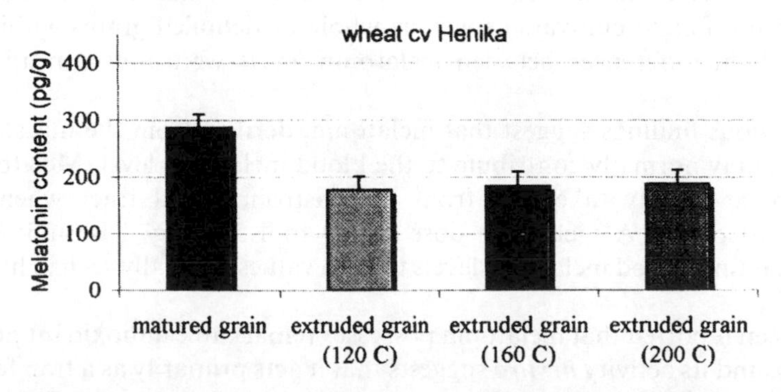

Figure 1 *Concentration of the radioimmunoassayable melatonin levels in wheat grain before and after extrusion cooking. Results are expressed as the mean ± SEM*

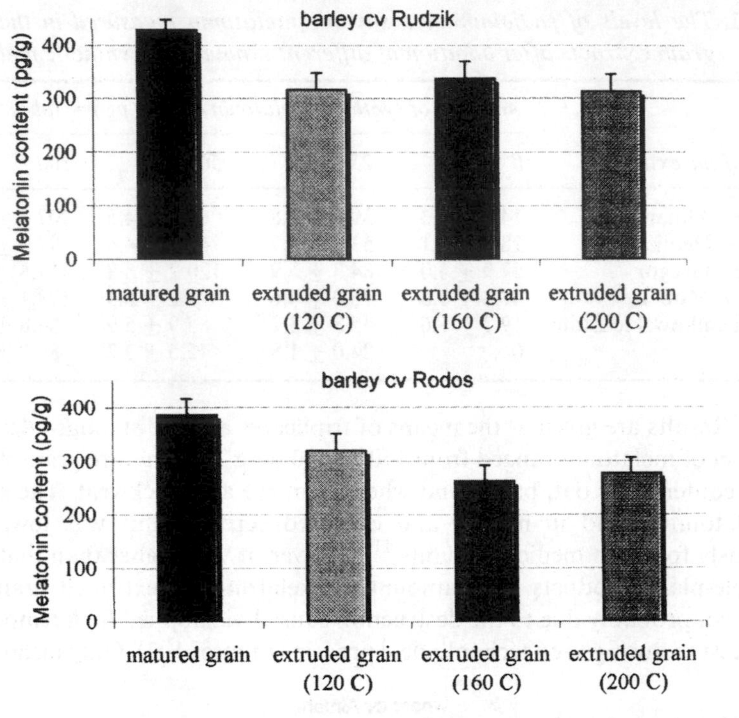

Figure 2 *Concentration of the radioimmunoassayable melatonin levels in barley grain before and after extrusion cooking. Results are expressed as the mean* $\pm$ *SEM*

the dehulled buckwheat grain was resistant to the extrusion process, giving similar amounts before and after extrusion. The extrusion cooking was applied because it is a versatile process that improves organoleptic and nutritional qualities in foods. The cost-to-benefit ratio of extrusion technology gives producers, processors, and consumers more choices by increasing the variety of ingredients used in cereal-based products.[17] In this work we investigated also the levels of melatonin in various mature cereal grains harvested in different years within different cultivars as well as whole or dehulled grains within one cultivar. Slight differences between melatonin levels were noted in all cases (Table 2).

The previous findings suggest that melatonin, derived from the ingestion of foodstuffs, may normally contribute to the blood melatonin level. Melatonin is known to be rapidly taken up from the gastrointestinal tract when it is administered orally. A melatonin dose of 1.0 to 1.5 $\mu g\ kg^{-1}$ is sufficient to raise the daytime blood melatonin levels to high values normally seen in humans at night.[18]

It has been reported that melatonin possesses remarkable antioxidant properties *in vivo*, and its activity *in vitro* suggests that it acts primarily as a trap for free radicals with an $^{\bullet}OH$ structure. Melatonin appears to supply $NADPH_2$ to the cell which is required for regenerating oxidised glutathione to reduced glu-

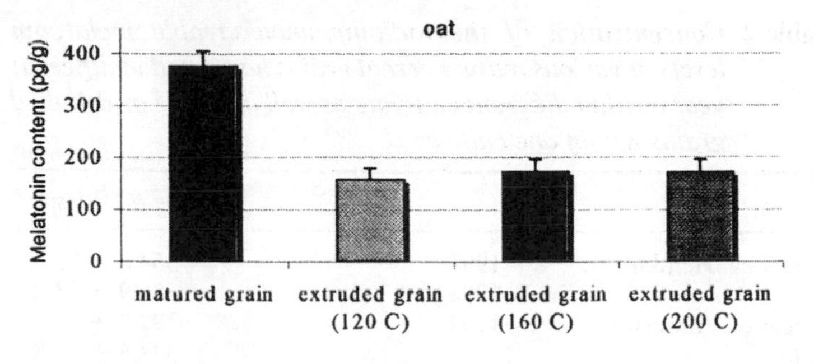

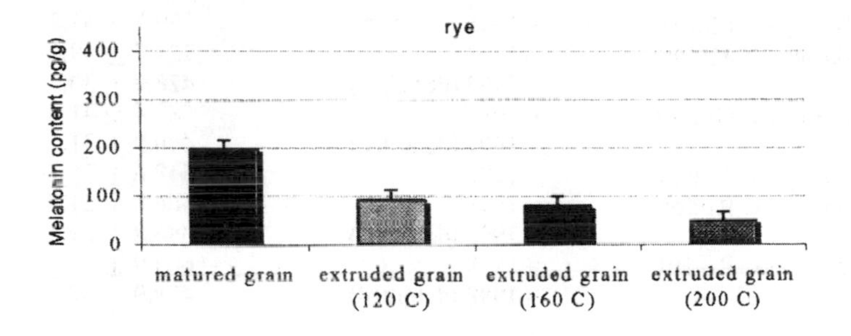

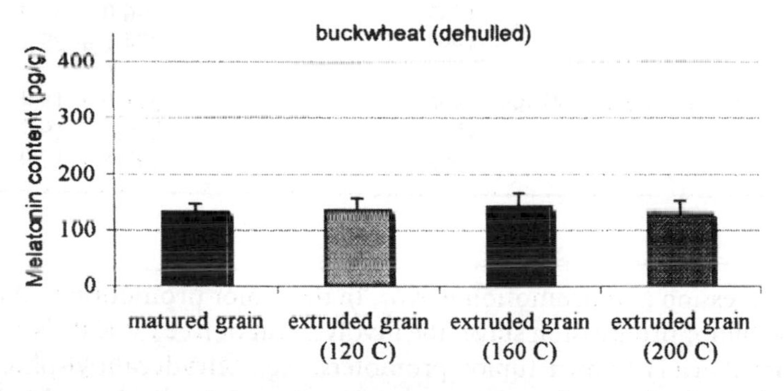

Figure 3 *Concentration of the radioimmunoassayable melatonin levels in oat, rye and buckwheat grain before and after extrusion cooking. Results are expressed as the mean ± SEM*

tathione by glutathione reductase (GR). Reduced glutathione is the cofactor required for glutathione peroxidase activity, which leads to the decreased formation of hydroxyl radicals by favouring the transformation of H_2O_2 to water.[3,12] As free radicals damage lipids, proteins, cell membranes and DNA, their removal could prevent development of certain chronic diseases, particularly cancer. Considering the potent antioxidant activity of melatonin and its synergistic interactions with antioxidants such as trolox, glutathione, ascorbate in aqueous system[5] and with α-tocopherol in a lipid bilayer,[6] it may be concluded that ingestion of foods selected for their high melatonin content may play a role in protection against radical mediated cellular damage as well as

Table 2 *Concentration of the radioimmunoassayable melatonin levels in various mature cereal grains harvested in different years within different cultivars as well as whole or dehulled grains within one cultivar*

Species		*pg g^{-1} tissue*
Wheat cv. Henika	– 1997	354.9 ± 26.3
	– 1998	286.9 ± 23.2
Wheat cv. Almarii	– 1997	392.2 ± 22.4
1998		328.8 ± 25.8
Barley cv. Maresi	– 1996	589.2 ± 31.8
Barley cv. Mobek	– 1997	724.7 ± 49.6
	– 1997 (dehulled)	**428.0** ± 30.7
Barley cv. Gregor	– 1997	589.4 ± 41.1
	– 1997 (dehulled)	**440.9** ± 31.5
Barley cv. Rodos	– 1996	448.6 ± 25.1
Barley cv. Rodos	– 1998	566.8 ± 21.2
	– 1998 (dehulled)	**386.8** ± 29.4
Barley cv. Rudzik	– 1998	604.9 ± 36.7
	– 1998 (dehulled)	**429.9** ± 32.1
Oat cv. Sławko	– 1996	323.2 ± 19.2
	– 1997	496.0 ± 34.9
	– 1998	374.1 ± 28.7
Rye cv. Dańkowskie Złote	– 1996	311.0 ± 18.4
	– 1997	240.5 ± 19.7
	– 1998	197.5 ± 17.9

cancer progression and promotion *in vivo*. In the tumor promotion and perhaps also in the tumor progression stage, the involvement of free radicals is suggested by the fact that (1) known tumor promoters, *e.g.*, tetradecanoyl-phorbol-13-acetate (TPA) and phorbol esters, stimulate oxygen radical production, (2) generators of free radicals (*e.g.*, superoxide radicals formed extracellularly by xanthine/xanthine oxidase or by human neutrophils) mimic the action of tumor promoters (*e.g.*, TPA), and (3) tumor promoters cause the inhibition of antioxidant enzyme (*e.g.*, SOD, CAT, GPX, GSSG-R) activities coupled with increased production of oxidized GSH.[2] Melatonin is considered to be a natural oncostatic agent for animals and man.[14,15]

4 Conclusion

The determination of melatonin in mature and extruded cereal grains studied in this work coupled with the observations that melatonin is a potent free radical scavenger require further investigation in relation to the significance of melatonin containing food or diets in the prevention of cancer in humans.

Acknowledgment

A research grant No. 5 P06G 024 13 from the Polish State Committee for Scientific Research is gratefully acknowledged.

5 References

1 Weisburger J.H. Nutritional approach to cancer prevention with emphasis on vitamins, antioxidants and carotenoids. *Am. J. Clin. Nutr.*, **53**, 226S–232S, 1991.
2 Sun Y. Free radicals, antioxidant enzymes, and carcinogenesis. *Free Rad. Biol. Med.*, **8**, 583, 1990.
3 Reiter R.J., Tan D., Cabrera J., D'Arpa D., Sainz R.M., Mayo J.C., Ramos S. The oxidant/antioxidant network: role of melatonin. *Biol. Signals Recept.*, **8**, 56–63, 1999.
4 Hattori A., Migitaka H., Iigo M., Itoh M., Yamamoto K., Ohtani-Kaneko R., Hara M., Suzuki T., Reiter R.J. Identification of melatonin in plants and its effects on plasma melatonin levels and binding to melatonin receptors in vertebrates. *Biochem. Mol. Biol. Int.*, **35**, 627–634, 1995.
5 Poeggeler B., Reiter R.J., Hardeland R., Sewerinek E., Melchiorri D., Barlow-Walden L.R. Melatonin, a mediator of electron transfer and repair reactions, acts synergistically with the chain-breaking antioxidants ascorbate, trolox and glutathione. *Neuroendocrinol. Lett.*, **17**, 87–92, 1995.
6 Livrea M.A., Luisa T., D'Arpa D., Morreale M. Reaction of melatonin with lipoperoxyl radicals in phospholipid bilayers. *Free Rad. Biol. Med.*, **23**(5), 706–711, 1997.
7 Doll R., Peto R. The causes of cancer: quantitative estimates of avoidable risks of cancer in the United States today. *J. Natl. Cancer Inst.*, **66**, 1191–1308, 1981.
8 Malone W.F. Studies evaluating antioxidants and β-carotene as chemopreventives. *Am. J. Clin. Nutr.*, **53**, 305S–313S, 1991.
9 Reiter R.J. Pineal function during aging: attenuation of the melatonin rhythm and its neurobiological consequences. *Acta Neurobiol. Exp.*, **54**, 31S–39S, 1994.
10 Dubbels R., Reiter R.J., Klenke E., Goebel A., Schnakenberg E., Ehlers C., Schiwara H.W., Schloot W. Melatonin in edible plants identified by radioimmunoassay and by high performance liquid chromatography-mass spectometry. *J. Pineal Res.*, **18**, 28–31, 1995.
11 Murch S.J., Simmons C.B., Saxena P.K. Melatonin in feverfew and other medicinal plants. *The Lancet*, **350**(9091), 1598, 1997.
12 Pierrefiche G., Laborit H. Oxygen free radiclas, melatonin and aging. *Exp. Gerontol.*, **30**(3/4), 213–227, 1995.
13 Fraser S., Cowen P., Franklin M., Franey C., Arendt J. Direct radioimmunoassay for melatonin in plasma. *Clin. Chem.*, **29**(2), 396–397, 1983.
14 Hill S.M., Blask D.E. Effects of the pineal hormone melatonin on proliferation and morphological characteristics of human breast cancer cells (MCF-7) in culture. *Cancer Res.*, **48**, 6121–6126, 1988.
15 Blask D.E. Melatonin in oncology. In: 'Melatonin', H.S. Yu, R.J. Reiter, eds. CRC Press, Boca Raton: 447–475, 1993.
16 Shida C.S., Castrucci A.M.L., Larny-Freund M.T. High melatonin solubility in aqueous medium. *J. Pineal Res.*, **16**, 198–201, 1994.
17 O'Connor C. Survey of the present extrusion cooking techniques in the food and confectionery. Pages 1–15 in: 'Extrusion Technology for the Food Industry'. Elsevier Applied Science. New York, 1987.
18 Vakkuri O., Leppaluto J., Kauppila A. Oral administration and distribution of melatonin in human serum, saliva and urine. *Life Sci.*, **37**, 489–495, 1985.

5.13

Chemopreventive Activity of Heat-processed Ginseng

Young-Sam Keum, Seong Su Han, Jung-Hwan Kim, Jeong-Hill Park and Young-Joon Surh*

COLLEGE OF PHARMACY, SEOUL NATIONAL UNIVERSITY, SEOUL 151–742, SOUTH KOREA

1 Introduction

Considerable attention has recently been focussed on dietary or medicinal phytochemicals that possess cancer chemopreventive properties. *Panax ginseng* C.A. Meyer has long been used in traditional oriental medicine. According to recent epidemiologic studies conducted in Korea, ginseng consumption reduced the risk of cancers of stomach, esophagus, colon, and lung.[1,2] A wide array of ginsenosides have been purified and many of them have been tested for their anticarcinogenic or antigenotoxic potential.[3–6]

Heat-treatment of ginseng at temperature higher than that applied to the conventional preparation of red ginseng substantially increases the antioxidant and other biological activities.[7] In the light of the close association between oxidative tissue damage and tumor promotion, we have evaluated anti-tumor promotional as well as antioxidant activities of heat-processed ginseng.

2 Inhibition of TPA-stimulated Superoxide Production in Human Promyelocytic Leukemia (HL-60) Cells

As an initial approach to determine the anti-tumor promotional activity of heat-processed ginseng, its effect on the TPA-stimulated production of superoxide in HL-60 cells was examined. As illustrated in Figure 1, heat-processed ginseng displayed a greater inhibitory effect on TPA-stimulated production of superoxide than did the white ginseng in differentiated HL-60 cells. The anti-tumor promotional activity of heat-processed ginseng was further verified by using a two stage mouse skin carcinogenesis model. Thus, the methanol extract of heat-

* Corresponding author: surh@plaza.snu.ac.kr

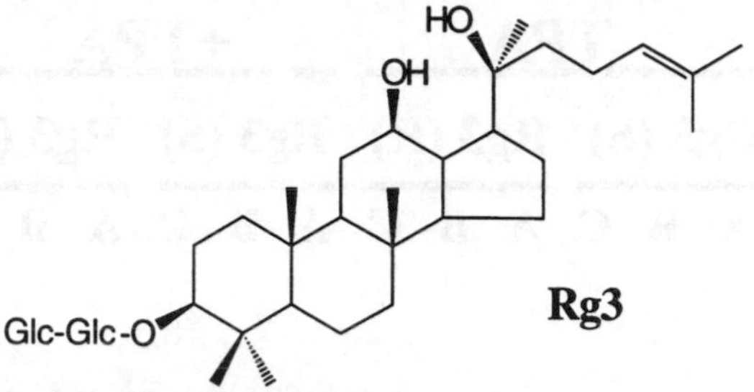

Figure 1 *Structure of Rg_3 derived from heat-processed ginseng*

treated ginseng markedly attenuated promotion of skin papillomagenesis in female ICR mice and also inhibited TPA-induced activation of epidermal ornithine decarboxylase and its mRNA expression.[8]

3 Suppression of TPA-induced Activation of the Nuclear Factor Kappa-B (NF-κB)

NF-κB is a ubiquitous eukaryotic transcription factor that is known to regulate expression of genes encoding proinflammatory cytokines, cell adhesion molecules, growth factors, *etc.* NF-κB is normally present in the cytoplasm of eukaryotic cells as an inactive complex with the inhibitory protein, I-κB. When

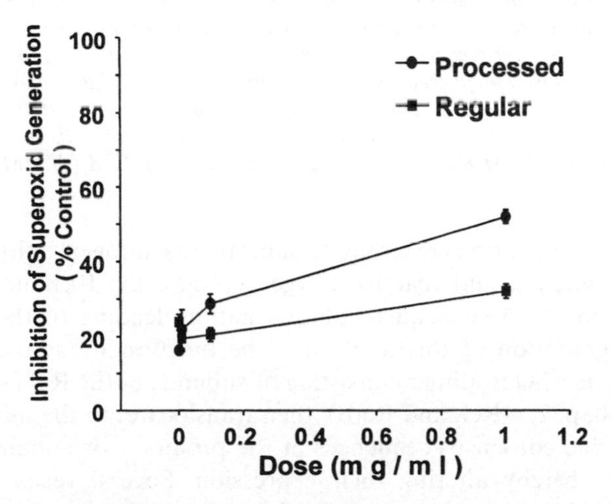

Figure 2 *Comparison of inhibitory effects of heat-processed and untreated ginseng on TPA-stimulated superoxide production in differentiated HL-60 cells. HL-60 cells differentiated by incubation with 1.3% DMSO at 37 °C for 6 days were preincubated for 10 min with the indicated amounts of the methanol extract of heat-processed (circle) or untreated regular (square) ginseng. After addition of TPA (8 μM) and cyt. C (60 μM), the mixtures were further incubated for 30 min and superoxide generation was determined by measuring the reduced cyt. C at 550 nm*

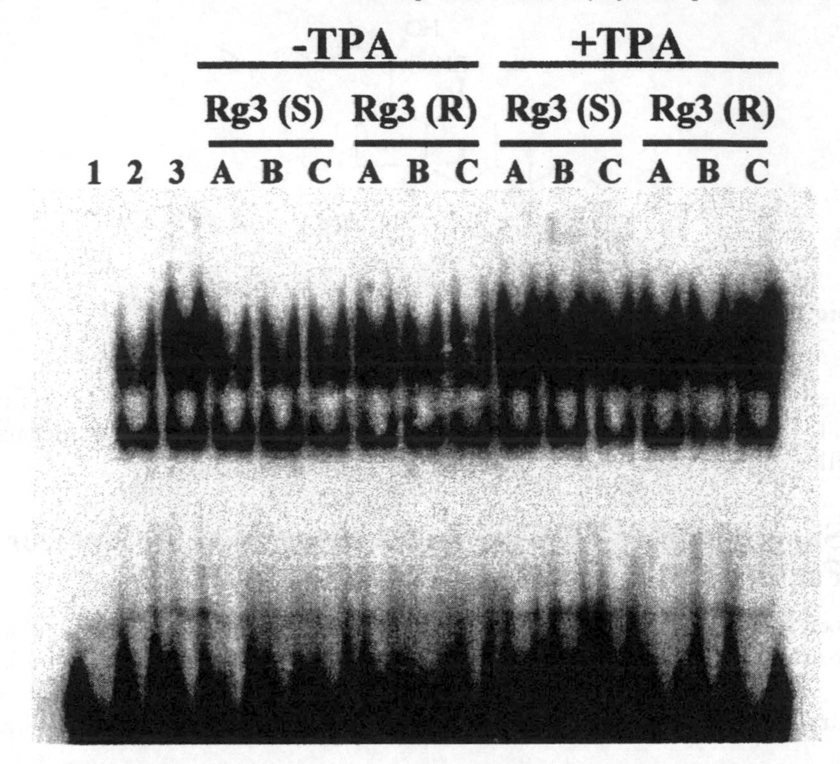

Figure 3 *Effects of enantiomers of Rg_3 on TPA-induced NF-κB activation in HL-60 cells. Cells were treated with each enantiomer 0.5 h prior to 10 μM TPA. Nuclear extractions were carried out 1 h after the incubation at 37 °C and NF-κB activation was measured by electrophoretic mobility shift assay (EMSA). Lane 1: probe only; lane 2, DMSO-treated control; lane 3, treated with TPA alone. Lanes A,B, and C indicate nuclear extracts from HL-60 cells treated with 0.1 μM, 1 μM and 10 μM Rg_3 in the absence or presence of TPA (10 nM)*

cells are exposed to various external stimuli, such as mitogens, ultraviolet, TNF-α, radiation, viruses, and reactive oxygen species, the I-κB undergoes rapid phosphorylation with subsequent ubiquitination, leading to the proteosome-mediated degradation of this inhibitor. The functionally active NF-κB, that exists mainly as a heterodimer consisting of subunits of the Rel family (*e.g.*, Rel A or p65, p50, p52, c-Rel, and RelB), then translocates to the nucleus, where it binds to specific consensus sequences in the promoter or enhancer regions of target genes, thereby altering their expression. Several recent reports have demonstrated the elevated nuclear Rel/NF-κB activity in a variety of tumor cells,[9,10] which implies the involvement of this ubiquitous transcription factor in cell proliferation, malignant transformation, and tumor development. In line with this notion, constitutive NF-κB p50-RelA prevented Hodgkin's lymphoma cells from undergoing apoptotic death.[11] Conversely, suppression of constitutive expression of NF-κB/Rel in B cell lymphoma led to induction of apoptosis.[12] Based on these findings and other lines of evidence, it is conceivable that

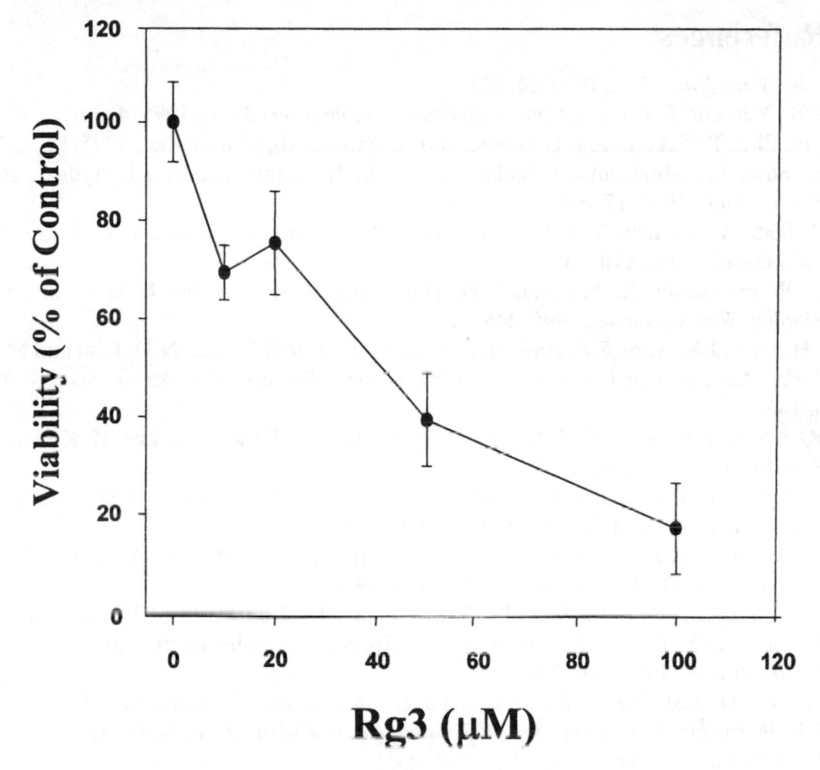

Figure 4 *Effect of Rg_3 on viability of HL-60 cells. Cells were exposed to indicated amounts of the compound for 6 h. Cell viability was determined by the MTT dye reduction assay*

NF-κB alters the expression of genes that contribute to the tumor cell survival, possibly by blocking apoptosis.[10,13] In the present study, we found that Rg_3 (structure shown in Figure 2), an active ingredient of the heat-processed ginseng, inhibited the TPA-induced activation of NF-κB in HL-60 cells as determined by the electrophoretic mobility shift assay (EMSA) using the nuclear extract from these cells (Figure 3). There were no striking differences between (*R*)- and (*S*)-enantiomers of Rg_3 in terms of suppressing TPA-stimulated NF-κB activation. In another experiment, Rg_3 reduced the viability of cultured HL-60 cells (Figure 4). Further investigation should follow to determine whether the observed anti-proliferative activity of this ginsenoside is mediated via induction of apoptosis.

Acknowledgements

This study was supported by a grant (1998 and 1999) from the Korean Research Foundation through the Research Institute of Pharmaceutical Sciences and also in part by a grant (1998) from the Korean Society of Ginseng.

4 References

1 T.K. Yun, *Nutr. Rev.*, 1996, **54**, S71.
2 T.K. Yun and S.Y. Choi, *Cancer Epidemiol. Biomarkers Prev.*, 1995, **4**, 401.
3 J.H. Zhu, T. Takeshita, I. Kitagawa and K. Morimoto, *Cancer Res.*, 1995, **55**, 1221.
4 K. Sato, M. Mochizuki, I. Saiki, Y.C. Yoo, K. Samukawa and I. Azuma, *Biol. Pharm. Bull.*, 1994, **17**, 635.
5 T. Tode, Y. Kikuchi, T. Kita, E. Hirata, E. Imaizumi and I. Nagata, *J. Cancer Res. Clin. Oncol.*, 1993, **120**, 24.
6 C. Wakayabashi, K. Murakami, H. Hasegawa, J. Murata and I. Saiki, *Biochem. Biophys. Res. Commun.*, 1998, **246**, 725.
7 J.H. Park, J.M. Kim, S.B. Han, N.Y. Kim, Y.-J. Surh, S.K. Lee, N.D. Kim and M.K. Park, 'Advances in Ginseng Research', Korean Society of Ginseng, Seoul, 1998, p. 146.
8 Y.-S. Keum, K.-K. Park, J.-M. Lee, K.-S. Chun, J.H. Park, S.-K. Lee, H. Kwon and Y.-J. Surh, *Cancer Lett.*, in press.
9 M.A. Sovak, R.E. Bellas, D.W. Kim, G.J. Zanieski, A.E. Rogers, A.M. Traish and G.E. Sonenshein, *J. Clin. Invest.*, 1997, **100**, 2952.
10 M.A. Sovak, R.E. Bellas, D.W. Kim, G.J. Zanieski, A.E. Rogers, A.M. Traish and G.E. Sonenshein, *J. Clin. Invest.*, 1997, **100**, 2952.
11 R.C. Bargou, F. Emmerich, D. Krappmann, K. Bommert, M.Y. Mapara, W. Arnold, H.D. Royer, E. Grinstein, A. Greiner, C. Scheidereit and B. Dorken, *J. Clin. Invest.*, 1997, **100**, 2961.
12 M. Wu, H. Lee, R.E. Bellas, S.L. Schauer, M.. Arsura, D. Katz, M.J. FitzGerald, T.L. Rothstein, D.H. Sherr and G.E. Sonenshein, *EMBO J.*, 1996, **15**, 4682.
13 T.J. Gilmore, *J. Clin. Invest.*, 1997, **100**, 2935.

5.14

Minimal Processing of Fruit and Vegetables: Influence on Concentration and Activity of Some Naturally Occurring Antioxidants in Orange Derivatives

M.C. Nicoli,[1] A. Piga,[1] V. Vacca,[1] F. Gambella,[1] S. D'Aquino[2] and M. Agabbio[1]

[1] DIPARTIMENTO DI SCIENZE AMBIENTALI AGRARIE E BIOTECNOLOGIE AGRO-ALIMENTARI, UNIVERSITÀ DEGLI STUDI DI SASSARI, VIALE ITALIA 39, 07100 SASSARI, ITALY
[2] ISTITUTO PER LA FISIOLOGIA DELLA MATURAZIONE E DELLA CONSERVAZIONE DEL FRUTTO DELLE SPECIE ARBOREE MEDITERRANEE, CNR, VIA DEI MILLE 48, 07100 SASSARI, ITALY

1 Introduction

Nutritional factors are widely proved to be critical for human health. Overwhelming, evidence from epidemiological studies showed that diets rich in fruit and vegetables are associated with a reduced risk of degenerative diseases.[1] This is attributed to the fact that these foods may provide an optimal mix of phytochemicals, such as antioxidants and their precursors.[2] However, it is widely recognised that the health promoting capacity of fruit and vegetables strictly depends on their technological history. Processing is expected to affect content, activity and bio-availability of naturally occurring antioxidants. Although some experimental evidence has recently demonstrated that processing may have many effects, not always resulting in a loss of the health promoting capacity of fruit and vegetables, uncertainty still exists about the effective incidence of the various technological steps.[3] This aspect, which is generally neglected or scarcely considered in present nutritional and epidemiological studies, is of great importance, considering that only a small amount of fruit and vegetables are consumed as fresh, whilst most of them need to be processed for safety, quality and economic reasons.[4] Thus, investigation on the effects of processing on the activity of naturally occurring antioxidants is a key factor in order to find out the best technological conditions for preserving the above cited beneficial properties and to achieve a correct interpretation of data on dietary habits and human health.

In the present investigation fresh and pasteurised orange juices, chosen by virtue of their high content in naturally occurring antioxidants and their widespread consumption, were considered. The changes in ascorbic acid concentration and in the overall antioxidant properties during juice preparation and storage, the latter carried out under different temperature conditions, were studied.

2 Materials and Methods

The study was carried out using fresh Italian oranges (*Citrus sinensis* L. cv. Salustiana) which were squeezed and immediately bottled in screw capped flasks in presence of air. Samples of fresh orange juice were than stored under refrigerated conditions (4 °C). Additional samples were subjected to pasteurisation in a water-bath at 90 °C for 9 minutes and than stored at 20 °C. Ascorbic acid was measured following the AOAC methodology.[5] The antioxidant activity of the aqueous phase of the samples was assessed following the bleaching rate of a stable polar radical $DPPH^{\bullet}$, as previously described.[6] The redox potential was measured using a platinum indicating electrode and a Ag/AgCl, Cl^{-}_{sat} reference electrode connected with a voltmeter. Total polyphenol content was assessed following Folin-Ciocalteu methodology.

3 Results and Discussion

Table 1 shows changes in ascorbic acid concentration measured in fresh orange juice samples during chilled storage. Although samples were bottled in presence of air, a moderate loss in ascorbic acid which did not exceed 12% was detected up to 15 day storage.

In Figure 1 the changes in the overall antioxidant capacity of the aqueous phase of the fresh orange juice as affected by storage is reported. It is interesting to observe that despite the slight reduction in ascorbic acid concentration, an

Table 1 *Changes in ascorbic acid concentration (expressed as mg g^{-1} of dry matter) of fresh orange juice during storage at 4 °C*

Storage time (days)	*Ascorbic acid (mg g_{dm}^{-1})*
0	5.03 ± 0.07
1	4.7 1 ± 0.03
4	4.80 ± 0.03
7	4.50 ± 0.02
11	4.52 ± 0.02
15	4.44 ± 0.03

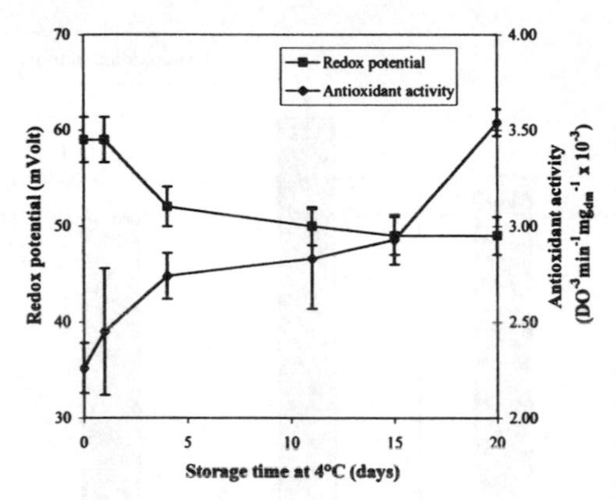

Figure 1 *Evolution of the antioxidant capacity and of the redox potential of fresh orange juice during chilled storage*

increase in the overall antioxidant properties of the product was detected. These results were also confirmed by the progressive reduction of the redox potential, indicating an increase in the reducing properties of the product.

Although ascorbic acid is the predominant identifiable antioxidant in orange juice, it is likely that other antioxidants such as polyphenols may play a significant contributory role. In particular, it has been recently observed that polyphenols constituents, by virtue of their antioxidant capacity, exhibit an ascorbate-sparing effect.[7] The progressive oxidation of polyphenols, whose concentration in the samples was found to be of 8.2 mg g^{-1} of dry matter, can explain the increase in the antioxidant properties of the fresh orange juice within 15 days of cold storage.

In fact, it has been recently stated that some polyphenols can exhibit, for intermediate oxidation level, higher radical scavenging efficiency than the non-oxidised ones. However, as the oxidation proceeds to the final stage, a progressive further loss in the antioxidant properties has been detected, due to the decreased ability of polyphenols, in their polymeric oxidised structure, to donate a hydrogen atom and/or to support an unpaired electron.[8–10] The changes in the antioxidant properties as well as in ascorbic acid concentration of orange juice samples subjected to pasteurisation and further storage at room temperature are shown in Figure 2.

It can be noted that, despite the dramatic decrease in ascorbic acid concentration, a moderate reduction in the antioxidant capacity, which did not exceed 30%, was detected. Results suggest that different and opposite events could be involved in determining the observed moderate changes in the antioxidant activity of the pasteurised orange juice. Considering the quantitative predominant role of ascorbic acid in determining the overall antioxidant capacity of the orange juice, it is likely that the expected dramatic loss in the antioxidant capacity, due to ascorbic acid thermal degradation, could be counterbalanced by the increased antioxidant ability of some partially oxidised polyphenol.

In conclusion, the changes in the antioxidant properties of orange juice

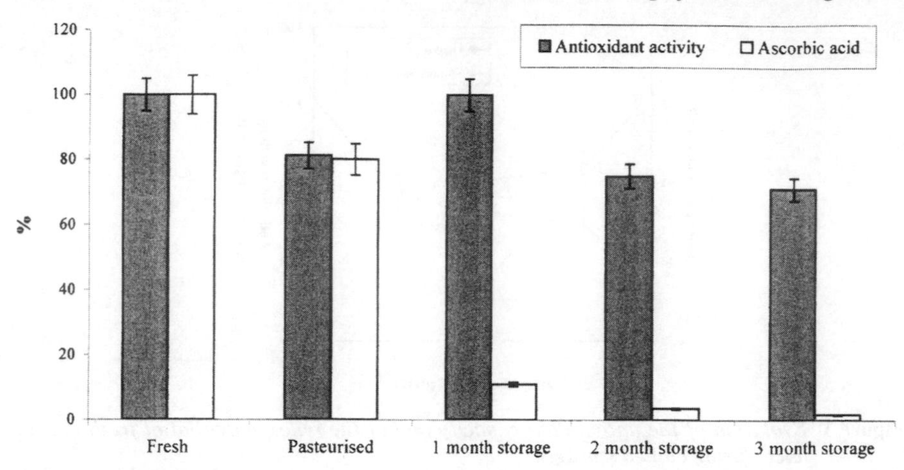

Figure 2 *Changes in ascorbic acid concentration and evolution of the antioxidant capacity of orange juice samples as affected by pasteurisation and storage. Data are expressed as a percentage referred to the fresh orange juice*

subjected to different technological conditions were found to be scarcely related to the content in ascorbic acid. Results would confirm the ascorbate-sparing effect of other minor naturally occurring antioxidants and suggest the role of these compounds in the maintenance of high values of antioxidant capacity, even after intense technological treatments. According to recent literature data, these effects could be mainly attributed to the polyphenol constituents.

4 References

1 World Cancer Research Fund and American Institute for Cancer Research, 'Food Nutrition and the Prevention of Cancer: a Global Perspective', 1997.

2 S. Southon, 'European Research towards Safer and Better Foods', Druckerei Gasser, Karlsrhue, 1998, p. 159.

3 M.C. Nicoli, M., Anese, M. Parpinel, S. Franceschi and C.R. Lerici, *Cancer Lett.*, 1997, **114**, 71.

4 H. Greenfield and D.A.T. Southgate, 'Food Composition Data. Production, Management and Use', Elsevier Applied Science, 1992.

5 AOAC, 'Official Methods of Analysis', Association of Official Analytical Chemists International, 1990.

6 L. Manzocco, M. Anese and M.C. Nicoli, *Lebensm.-Wiss. Technol.*, 1998, **31**, 694.

7 N.J. Miller and C.A. Rice-Evans, *Food Chem.*, 1997, **60**, 331.

8 H.S. Cheigh, S.H. Um and C.Y. Lee, *ACS Sym. Series*, 1995, No. 600, p. 200.

9 N. Saint-Cricq de Gaulejac, C. Provost and N.J. Vivas, *J. Agric. Food Chem.*, 1999, **47**, 425.

10 L. Manzocco, S. Calligaris, D. Mastrocola and M.C. Nicoli, this volume.

Section 6

Defence Systems: Cellular Differentiation and Apoptosis

6.1

Diet/Gene Interactions: Food and Botanical Sources of Cyclooxygenase Inhibitors for the Prevention of Colon Cancer

Michael J. Wargovich

DEPARTMENT OF PATHOLOGY, UNIVERSITY OF SOUTH CAROLINA SCHOOL OF MEDICINE AND SOUTH CAROLINA CANCER CENTER, SEVEN RICHLAND MEDICAL PARK DRIVE, COLUMBIA, SC 29203, USA

1 Introduction

By some estimates at least half of the pharmaceuticals used in the treatment of cancer are derived of plant origin. As we understand with greater precision the molecular events underlying the causation of colon cancer, the possibility of targeting critical pathways using botanically-derived cancer preventive agents becomes more realistic. One important pathway in colon cancer is the process of inflammation. Anti-inflammatory drugs show strong chemopreventive activity in animal models. Some plants are excellent sources of anti-inflammatory compounds. Such phytochemicals may be natural 'NSAIDs' (non-steroidal anti-inflammatory drugs), acting like aspirin and related drugs for the prevention of colon cancer and other solid tumors. A number of recent case-control studies have suggested that daily or 'continual' use of aspirin substantially reduces the risk of mortality from colon cancer, which by some estimates may be 50% lower than risk for colon cancer in the general population. Several large clinical trials are now in progress in the United States and Europe to assess the chemopreventive efficacy of aspirin and other NSAIDs in the prevention of premalignant adenomas of the colon.[1] Suppression of recurrence of colonic adenomas will strongly support the use of NSAIDs in the prevention of colon cancer. However, the use of NSAIDs in the general public is attenuated by possible serious side effects including enhanced risk for ulcers in the intestine.[2] Current consensus reserves the broad use of NSAIDs as colon preventives only for the highest risk populations. Thus, it seems reasonable and timely to conduct a search for safer and more tolerable NSAID-like compounds. This search extends to natural products in foods, herbals, and botanicals as sources of anti-inflammatory compounds.

2 Cyclooxygenases, Prostaglandins, and Colon Cancer

Cyclooxygenase is the enzyme responsible for the conversion of dietary arachidonic acid to a number of bioactive molecules including prostaglandins, thromboxanes, and leukotrienes. These molecules have wide-ranging effects in the body including the regulation of inflammation, renal blood flow, and cytoprotection of the stomach lining.[3]

Two isoforms of cyclooxygenase exist, encoding separate genes. Cyclooxygenase-1 (COX 1) is the constitutive form, which is expressed in many cell types and is thought to function in normal maintenance functions such as platelet aggregation and gastric cytoprotection. Cyclooxygenase-2 (COX-2) is the inducible form, which is thought to be proinflammatory, and is stimulated by mitogens and tumor promoters.[4] COX-2 is, however, expressed in kidney, brain, and pancreas.[5] The intense interest in the COX proteins stems from a rash of recent epidemiologic studies associating a reduction in risk for colorectal cancer in populations, who for other medical reasons, use aspirin or other non-steroidal anti-inflammatory drugs (NSAIDs) on a regular basis. Table 1 lists the current epidemiologic evidence for protection against colorectal cancer by NSAID use.[6–16]

The population-based evidence is overwhelmingly one-sided; consumption of aspirin on a continual basis (daily or weekly) diminishes colon cancer risk by approximately 50%. Very recent studies suggest that non-aspirin NSAIDs also confer protection and these early conclusions may soon be confirmed as the

Table 1 *Epidemiology of NSAID use and risk for colon cancer*

Investigator	*Location*	*Type of study*	*NSAID use*	*Result RR + 95% CI*
Giovannucci	USA	Case-control	Aspirin > 20y	0.56 (0.36–0.90)
Gridley	Sweden	Cohort	Aspirin	0.68 (0.6–0.98)
Kune	Australia	Case-control	Aspirin	0.53 (0.35–0.80)
Peleg	USA	Case-control	Aspirin	0.52 (0.30–0.91)
Reeves	USA	Case-control	Non-aspirin NSAIDS	0.43 (0.20–0.88)
Rosenberg	USA	Case-control	Aspirin	0.70 (0.5–0.8)
Schreinemachers	USA	Cohort	Aspirin & Other NSAIDS	0.35 (0.17–0.73)
Smalley	USA	Cohort	Non-aspirin NSAIDS	0.61 (0.48–0.77)
Suh	USA	Cohort	Aspirin	0.44 (0.18–1.10)
Sturmer	USA	Clinical Trial	Aspirin	1.07 (0.75–1.53)
Thun	USA	Cohort	Aspirin	0.60 (0.4–0.89)

length of time people are exposed to non-aspirin NSAIDs increases.[13] There are, however, risks associated with long-term use of aspirin and other NSAIDs. These toxicities revolve around the housekeeping functions of COX-1, *i.e.*, renal function, cytoprotection of the gastric lining, and platelet aggregation. Overuse, or sensitivity to aspirin and other NSAIDS manifests itself in renal impairment, ulceration of the stomach lining, or bleeding.[17] The advent of COX-2 selective drugs may largely reduce the side effects of NSAID use, but at the current time highly specific COX-2 drugs with absolute evidence of a lack of side effects are not yet on the market.[5,18]

From the mechanistic point of view a conundrum exists concerning the fact that most known NSAIDs with side effects are potent inhibitors of colon cancer in animal models. For several years our laboratory has worked with the National Cancer Institute to screen potential chemopreventives for colon cancer. Nearly every over the counter or prescription NSAID we have tested has been effective in inhibiting the development of aberrant crypt foci, early progenitor lesions for colon cancer in the rat colon.[19] These include aspirin, ibuprofen, ketoprofen, and naproxen, all available in the USA over the counter. Also active in preventing colon cancer are the more powerful NSAIDs, flurbiprofen, sulindac, indomethacin, and piroxicam. Our data allow us to conclude that chemopreventive efficacy is highly related to propensity for the NSAID to be ulcerogenic. Given that NSAIDs do have risk associated with their continued use, we pose the question of whether herbals or botanicals exist that have NSAID-like activity while hopefully showing a more promising safety profile.

3 Herbal and Botanical Sources of Anti-inflammatory Cancer Chemopreventive Agents

It is not surprising that anti-inflammatory compounds exist in plants. Aspirin itself is a synthetic derivative of a plant compound, salicylic acid, known to provide analgesic relief.[20] Herbs and botanicals are sources of some exceptionally strong anti-inflammatory compounds and many of these have proven to inhibit tumorigenesis in animal models. The polyphenolic antioxidants in tea inhibit several forms of cancer in animals and appear to be especially active toward tumors of cutaneous origin, notably skin cancer.[21] The tea polyphenols appear to have a unique niche in the chemoprevention world, as these compounds appear to be inhibitory at concentrations that are found in tea infusions drunk by Asian populations that are at reduced risk for cancer[22]. Other anti-inflammatory compounds in botanicals with reported cancer chemopreventive activity include: resveratrol from grape seed extract, curcumin from the spice turmeric, ursolic acid in the spice rosemary, several organosulfur compounds in garlic, and the onion flavonoid quercetin.[23,24,25] In our laboratory we have examined several of these botanical compounds for efficacy in inhibiting aberrant crypt foci, progenitor lesions for colon cancer in the rat colon.[26,27] As shown in Figure 1, supplementation of a semisynthetic

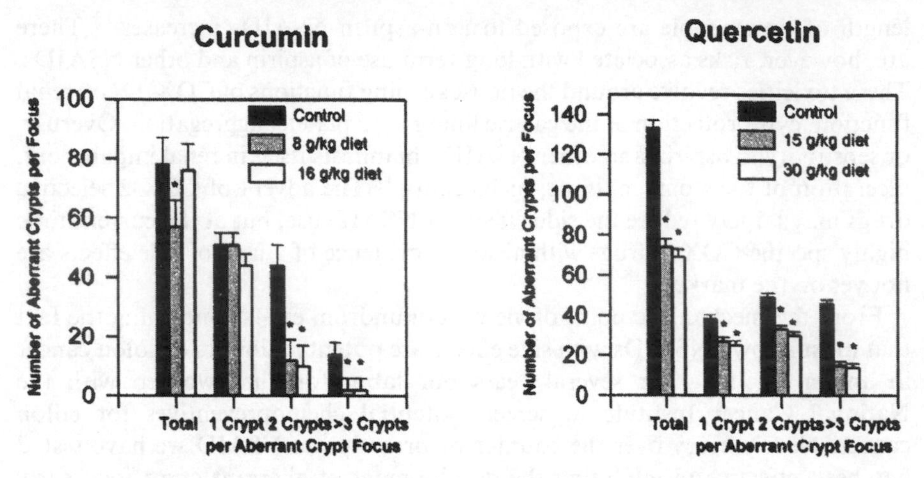

Figure 1 *Curcumin and quercetin inhibit azoxymethane-induced aberrant crypt foci in the rat colon. Either compound was fed in the diet at the doses indicated for 4 weeks. Black bars represent values for rats with no test agent in the diet*

diet with graded doses of either curcumin or quercetin reduced aberrant crypt formation in the rat colon. Quercetin appears to be the more potent of the two compounds in reducing the total incidence of aberrant crypt foci compared with controls receiving neither compound in the diet. Both agents seem to be effective in reducing the number of foci containing three or more aberrant crypts. Foci with multiple crypts are more likely to progress to colonic adenomas and cancers.[28] Other botanicals we have tested in this assay are also moderate to strong inhibitors of colonic aberrant crypt foci. These include resveratrol, the tea catechin, epigallocatechin gallate, rutin from onions, dialkyl sulfides from garlic, and ferulic acid, found in many plant foods. Some botanicals were also found to be ineffective in the assay including silymarin from milk thistle, esculetin, and nerolidol. Many of the botanicals demonstrating strong inhibitory effect toward aberrant crypt formation also have been shown to inhibit tumorigenesis in the full-length carcinogenesis protocols.

Research that remains to be done includes determining whether herbal and botanical compounds with NSAID-like activity have effects on specific COX isoforms. Prostaglandin-independent mechanisms may also be involved. Several NSAID drugs are also ligands for peroxisome-proliferator activated receptors (PPARs). PPARs are a group of nuclear receptor protein transcription factors that, when activated, induce genes involved in differentiation.[29] PPAR γ, the predominant form found in the colon, is responsive to certain NSAIDs. The unanswered question is whether certain herbals or botanicals function directly as ligands for PPARs, or whether reduction in the production of certain prostanoids (also ligands for PPARs) influences this transcription signaling system.

4 Summary

It is clear that some herbals and botanicals contain anti-inflammatory compounds that potentially will prevent colon cancer, if prostaglandins are important mediators of colon tumor growth and progression. Important future research should focus on the identity of phytochemical NSAIDs and determine both their relative potency and specificity for the COX-1 and COX-2 enzymes. Further research is needed to define if PPAR activation represents a second pathway (that may not be totally independent of prostanoids) by which such natural compounds inhibit colon cancer.

5 References

1 E. R. Greenberg, J. A. Baron, D. H. Freeman, Jr., J. S. Mandel and R. Haile, *JNCI*, 1993, **85**, 912.
2 D. Y. Graham and J. L. Smith, *Ann. Intern. Med.*, 1986, **104**, 390.
3 L. J. Marnett, *Cancer Res.*, 1992, **52**, 5575.
4 R. N. DuBois and W. E. Smalley, *J. Gastroenterol.*, 1996, **31**, 898.
5 L. Schachna and P. F. Ryan, *Med. J. Aust.*, 1999, **171**, 175.
6 E. Giovannucci, K. M. Egan, D. J. Hunter, M. J. Stampfer, G. A. Colditz, W. C. Willett and F. E. Speizer, *N. Engl. J. Med.*, 1995, **333**, 609.
7 G. Gridley, J. K. McLaughlin, A. Ekbom, L. Klareskog, H. O. Adami, D. G. Hacker, R. Hoover and J. F. Fraumeni, Jr., *JNCI*, 1993, **85**, 307.
8 G. A. Kune, S. Kune and L. F. Watson, *Cancer Res.*, 1988, **48**, 4399.
9 I. I. Peleg, H. T. Maibach, S. H. Brown and C. M. Wilcox, *Arch. Intern. Med.*, 1994, **154**, 394.
10 M. J. Reeves, P. A. Newcomb, A. Trentham-Dietz, B. E. Storer and P. L. Remington, *Cancer Epi. Biomarkers Prev.*, 1996, **5**, 955.
11 L. Rosenberg, C. Louik and S. Shapiro, *Cancer*, 1998, **82**, 2326.
12 D. M. Schreinemachers and R. B. Everson, *Epidemiol.*, 1994, **5**, 138.
13 W. Smalley, W. A. Ray, J. Daugherty and M. R. Griffin, *Arch. Intern. Med.*, 1999, **159**, 161.
14 O. Suh, C. Mettlin and N. J. Petrelli, *Cancer*, 1993, **72**, 1171.
15 T. Sturmer, R. J. Glynn, I. M. Lee, J. E. Manson, J. E. Buring and C. H. Hennekens, *Ann. Intern. Med.*, 1998, **128**, 713.
16 M. J. Thun, E. E. Calle, M. M. Namboodiri, W. D. Flanders, R. J. Coates, T. Byers, P. Boffetta, L. Garfinkel and C. W. Heath, Jr., *JNCI*, 1992, **84**, 1491.
17 C. Gustafson-Svard, I. Lilja, O. Hallbook and R. Sjodahl, *Ann. Med.*, 1997, **29**, 247.
18 K. M. Sheehan, K. Sheahan, D. P. O'Donoghue, F. MacSweeney, R. M. Conroy, D. J. Fitzgerald and F. E. Murray, *JAMA*, 1999, **282**, 1254.
19 M. J. Wargovich, C. D. Chen, C. Harris, E. Yang and M. Velasco, *Int. J. Cancer*, 1994, **59**, 457.
20 G. Weissman, *Sci. Am.*, 1991, **264**, 84.
21 L. Kohlmeier, K. G. Weterings, S. Steck and F. J. Kok, *Nutr. Cancer*, 1997, **27**, 1.
22 I. E. Dreosti, M. J. Wargovich and C. S. Yang, *Crit. Rev. Food Sci. Nutr.*, 1997, **37**, 761.
23 M. Jang, L. Cai, G. O. Udeani, K. V. Slowing, C. F. Thomas, C. W. Beecher, H. H. Fong, N. R. Farnsworth, A. D. Kinghorn, R. G. Mehta, R. C. Moon and J. M. Pezzuto, *Science*, 1997, **275**, 218.

24 M. Nagabhushan and S. V. Bhide, *J. Am. Coll. Nutr.*, 1992, **11**, 192.
25 A. Najid, A. Simon, J. Cook, H. C. Rabinovitch, C. Delage, A. J. Chulia and M. Rigaud, *FEBS Lett.*, 1992, **299**, 213.
26 E. E. Deschner, J. F. Ruperto, G. Y. Wong and H. L. Newmark, *Nutr. Cancer*, 1993, **20**, 199.
27 M. J. Wargovich, 'Cancer chemoprevention', CRC Press, 1992, p. 195.
28 T. Takayama, S. Katsuki, Y. Takahashi, M. Ohi, S. Nojiri, S. Sakamaki, J. Kato, K. Kogawa, H. Miyake and Y. Niitsu. *N. Engl. J. Med.*, 1998, **339**, 1277.
29 L. Gelman, J. C. Fruchart and J. Auwerx, *Cell Mol. Life Sci.*, 1999, **55**, 932.

6.2

Cell Proliferation and Apoptosis in the Liver of Mice with Different Genetic Susceptibility to Tumor Induction and Its Control by Food and TGF-β1

M. Chabicovsky, U. Wastl, K. Hufnagl, W. Bursch and R. Schulte-Hermann

INSTITUTE OF TUMOR BIOLOGY AND CANCER RESEARCH, UNIVERSITY OF VIENNA, VIENNA, AUSTRIA

1 Introduction

Mouse liver tumor response is a routinely used endpoint in studies on the carcinogenicity of numerous chemicals. However, the evaluation of data for risk assessment is complicated by the different susceptibilities of mouse strains in developing spontaneous and chemically induced liver tumors: sensitive strains like C3H/He and B6C3F1 are characterized by high incidences of spontaneous liver tumors (60% in male C3H/He, 30–50% in male B6C3F1 at 70 weeks of age), while less than 1% if two years old C57Bl/6J mice are affected.[1, 2] Exploring the mechanisms causing this strain difference may provide opportunities to elucidate the genetic and cellular basis of predisposition to carcinogenesis. In this respect we looked for the cell proliferation and death rate of hepatocytes in normal and regressing hyperplastic livers as well as on its control by food and TGF-β1 (Transforming Growth Factor beta 1).

2 Material and Methods

Male C57Bl/6J and C3H/He mice were housed individually under standardized conditions (controlled temperature and humidity, 12 h light from 10 p.m. to 10 a.m., food and tap water *ad libitum*). From 14 weeks of age mice were continuously fed a diet containing an adjusted concentration (0.05–0.07%) of PB (5-ethyl-5-phenylbarbituric acid) to allow a daily intake of approximately 90 mg kg^{-1} of body weight for 7 days. Subsequently some mice received a standard diet for 14 days. In addition to this *ad libitum* fed withdrawal group, other mice were subjected to 40% food restriction (60% of the *ad libitum* intake) upon cessation of PB treatment and were sacrificed 2, 4, 7 and 14 days after onset of food restriction. Other PB treated mice received a single intravenous injection of mature TGF-β1, kindly provided by Oncogene (Seattle, WA) at a dose of 56 μg TGF-β1 kg^{-1} of body weight, 45 hours after cessation of dietary PB. Mice were sacrificed either 3 hours or 51 hours after the injection.

At sacrifice the liver was removed, weighed, fixed in Carnoy's solution, embedded in paraffin, and 5 μm sections were cut and stained with hematoxilin and eosin. Apoptotic cells were detected by morphological features: condensation of cytoplasm and chromatin at the nuclear membrane, and fragmentation of nucleus and cytoplasm.

3 Results

The strain dependent effect of PB treatment on the increase in relative liver weight (*i.e.* liver to body weight ratio) tendentiously correlates with the different susceptibility to the tumor promotion by PB: in the susceptible C3H/He strain the relative liver weight was 49% higher than that of controls ($p < 0.01$) as compared to the resistant C57Bl/6J strain, where it was only 37% ($p < 0.01$) increased (Table 1).

Cessation of PB treatment led to a rapid regression of liver hypertrophy: liver mass, hepatocyte volume and hepatic protein content returned to control levels (data not shown). However, apoptotic incidences were not significantly increased in either of the strains (Table 1).

Table 1 *Tumor incidence, liver growth and apoptosis in C3H/He and C57Bl/6J mice*

			C3H/He	*C57Bl/6J*
A	tumor incidence NDEA = > PB[a]		high	low
	PB induced liver growth (relative liver weight)[b]		49%	37%
B	apoptosis in livers of untreated mice		0	0
	apoptosis in regressing livers		0	0.16 ± 0.24
	apoptosis induced by food restriction:	2 days	0.08 ± 0.16	0.07 ± 0.10
		4 days	0.07 ± 0.12	0
		7 days	0.03 ± 0.05	0
		14 days	0	0.07 ± 0.14
	apoptosis induced by TGF-β1:	3 h after TGF-β1	0.02 ± 0.04	0.05 ± 0.08
		51 h after TGF-β1	0	0

Table 1A: Tumor incidence and liver growth induced by PB. [a] At five weeks of age mice received a single i.p. injection of NDEA (*N*-nitrosodiethylamin; 90 mg kg^{-1} of body weight). From 7 weeks mice were treated with dietary PB (daily intake: 90 mg kg^{-1} of body weight) and killed 20, 40, 52, 76, or 92 weeks after initiation with NDEA. Macroscopically visible liver lesions were sampled individually. [b] Values are % changes from the corresponding control value.
1B: Apoptotic incidences in the liver of different mouse strains during regression after PB withdrawal, and induced by food restriction and TGF-β1 injection. Mean and SD of apoptotic incidences (4–8 animals/group; at least 1000 hepatocytes per animal).

Food restriction to 60% of the *ad libitum* diet during the withdrawal period led to a distinct loss of body weight, liver weight and hepatic protein content in both strains, indicating hypotrophy of the liver, which is also demonstrated by increased numbers of hepatocytes per visual field (data not shown). But histological analysis failed to detect any significant increase of apoptotic bodies at all timepoints investigated (Table 1). Also the hepatic DNA content remained at elevated levels similar to the *ad libitum* fed groups (data not shown).

The regressing mouse liver (*i.e.* 45 hours after cessation of PB) did not respond to an i.v. injection of TGF-β1 with changes in relative liver weight, hepatic protein and DNA content (data not shown) or increased incidences of apoptosis (Table 1).

4 Discussion

The present study could establish a strain difference in liver growth after PB treatment: the tumor susceptible strain (C3H/He) responds to the tumor promoter with more extensive growth as compared to the tumor resistant strain (C57Bl/6J). This result correlates positively with the different tumor promoting effect of PB found within these two strains.[2–5] But the PB induced liver hyperplasia did not regress after cessation of the growth stimulus and moreover was not affected by 14 days of 40% food restriction. Thus, an elevated DNA content persisted and was not removed through apoptosis, which is in contrast to observations in rats. In rats liver hyperplasia regresses by an increased apoptotic rate within the first two days after PB withdrawal.[6,7] Similarly, food restricted rats loose liver mass by reduction of hepatic DNA content *via* increased apoptosis.[8] Furthermore, TGF-β1, which has been shown to induce apoptosis in primary rat hepatocytes and in the hyperplastic rat liver by acting synergistically with CPA withdrawal,[9,10] did not result in increased apoptotic incidences in mouse liver.

These differences between rats and mice rises the question of the significance of apoptosis in the development of liver tumors in mice. Observations in rats show that apoptosis acts as a defense mechanism against cancerogenesis by the preferential suicide of preneoplastic cells,[8] and it has to be examined to see whether this is true also for hepatocarcinogenesis in mice.

5 References

1 P. Grasso and J. Hardy, in: P. M. Newberne and W. H. Butler. 'Mouse Hepatic Neoplasia', Elsevier, Amsterdam, 1975, p. 111.
2 F. F. Becker, *Cancer Res.*, 1985, **45**, 768.
3 B. A. Diwan, J. M. Rice, M. Ohshima and J. M. Ward, *Carcinogenesis*, 1986, **7**, 215.
4 A. Diwan, J. M. Rice and J. M. Ward, in: D. E. Stevenson, R. M. McClain, J. A. Popp, T. J. Slaga, J. M. Ward and H. C. Pitot. 'Progress in Clinical and Biological Research. Mouse liver carcinogenesis: mechanisms and species comparison', New York, 1990, Vol. 331, p. 69.

5 J. G. Evans, M. A. Collins, B. G. Lake and W. H. Butler, *Toxicol. Pathol.*, 1992, **20**, 585.
6 W. Bursch, B. Lauer, I. Timmermann-Trosiener, G. Barthel, J. Schuppler and R. Schulte-Hermann, *Carcinogenesis*, 1984, **5**, 453.
7 W. Bursch, B. Düsterberg and R. Schulte-Hermann, *Arch. Toxicol.*, 1986, **59**, 221.
8 B. Grasl-Kraupp, W. Bursch, B. Ruttkay-Nedecky, A. Wagner, B. Lauer and R. Schulte-Hermann, *Proc. Natl. Acad. Sci. USA*, 1994, **91**, 9995.
9 F. Oberhammer, M. Pavelka, S. Sharma, R. Tiefenbacher, A. F. Purchio, W. Bursch and R. Schulte-Hermann, *Proc. Natl. Acad. Sci. USA*, 1992, **89**, 5408.
10 F. Oberhammer, W. Bursch, R. Tiefenbacher, G. Fröschl, M. Pavelka, T. Purchio and R. Schulte-Hermann, *Hepatology*, 1993, **18**, 1238.

6.3

Proliferative Actions of Dietary Lectins on Gastrointestinal Epithelia

Anthony J. FitzGerald,[1,*] Mark Jordinson,[1] Jonathan M. Rhodes[2] and Robert A. Goodlad[3]

[1] DEPARTMENT OF HISTOPATHOLOGY, IMPERIAL COLLEGE SCHOOL OF MEDICINE, DIVISION OF INVESTIGATIVE SCIENCES, HAMMERSMITH HOSPITAL, DU CANE ROAD, LONDON W12 ONN, UK
[2] DEPARTMENT OF MEDICINE, UNIVERSITY OF LIVERPOOL, LIVERPOOL, UK
[3] HISTOPATHOLOGY UNIT, IMPERIAL CANCER RESEARCH FUND, 44 LINCOLN'S INN FIELDS, LONDON, UK

1 Introduction

Lectins are highly specific carbohydrate binding proteins or glycoproteins that are ubiquitous in nature and prevalent in our diets.[1] Many lectins resist digestion and pass through the gastrointestinal tract where they remain active and several studies have demonstrated biological activity in the small intestine[2] and in the colon.[3] The normal human diet contains many sources of lectins which bind to specific carbohydrate motifs on the cell surface.[1,4] Lectins were chosen to represent groups with different carbohydrate binding. Concanavalin A (Con-A), the lectin present in Jack beans, binds to mannose and glucose.[5] Wheat germ agglutinin (WGA) binds to *N*-acetylglucosamine and sialic acid and the lectin present in red kidney beans, phytohemagglutinin, (PHA) bind strongly to complex carbohydrate groups.[6,7] Peanut agglutinin (PNA) binds to Gal β_{1-3}Gal NAc and stimulates proliferation of human colonic epithelium, and colonic carcinoma cell lines.[8–10] ABA (*Agaricus bisporus* agglutinin), from the edible common mushroom, binds to the same cell surface glycoprotein as PNA, but inhibits proliferation (by up to 80% in some cases) in a variety of epithelial cells, such as HT29, Caco-2, MCF-7 and Rama-27 without any signs of cytotoxicity.[11] We therefore tested the hypothesis whether certain lectins are

* Correspondence to: A.Fitzgerald@icrf.icnet.uk

able to reverse the atrophy seen in our model of TPN-fed rats because they diminish proliferation of gastrointestinal epithelia to a basal level, and if they can stimulate proliferation.

2 Materials and Methods

Chemicals were purchased from Sigma (Poole, Dorset) unless otherwise stated. Lectins; Peanut agglutinin was supplied by Prof. J. Rhodes, (University Hospital, Liverpool, UK), concanavalin A, wheatgerm agglutinin from EY Laboratories (Leicestershire UK), phytohemagglutinin was produced by Prof. A. Pusztai (Rowett Research Institute, Bucksburn, Aberdeen, UK).

Preparation of rats. Male rats were anaesthetised with hypnorm and diazepam. A silastic cannula was tied into the right external jugular vein. A second cannula was inserted into the squamous portion of the stomach. Both cannulae were connected through a stainless steel skin button and tethered to a two-channel fluid swivel joint (Linton Instrumentation, Norfolk). The TPN diet was delivered to the rats via the right external jugular vein at a rate of 60 ml/rat/day, giving 1.8 g N, 6.0 g lipid, 8.5 g glucose and 1047 kJ kg^{-1} day^{-1}.[12]

Experimental protocol. Groups of rats (n = 6 per group) received TPN for 2 days alone. Beginning on the third day lectins were infused into the rat stomachs with vehicle (saline 0.5 ml) as a once daily bolus dose of lectin (25 mg day^{-1}). On the final day rats were injected with vincristine sulphate, 1 mg kg^{-1}, intra-peritoneal (David Bull Laboratories, Warwick, UK). One hour later a final dose of lectin was administered. One hour later they were killed with pentobarbitone; blood was then taken by cardiac puncture for measurements of plasma hormones by radioimmunoassay. Samples of the small intestine and colon were fixed in Carnoy's fluid and stored in 70% (v/v) ethanol.

Assessment of proliferation by metaphase arrest. Pieces of fixed tissue were hydrated, hydrolysed in 1 M HCl at 60 °C for 10 minutes and then stained in Schiff's reagent and the crypts displayed by micro-dissection.[13] The numbers of arrested metaphases in 10 small intestinal crypts and 20 colonic crypts were counted in each specimen. The number of cells arrested in metaphase was used in the subsequent analyses.

Statistical analysis. All results are presented as the group mean ± standard error of that mean. Data was tested as appropriate by two-tailed t-test or by analysis of variance.

3 Results

3.1 Effect of Lectins on Organ Weights

The effect of intragastric administration of lectins upon the wet weight of gastrointestinal organs is shown in Figure 1. PHA caused a 27% increase ($P < 0.001$) in the wet weight of the small intestine while both WGA and PNA increased colonic weight by 38% ($P < 0.01$) and 53% ($P < 0.001$) respectively. The wet weight of the pancreas was increased by all of the lectins as follows,

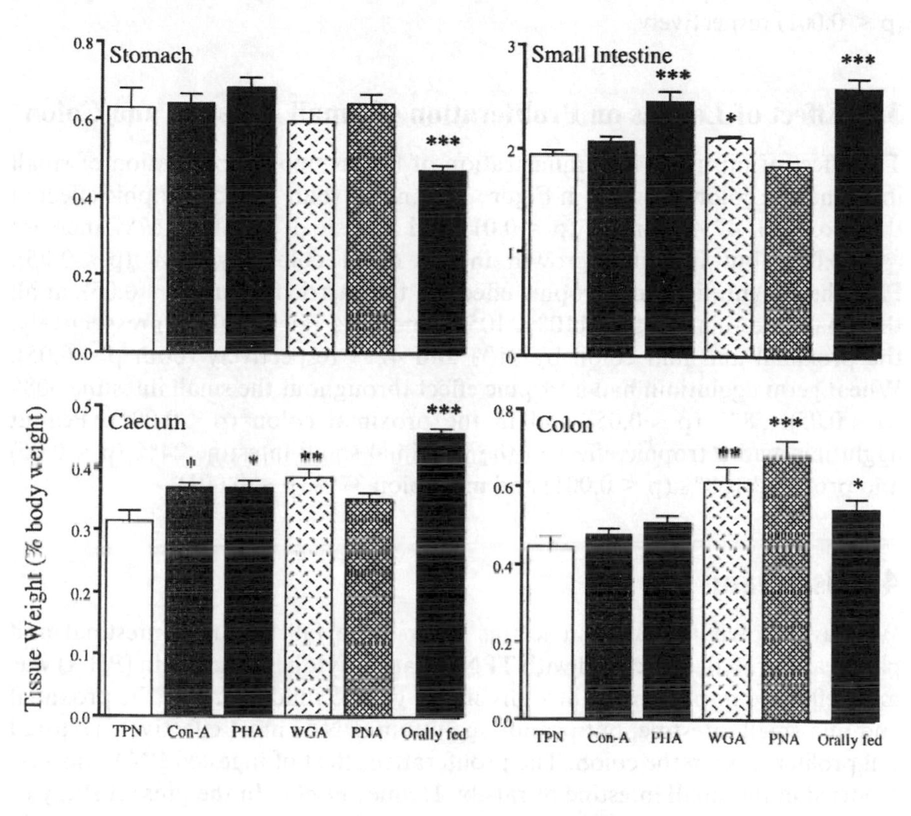

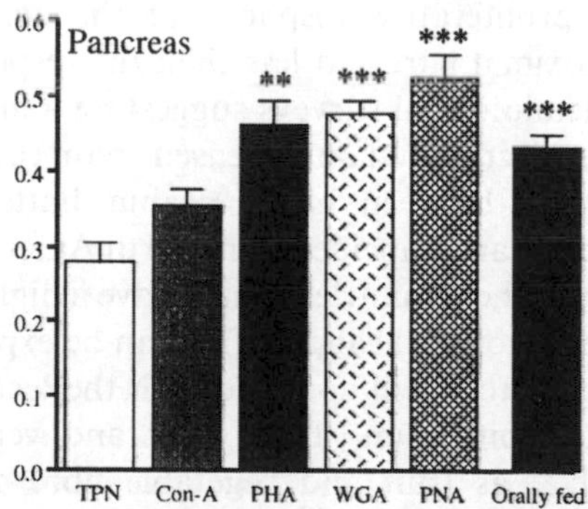

Figure 1 *Effects of intragastric administration of lectins on the wet weight of gastrointestinal organs of rats receiving total parenteral nutrition (TPN). Con-A = concanavalin-A, PHA = phytohemagglutinin, WGA = wheatgerm agglutinin and PNA = peanut agglutinin. * = $p < 0.05$, ** = $p < 0.01$ and *** = $p < 0.001$ vs. TPN alone*

PHA by 66% ($p < 0.01$), WGA by 72% ($p < 0.001$) and PNA by 89% ($p < 0.001$) respectively.

3.2 Effect of Lectins on Proliferation of Small Intestine and Colon

The effect of intragastric administration of lectins upon proliferation of small intestine and colon is shown in Figure 2. Concanavalin A had a trophic effect in the proximal, 47% increase ($p < 0.01$) and mid small intestine, 67% increase ($p < 0.001$) but inhibited growth in the distal colon by 48% ($p < 0.05$). Phytohemaggutinin had a trophic effect on the gastric fundus ($p < 0.05$), at all the small intestinal sites by 110%, 105% and 99% (all $p < 0.001$) respectively, the proximal and mid colon by 81% and 42% respectively (both $p < 0.05$). Wheat germ agglutinin had a trophic effect throughout the small intestine 50% ($p < 0.05$), 58% ($p < 0.05$) and in the proximal colon ($p < 0.001$). Peanut agglutinin had a trophic effect in the proximal small intestine 24% ($p < 0.05$) and proximal 168% ($p < 0.001$) and mid colon 37% ($p < 0.001$).

4 Discussion

We have shown that certain lectins reverse the fall in gastrointestinal and pancreatic growth associated with TPN in rats. Phytohemagglutinin (PHA) was most effective in preventing atrophy in the gastric fundus and in the proximal and mid small intestine, but peanut agglutinin (PNA) most effectively restored cell proliferation in the colon. The proliferative effect of ingested PNA was first reported in the small intestine of rats by Henney *et al.*[14] In the present study of TPN-fed rats the proliferative response of the small intestine was weak, restricted to the proximal part and less than the response of the colon. There have been few epidemiological surveys suggesting a link between the effects of consumption of these lectins and an increased colon cancer risk. One such study showed a relationship between heavy peanut butter consumption and an increased risk of rectal cancer incidence in North America.[15] However in certain regions of Africa and other countries, which have a high intake of peanuts, there is a low incidence of colorectal cancer. This can be explained by the hypothesis that diets high in galactose will compete with the lectin for potential binding sites on the colonic epithelium. Thus, fruit and vegetables may indeed be beneficial to health,[16] as fruit and vegetable fibre contains a much higher galactose content than cereal fibre.[17]

Further studies are necessary to investigate the proliferative effects of various lectins on normal gastrointestinal epithelium and in animals models of carcinogenesis. These lectins may also interact with dietary fibres containing galactose so a multi-factorial study is therefore planned using two types of dietary fibre to investigate this interaction. This study will have implications on the relationship between these lectins and the consumption of fruit and vegetable fibres as opposed to cereal fibres on human health. As these lectins are capable of having significant effects on epithelial proliferation, they may have important implica-

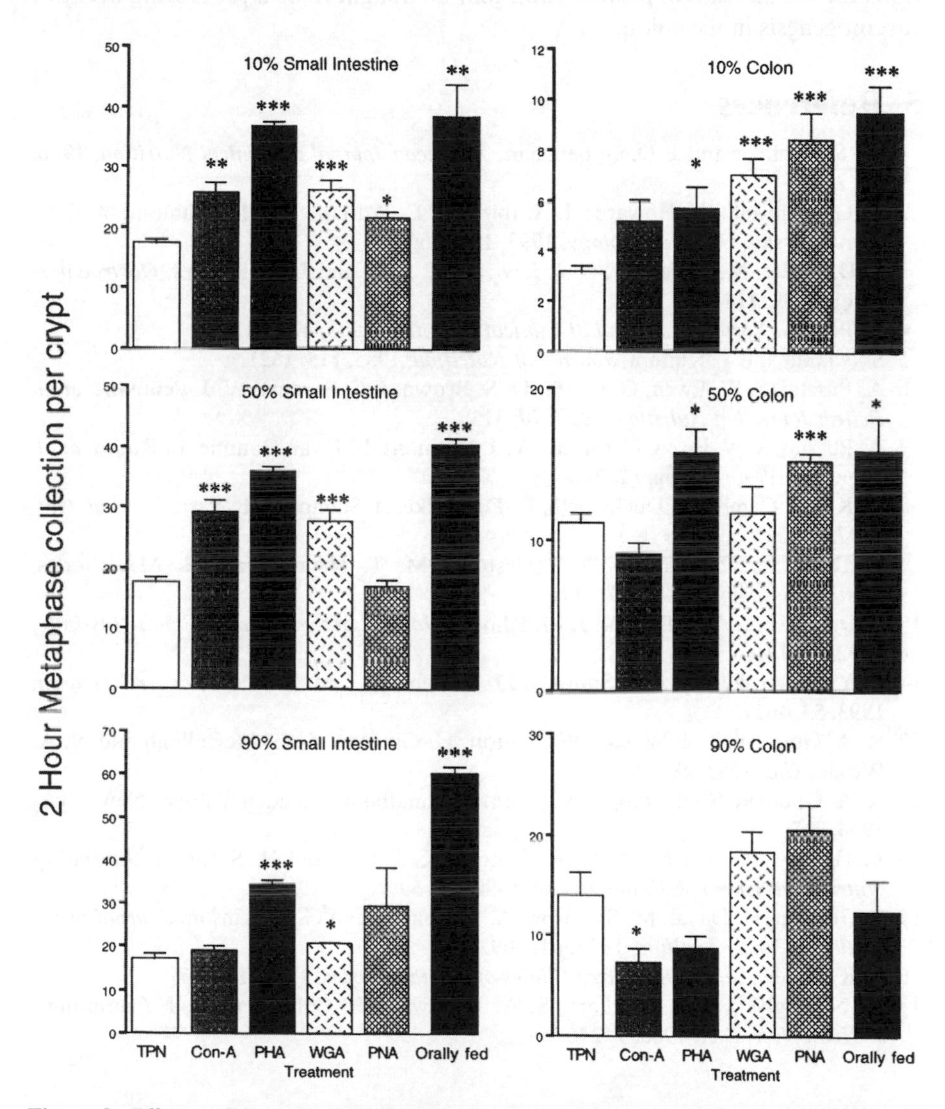

Figure 2 *Effects of intragastric administration of lectins on proliferation of the small intestine and colon measured by metaphase arrest in rats receiving total parenteral nutrition (TPN). Con-A = concanavalin-A, PHA = phytohemagglutinin, WGA = wheatgerm agglutinin and PNA = peanut agglutinin. Orally fed animals received rat chow.* * = $p < 0.05$, ** = $p < 0.01$ *and* *** = $p < 0.001$ *vs. TPN alone*

tions for the increase in proliferation that is thought to be a promoting event of carcinogenesis in the colon.

5 References

1 M. S. Nachbar and J. D. Oppenheim, *American Journal of Clinical Nutrition*, 1980, **33**, 2338.
2 J. G. Banwell, R. Howard, I. Kabir, T. E. Adrian, R. H. Diamond and C. Abramowsky, *Gastroenterology*, 1993, **104**, 1669.
3 S. D. Ryder, M. R. Jacyna, A. J. Levi, P. M. Rizzi and J. M. Rhodes, *Gastroenterology*, 1998, **114**, 44.
4 A. Pusztai, *European Journal of Clinical Nutrition*, 1993, **47**, 691.
5 S. Nakata and T. Kimura, *Journal of Nutrition*, 1985, **115**, 1621.
6 A. Pusztai, S. W. Ewen, G. Grant, D. S. Brown, J. C. Stewart, W. J. Peumans, *et al. British Journal of Nutrition*, 1993, **70**, 313.
7 A. Pusztai, S. W. Ewen, G. Grant, W. J. Peumans, E. J. van Damme, L. Rubio, *et al. Digestion*, 1990, **46 Suppl 2**, 308.
8 R. Kiss, I. Camby, C. Duckworth, R. De Decker, I. Salmon, J. L. Pasteels, *et al. Gut*, 1997, **40**, 253.
9 S. D. Ryder, N. Parker, D. Ecclestone, M. T. Haqqani and J. M. Rhodes, *Gastroenterology*, 1994, **106**, 117.
10 S. D. Ryder, J. A. Smith and J. M. Rhodes, *Journal of the National Cancer Institute*, 1992, **84**, 1410.
11 L. Yu, D. G. Fernig, J. A. Smith, J. D. Milton and J. M. Rhodes, *Cancer Research*, 1993, **53**, 4627.
12 R. A. Goodlad, T. J. Wilson, W. Lenton, H. Gregory, K. G. McCullagh and N. A. Wright, *Gut*, 1987, **28**, 573.
13 R. A. Goodlad, 'Cell biology: A laboratory handbook.' Academic Press, New York, 1994, **2**, 205.
14 L. Henney, E. M. Ahmed, D. E. George, K. J. Kao and H. S. Sitren, *Journal of Nutritional Science & Vitaminology*, 1990, **36**, 599.
15 S. Graham, H. Dayal, M. Swanson, A. Mittelman and G. Wilkinson, *Journal of the National Cancer Institute*, 1978, **61**, 709.
16 L. R. Jacobs, *Gastroenterology Clinics of North America*, 1988, **17**, 747.
17 H. N. Englyst, S. A. Bingham, S. A. Runswick, E. Collinson and J. Cummings, *J. Hum. Nutr. Diet*, 1988, **1**, 246.

6.4

The Significance of *N*-Nitrosamines in the Histological Rebuilding of Gastric Mucosa in Patients with a Gastric Ulcer

M. Schlegel-Zawadzka,[1] Z. Kopanski,[2] A. Plaszczak,[2]
A. Bruchnalska[2] and T. Wojewoda[2]

[1] DEPARTMENT OF FOOD CHEMISTRY AND NUTRITION, JAGIELLONIAN UNIVERSITY, KRAKOW, POLAND
[2] MILITARY CLINICAL HOSPITAL, KRAKOW, POLAND

1 Summary

The analysis included 48 patients with a gastric ulcer in whom the concentration of *N*-nitrosamines in the gastric juice was chromatographically determined. The results of the analytical tests were referred to the type of pathological changes in the mucosa of the stomach.

It was confirmed that *N*-nitrosamines are a factor involved in the formation of pathological alterations in that organ but in an uneven way (the relative factors are given in brackets). They most strongly influence the generation of dysplasia (1), affect the presence of the normal mucosa (0.2), influence inflammatory changes (0.15) and atrophic alterations (0.1).

2 Introduction

N-nitrosamines have been known for over a century. However, particular attention was only paid to them after 1956 as a result of the observation by Magge and Barnes of the carcinogenic properties of *N*-nitrosodimethylamine.[1] The increase of their concentration in the gastric juice of man is linked with an increase in the concentration of bacteria capable of reducing nitrates and nitrites.

Gastric ulcers are frequently associated with alterations of the gastric microflora, including the development of the infection by *Helicobacter pylori*, which in the opinion of some authors provokes an increase in the concentration of *N*-nitrosamines.[2] In this situation, the question may arise as to what extent the increase of the concentration in the gastric juice of *N*-nitrosamines is correlated with the growth of the pathology of the mucosa in the gastric ulcer. We have studied this problem and present our findings in the present communication.

3 Materials and Method

The analysis included 48 patients (35 men and 13 women) aged 27 to 65 years in whom a gastric peptic ulcer had been confirmed. All the patients were subjected to an endoscopic test with histological evaluation of segments of the gastric mucosa as well as to a chromatographic analysis of the concentration in the gastric juice of *N*-nitrosamines.

In estimating the histopathological changes we used the cryteria given by Urban, Oechlert and the WHO.[3–5] We differentiated the normal mucosa (NM), inflammatory changes (IC), atrophic changes (AC) and dysplasia (D). The chromatographic determinations of the concentration in the gastric juice of *N*-nitrosamines (*N*-nitrosodiethanolamine, *N*-nitrosodimethylamine, *N*-nitrosomorpholine) were carried out using a high pressure liquid chromatograph (KONTRON) with a piston pump (model 420), with a 422 spectrophotometric detector. The measurements were made at a wavelength of 245 nm. The concentration of *N*-nitrosamines is expressed in μmol l^{-1}.

The results were processed using SAS for PC IBM rel. 6.03. The hypothesis of normality of distribution was verified with the W test. The strength of the link between the occurrence of the respective histological changes and the concentration of *N*-nitrosamines was studied with GLM (Repeated Measures) and expressed in F-value. All the decisions were taken on the critical level $p < 0.05$.

4 Results

Among the patients analysed, the concetration in the gastric juice of *N*-nitrosamines fluctuated from 0.025 to 1.890 μmol l^{-1}, averaging 0.268 ± 0.089 μmol l^{-1}. The changes of the concentration in the gastric juice of *N*-nitrosamines in relation to the pathology of the gastric mucosa of patients with an ulcer are presented in Table 1.

The results obtained indicate that in patients with a gastric ulcer the concentration in the gastric juice of *N*-nitrosamines statistically significantly influences the occurrence of all the histological changes studied. The strength with which the *N*-nitrosamines affect the pathological alterations of the gastric mucosa, however, differs. This is confirmed by the various values of the F statistic that were given by GLM for the respective histological changes. If those values are arranged in decreasing order: FD : FNM : FIC : FAC then the following ratios are obtained: $28.7:5.7:4.0:2.8 = 1:1/5:1/7:1/10$. That is, the changes of the concentration in the gastric juice of *N*-nitrosamines in patients with a gastric ulcer most strongly affect the occurrence of D (1.0), the presence of NM more weakly (0.2), the occurrence of IC yet more weakly (0.15) and the occurrence of AC most weakly of all (0.1).

Table 1 *Concentration in the gastric juice of N-nitrosamines in relation to the histological changes of the mucosa of the stomach of patients with an ulcer in that organ*

Type of histopathological change[a]	*Number of cases studied (n)*	*Concentration of N-nitrosamines in the gastric juice (average* ± SD) *(*μmol l^{-1})	*p*	*F-value*
Normal mucosa	18	0.168 ± 0.058	0.05	5.7
Inflammatory changes	38	0.198 ± 0.087	0.01	4.0
Atrophic changes	24	0.297 ± 0.064	0.001	2.8
Dysplasia	17	0.409 ± 0.074	0.001	28.4

[a] In the majority of patients more than one type of histopathological change occurred.

5 Discussion

In patients with a gastric peptic ulcer there is a more frequent occurrence of intestinal metaplasia and dysplasia than in the rest of the population. In the opinion of many authors, the appearance of these changes may already signal an increased danger of cancer.[6] Certainly, there exist many factors creating such a direction of the morphological alterations. One of them may be the rise of the concentration in the gastric juice of *N*-nitrosamines. Our own studies indicate that *N*-nitrosamines are a factor influencing the generation in an uneven way of particular histological changes of the mucosa of the stomach in patients with an ulcer in that organ.

They act most strongly on the production of dysplasia (1), more weakly on the presence of the normal mucosa (0.2), yet more weakly on inflammatory changes (0.15), and most weakly of all, on atrophic changes (0.1). Our observations have been partly confirmed in the studies by other authors.[2,6]

6 References

1 L. Fishbein, *Sci. Total Environ.*, 1979, **13**, 157.
2 B. Pignatelli, C. Malareille, A. Rogatko, A. Hautefeuille, P. Thuiller, B. Munoz and F. Bergor, *Eur. J. Cancer*, 1993, **14**, 2031.
3 W. Oehlert, 'Klinische Pathologie des Mages – Darm – Traktes-, in: 'Histologische Diagnose und Differentialdiagnose am gastroenterologischen Biopsiematerial', Schattauer-Verlag, Stuttgart, New York, 1978, p. 78.
4 K. Oota and L. H. Sobina, 'Histological Classification of Stomach and Oesophagus Tumours. International Classification of Tumours' Nr 18, WHO, Geneva, 1977.
5 A. Urban, 'Patologia przewodu pokarmowego. Podstawy morfologii pod red', J. Groniowskiego, S. Krusia, PZWL, Warszawa, 1993, p. 558.
6 M. J. Hill, 'Cancer of the Stomach', in: M. J. Hill (ed.) 'Microbes and Human Carcinogenesis', Edward Arnold, London, p. 36.

6.5

Inhibitory Effects of VES on the Growth of Human Squamous Gastric Carcinoma Cells

Kun Wu, Jian Guo and Yujuan Shan

PUBLIC HEALTH COLLEGE, HARBIN MEDICAL UNIVERSITY, HARBIN 150001 P.R. CHINA

1 Introduction

Vitamin E succinate (VES, RRR-alpha-tocopheryl succinate), a derivative of natural Vitamin E, inhibits the proliferation of various tumor cell types. Although the mechanisms of inhibition are not yet fully understood, cell cycle blockage,[1] cell growth inhibition,[2] differentiation promotion,[3] induction of apoptosis,[4] induced secretion and activation of potent epithelia cell growth inhibitor, TGF-beta, and enhanced expression of cell surface protein required for TGF-beta signaling[5] have been observed in the cells treated with VES. The inhibitory effects on tumor cells after VES treatment of human and rat neuroblastoma, rat neuroglioma, human prostatic cancer,[3] human breast cancer[6,7] have also been reported, but there is no information about the effects of VES on human gastric carcinoma cell. Gastric carcinoma is one of the most malignant tumors in China. In order to further study the mechanisms of VES-mediated inhibition of tumor cell growth, SGC-7901 cells were used as an *in vitro* model. Cell growth inhibition, apoptosis and DNA synthesis arrest were observed in SGC-7901 cells treated with VES, implicating the possibility of VES as chemopreventive and chemotherapeutic drug in the future.

2 Materials and Methods

Materials. VES and DAPI staining was purchased from Sigma Co. Ltd. RPMI-1640 media was purchased from Gibco Co. Ltd. *In situ* cell death detection kit was purchased from Boehringer Mannheim Co. Ltd. ^{3}H-TdR was supplied by the Chinese Science Institute.

Methods. Growth curve: SGC-7901 cells were subcultured in 24-well plates and counted daily for seven days. Growth curves were drawn and inhibition was calculated. *The characteristic changes of apoptotic cells*: The cells treated with separate doses of VES were harvested after 48 hours. Cell morphology was observed by electron microscopy. *Apoptosis detected by flow cytometry*: The cells were harvested after treatment of 24 and 48 hours; the samples were measured by FACScan flow cytometry at 585 nm. Apoptosis tested by *in situ* cell death detection kit (TUNEL). *[^{3}H-TdR] incorporation assay*: The cells in 24-well plates (2×10^4 ml^{-1}) were treated with [^{3}H-TdR] (ultimate concentration 0.5 μCi ml^{-1}) for 6 hours. The cells were collected onto the fiberglass filter using multicipital cytocollectors. Radiointensity was determined by liquid scintillation counter. *DAPI staining*: Both adherent and non-adherent cells were harvested and stained with DAPI, then viewed.

3 Results

The growth curve (Figure 1) showed that the cells were inhibited by 24.7%, 49.2%, 68.4% after 24 hours of treatment with 5, 10 and 20 μg ml^{-1} VES, respectively. By the end of the experiment, the percentage of inhibition had reached 100% in the groups of 10 and 20 μg ml^{-1} VES.

There were also some characteristic changes in morphology such as chromatin condensation and cell shrinkage in volume as viewed by the electron microscope and DAPI staining. The results of apoptosis detected with flow cytometry showed that the pre-G1 peaks could clearly be seen in the cells treated with VES at 10 and 20 μg ml^{-1} after 48 hours. The results in the TUNEL assay showed that VES could induce SGC-7901 cells to apoptosis by 12.48%, 57.63%, 89.96% at 5, 10 and 20 μg ml^{-1}, respectively. VES inhibited the DNA synthesis in dose-dependent and time-dependent manners. The DNA synthesis was significantly reduced by 35%, 45%, 98% after 24 h treatment at 20 μg ml^{-1} and 48 h at 10 and 20 μg ml^{-1} (Table 1).

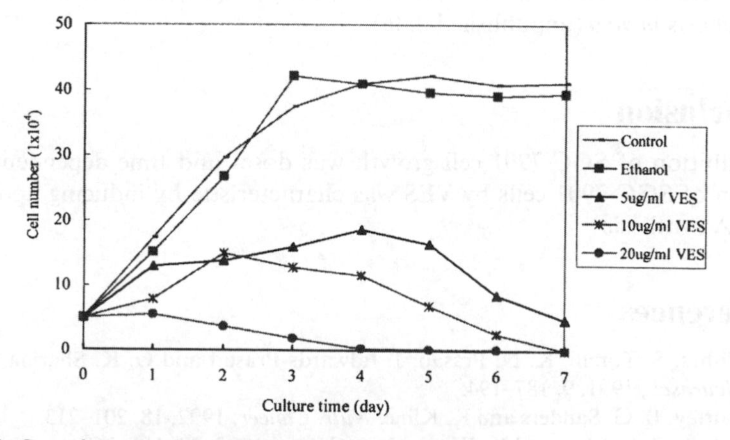

Figure 1 *Growth curve of SGC-7901 cell treated with VES*

Table 1 *Inhibitory effect of VES on DNA synthesis incorporated with ^{3}H-TdR*

	Radiointensity (CPM/ 5 × 10^5)	
Doses (μg ml^{-1})	*24 h*	*48 h*
0	2308 ± 368 (0)	5156 ± 492 (0)
5	2226 ± 375 (4)	4935 ± 585 (4)
10	2058 ± 386 (11)	2854 ± 360 (45)[a]
20	1492 ± 289 (35)[a]	80 ± 11 (98)[a]

[a] Compared with control group, P<0.01, Numbers in parentheses are % of inhibition.

4 Discussion

The data showed that VES induced more than 80% of SGC-7901 cells to undergo apoptosis after 48 hours treatment, both by counting the actual cell number and TUNEL. In addition, VES significantly inhibited DNA synthesis by 45% after 48 hours treatment. VES induced DNA synthesis arrest, by preventing SGC-7901 cells from progressing into the S-phase, ultimately blocking the proliferation of cells. A large number of reports postulate that VES evokes this effect, at least in part, via activation of TGF-beta, multifunctional cytokines regulating cell proliferation and differentiation.[8] However, it seems that a cell blocked cycle is not a prerequisite for apoptosis. The use of flow cytometry for both the detection of apoptosis and VES-induced cell cycle showed that VES-induced apoptotic cells could be cleaved from all phases of the cell cycle. However, VES has not been extensiviely studied for its *in vivo* antitumor properties. Using the hamster buccal pouch as a animal model showed that injection of VES directly into the tumor bearing buccal pouch caused chemically induced oral epidermoid carcinomas to regress.[9] The research in our laboratory demonstrated that VES inhibited B(a)P-induced forestomach tumorigenesis *in vivo* (unpublished data).

5 Conclusion

VES inhibition of SGC-7901 cell growth was dose- and time-dependent. The inhibition of SGC-7901 cells by VES was characteristic by inducing apoptosis and DNA synthesis.

6 References

1 R. J. Cohrs, S. Torelli, K. N. Prasad, J. Edwards-Prasad and O. K. Sharma, *Int. J. Devl. Neurosci.*, 1991, **9**, 187–194.
2 J. M. Turley, B. G. Sanders and K. Kline, *Nutr. Cancer*, 1992, **18**, 201–213.
3 K. Israel, B. G. Sanders and K. Kline, *Nutr. Cancer*, 1995, **24**, 161–169.

4 M. Qian, B. G. Sanders and K. Kline, *Nutr. Cancer*, 1996, **25**, 9–26.

5 J. M. Turley, S. Funakoshi, F. W. Ruscetti, J. Kasper, W. J. Murphy, D. L. Longo and M. C. Birchenall-Roberts, *Cell Growth Differ*., 1995, **6**, 655–663.

6 A. Charpentier, M. Simmons-Menchaca, J. Turley, B. Zhao, B. G. Sanders and K. Kline, *Nutr. Cancer*, 1993, **19**, 225–239.

7 A. Charpentier, M. Simmons-Menchaca, W. Yu, B. Zhao and M. Qian, *Nutr. Cancer*, 1996, **26**, 237–250.

8 J. M. Turley, B. G. Sanders and K. Kline, *Nutr. Cancer*, 1995, **23**, 43–54.

9 K. Shklar, J. Schwartz, D. P. Trickler and K. Niukian, *J. Nat. Cancer Inst*., 1987, **78**, 987–992.

6.6

Glutamine and Epithelial Cell Proliferation in the Intestine of Parenterally Fed Rats

N. Mandir and R.A. Goodlad*

DIVISION OF INVESTIGATIVE SCIENCE, IMPERIAL COLLEGE SCHOOL OF MEDICINE, HAMMERSMITH HOSPITAL, DUCANE ROAD, LONDON, AND HISTOPATHOLOGY UNIT, IMPERIAL CANCER RESEARCH FUND, 44 LINCOLN'S INN FIELDS, LONDON WC2A 3PX, UK

1 Introduction

Various nutrients can stimulate intestinal epithelial cell proliferation and there is evidence that the gut can become deficient in glutamine, which although a nonessential amino acid is one of the preferred respiratory fuels for enterocytes.[1,2] There are several reports of beneficial effects of additional glutamine.[3,4] We used our TPN feeding system to investigate the effects of glutamine on cell kinetics at various sites in the gut. We scored the number and distribution of (native) mitotic figures in crypt squashes[5] and also of bromodeoxyuridine labelled DNA sythesising cells.[6]

2 Materials and Methods

Rats were infused with a total parenteral nutrition (TPN) diet for six days.[7] Group 1 was infused with the basic TPN diet. Group 2 was given TPN supplemented with 2% alanine. Groups 3, 4 and 5 were given TPN supplemented with 0.5, 1.5 and 2% glutamine respectively. Group 6 was fed orally.

Rats were killed one hour after being given 50 mg kg^{-1} of bromodeoxyuridine. Samples of the small intestine and colon, defined by their percentage length, were fixed in Carnoy's fluid for three hours and then transferred to, and stored in, 70% ethanol. For the microdissection method a small piece of the tissue was stained by the Feulgen reaction, gently teased apart and the number of native mitoses in 20 intestinal crypts or colonic crypts counted and the

* Correspondence to: goodlad@icrf.icnet.uk

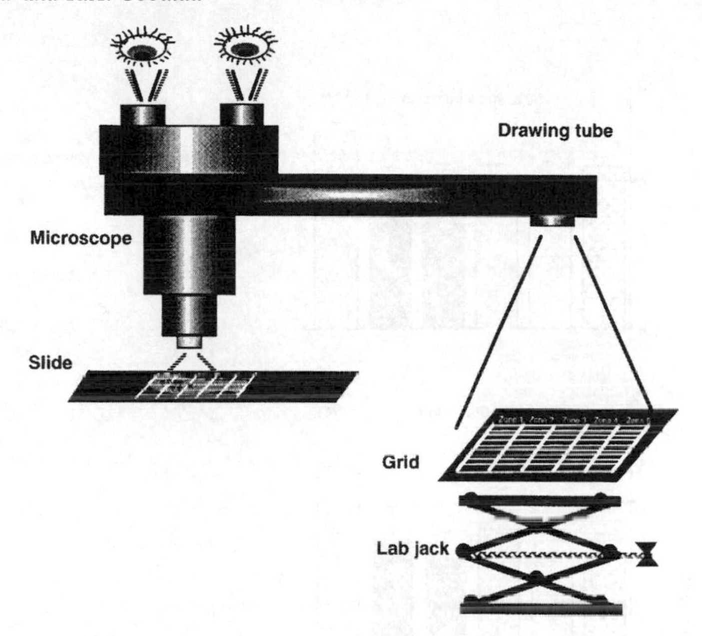

Figure 1 *The use of a drawing tube to superimpose and fit a grid to individual crypts. The grid is placed on a laboratory jack so that its size in the optical field can be adjusted. This standardised the crypts into five equal zones for the quantification of total mitoses and mitoses per zone*

distribution of mitoses in 5 zones determined by using a drawing tube to project an image of a five zone grid into the field of view of the microscope (Figure 1).

BrdU labelling was visualised by immunohistochemistry[8] and 30 well-oriented crypts were scored to determine the presence and location of labelled and mitotic cells and the crypt length.[6,9] The crypts were standardised, that is the distribution data per crypt was adjusted to the average crypt length.

3 Results

No effect of glutamine was seen on the weights of the major regions of the gut. The weights of the small intestine, caecum and colon were all significantly heavier in the orally fed group than in the TPN groups ($p < 0.001$).

Mitotic counts for the TPN groups were significantly lower than for the orally fed in both the small intestine and colon ($p < 0.001$). There was no effect of glutamine on mitoses in the small intestine (Figure 2). However, there was an indication of increased mitotic activity in the colon ($p = 0.03$ by two way analysis, with no interaction between glutamine and site).

There were no differences attributable to glutamine in the distribution of mitotic and labelled cells in histological sections in the small intestine. However, the orally fed group had more labelled cells per crypt column ($p = 0.05$) and had longer crypts ($p = 0.027$); thus the growth fraction of the orally fed rats was significantly reduced ($p = 0.024$). It is important to stress that there was no

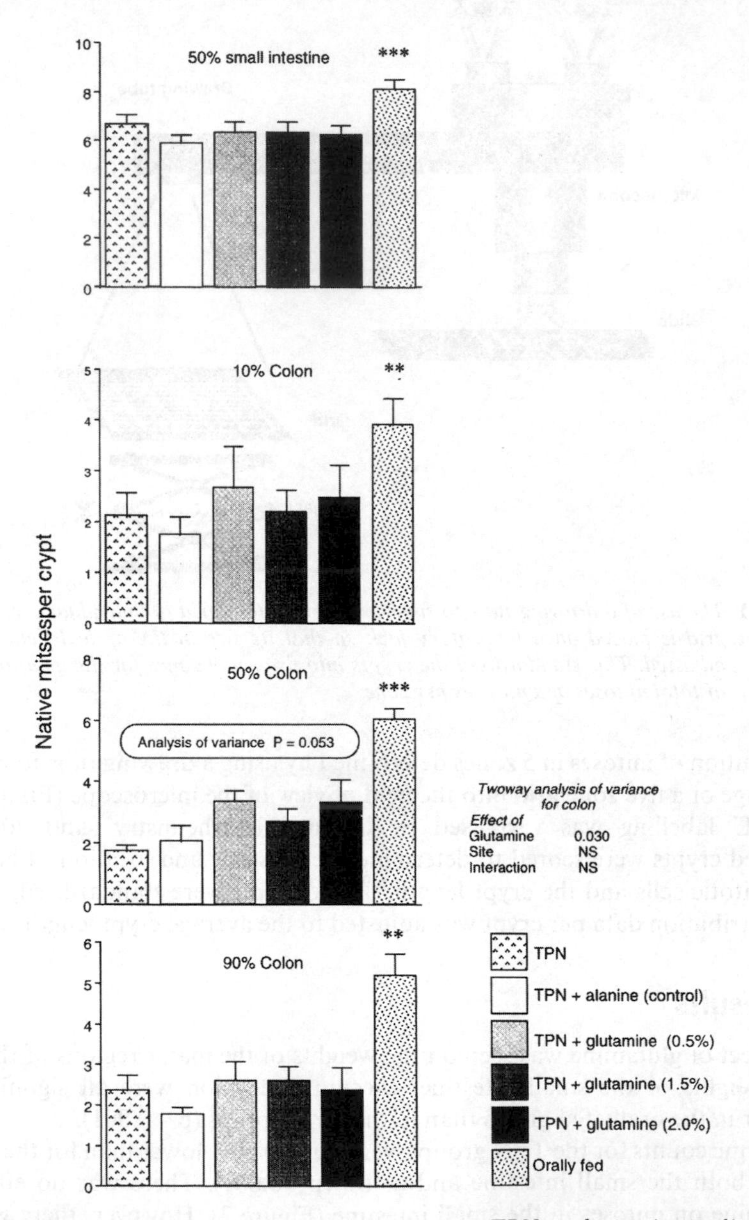

Figure 2 *Mitoses per crypt in rats (n = 6) fed TPN or TPN + alanine (control) or TPN plus 0.5, 1.5 or 2% glutamine. An orally fed (ad lib) group was also included*

difference in labelling index as both the nominator and denominator had changed together. In the colon, there were more mitoses and labelled cells per crypt column and this was reflected in a small but significant reduction in the growth fraction.

The data from histological sections compared very well with the distribution of mitotic figures scored by superimposing a grid over the individual micro-dissected crypts. The results of the zonal distribution of mitoses are presented in Figure 3.

In the small intestine both measures showed that there was no difference attributable to glutamine, and that the orally fed rats had more labelled cells in zones one and two and more mitotic figures in zones one, two, three and four.

In the colon there were no significant differences between groups for the labelling distributions; the mitotic data, however, showed several differences between the TPN and the orally fed in terms of absolute number of mitoses per zone.

There were no significant differences in the total number of labelled cells in the colon, but there was a large increase in the total number of mitoses scored in the orally fed ($p = 0.007$).

When the distribution data was converted to relative distributions, little effect was seen except for a slight increase in activity in zone 4 of the orally fed.

4 Discussion

Glutamine has been reported to be the preferred fuel for the small intestine with the short chain fatty acid, butyrate the preferred fuel for the large bowel (followed by glutamine then glucose).[2] However, one must consider, especially for butyrate, if this is a preference based on nutritional need or the need to metabolise/detoxify the substrate. Whilst glutamine had some effects on the colon, they were not pronounced. Far more dramatic effects of glutamine have been published; however, other papers have only shown weak or even no effects. Some of these differences may be due to the choice of models, especially whether the normal intestine or injured intestine is involved.[10] Our rats were not stressed and it is possible that the actions of glutamine may be more pronounced when stores of glutamine are depleted and energy requirements are increased following illness or injury.[4,11]

Also the site of action of glutamine may vary and we have focused on the effects on cell proliferation, which may not be the main action. Protein synthesis[12] or cell transport function[13] may be more important.

In this study we used two methods to study cell proliferation and we have demonstrated that the crypt squash technique can yield comparable results to scoring S phase label. As well as being far quicker, the method was also better able to detect the well known differences between orally fed and TPN rats and avoid the problems associated with both the nominator and denominator altering in concert. In addition, the method does not need subsequent re-analysis to 'standardised' crypts; to reiterate one should use 'the crypt the whole crypt and nothing but the crypt'.[14]

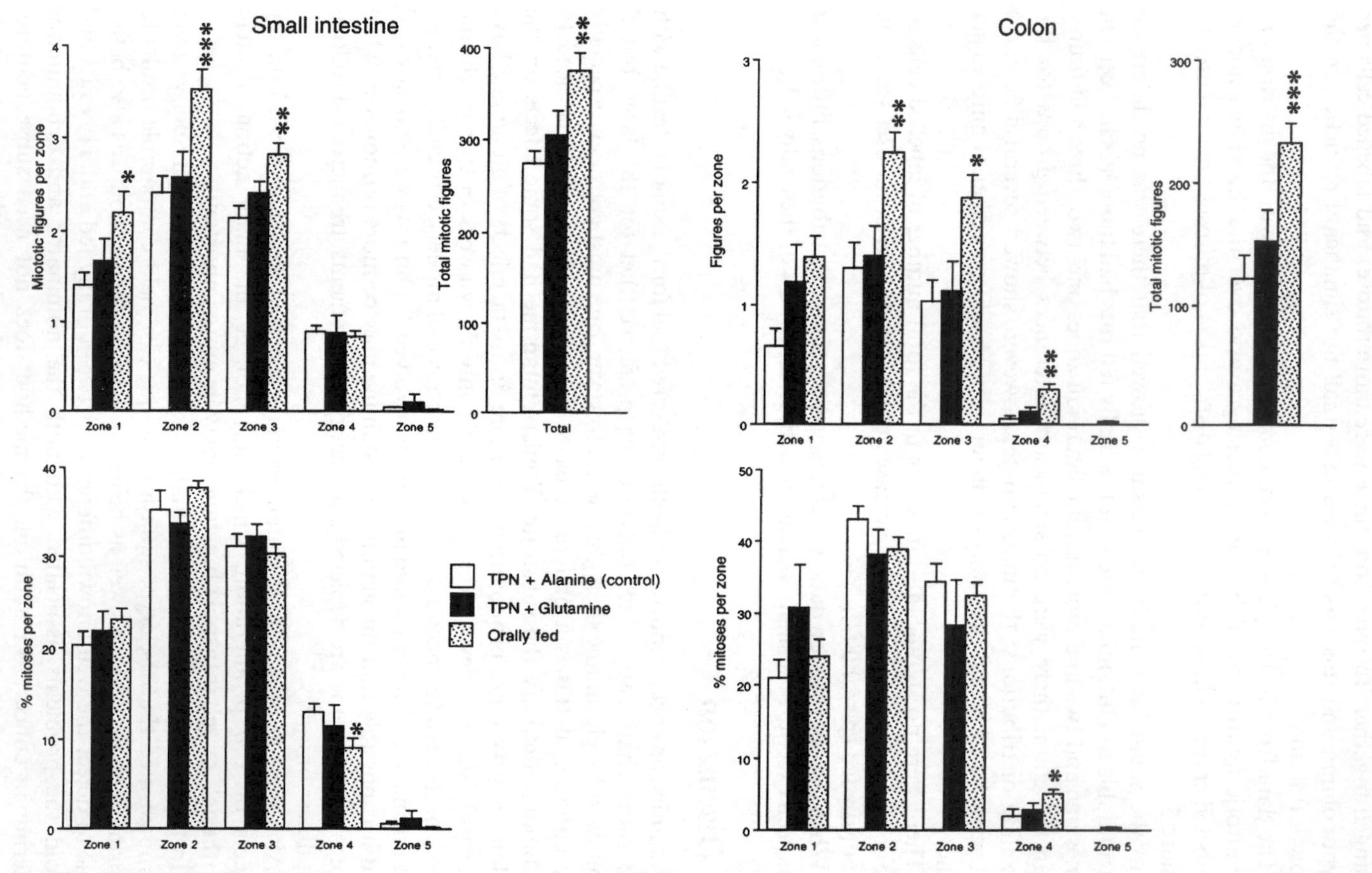

Figure 3 *The distribution of mitotic figures in the five zones of small and large bowel crypts, expressed as total number and as percentage distribution*

5 References

1 W. E. Roediger. Utilization of nutrients by isolated epithelial cells of the rat colon. *Gastroenterology*, 1982, **83**, 424–429.

2 M. A. Chapman, M. F. Grahn, P. Giamundo, P. R. O'Connell, D. Onwu, M. Hutton, *et al.* New technique to measure mucosal metabolism and its use to map substrate utilization in the healthy human large bowel. *British Journal of Surgery*, 1993, **80**, 445–449.

3 S. T. O'Dwyer, R. J. Smith, T. L. Hwang and D. W. Wilmore. Maintenance of small bowel mucosa with glutamine-enriched parenteral nutrition. *Journal of Parenteral & Enteral Nutrition*, 1989, **13**, 579–585.

4 E. Newsholme and G. Hardy. Supplementation of diets with nutritional pharmaceuticals. [Review] [16 refs]. *Nutrition* 1997, **13**, 837–839.

5 R. A. Goodlad. Microdissection-based techniques for the determination of cell proliferation in gastrointestinal epithelium: application to animal and human studies. In: Celis JE, ed. *Cell biology: A laboratory handbook*. New York: Academic Press, 1994, 205–216.

6 R. A. Goodlad, C. Y. Lee and N. A. Wright. Cell proliferation in the small intestine and colon of intravenously fed rats: effects of urogastrone-epidermal growth factor. *Cell Prolif.*, 1992, **25**, 393–404.

7 R. A. Goodlad, T. G. Wilson, W. Lenton, N. A. Wright, H. Gregory and K. G. McCullagh. Intravenous but not intragastric urogastrone-EGF is trophic to the intestinal epithelium of parenterally fed rats. *Gut*, 1987, **28**, 573–582.

8 S. Chwalinski, C. S. Potten and G. Evans. Double labelling with bromodeoxyuridine and 3H-thymidine of proliferating cells in small intestinal epithelium in steady state and after irradiation. *Cell Tiss. kinet.*, 1988, **21**, 317–329.

9 R. A. Goodlad and N. A. Wright. Epithelial kinetics, control and consequences of alterations in disease. In: Whitehead R, ed. *Gastrointestinal and oesophageal pathology*. Edinburgh: Churchill Livingstone, 1995: 97–116.

10 W. Scheppach, G. Dusel, T. Kuhn, C. Loges, H. Karch, H. P. Bartram, *et al.* Effect of L-glutamine and n-butyrate on the restitution of rat colonic mucosa after acid induced injury. *Gut*, 1996, **38**, 878–885.

11 G. Spaeth, T. Gottwald, W. Haas and M. Holmer. Glutamine peptide does not improve gut barrier function and mucosal immunity in total parenteral nutrition. *Journal of Parenteral & Enteral Nutrition*, 1993, **17**, 317–323.

12 T. Higashiguchi, P. O. Hasselgren, K. Wagner and J. E. Fischer. Effect of glutamine on protein synthesis in isolated intestinal epithelial cells. *Journal of Parenteral & Enteral Nutrition*, 1993, **17**, 307–314.

13 T. P. Sarac, A. S. Seydel, C. K. Ryan, P. Q. Bessey, J. H. Miller, W. W. Souba, *et al.* Sequential alterations in gut mucosal amino acid and glucose transport after 70% small bowel resection. *Surgery*, 1996, **120**, 503–508.

14 R. A. Goodlad. The whole crypt and nothing but the crypt. *European Journal of Gastroenterology and Hepatology*, 1992, **4**, 1035–1036.

6.7

Leptin, Starvation and the Mouse Gut

M. Chaudhary, A.J. FitzGerald, N. Mandir,[1] J.K. Howard, G.M. Lord, M.A. Ghatei, S.R. Bloom and R.A. Goodlad[1,*]

DIVISION OF INVESTIGATIVE SCIENCE, IMPERIAL COLLEGE SCHOOL OF MEDICINE, HAMMERSMITH HOSPITAL, DUCANE ROAD, LONDON W12 0NN, AND [1]HISTOPATHOLOGY UNIT, IMPERIAL CANCER RESEARCH FUND, 44 LINCOLN'S INN FIELDS, LONDON WC2A 3PX, UK

Leptin, a peptide hormone and the ob/ob gene product, is secreted by adipocytes, which can act as a satiety factor to regulate food intake. Leptin, however, has a large variety of other actions and there are some indications that one of these could be to stimulate mucosal growth.

Mice were either fed, starved for 48 h or starved and given leptin twice daily. Starvation led to a 20% decrease in body weight and in the weight of the intestine. Starvation also markedly inhibited intestinal epithelial cell proliferation. Leptin had little effect on the small intestine; however, in the hindgut it was associated with small but significant decreases in caecal weight, mitotic counts and colonic crypt area.

1 Introduction

Leptin, the protein product of the obese gene (ob) is synthesised in adipose tissue[1] and leptin deficiency leads to obesity, hyperphagia, hyperglycaemia, hyperinsulinaemia and insulin-resistance, hypothermia and infertility.[2]

While leptin is usually considered as a regulator of food intake, energy expenditure and consequently body weight, the leptin receptor is expressed in many tissues including the jejunum.[3] Leptin may thus have many other actions, perhaps related to its being a sensor of fat (and thus reproductive) status.[4] Leptin can reverse the immunosuppressive effects of starvation thus linking nutritional status to cellular immune function.[5] To test the hypothesis that leptin may also moderate intestinal adaptation to starvation,[6,7] we gave

* Correspondence to: goodlad@icrf.icnet.uk

leptin to starved mice and measured the proliferative and morphological response.

2 Methods

Three groups of 10 C57BL male mice (Harlan, Oxford) were used. The first group was fed *ad libitum*, the second group was starved for 48 hours and the third group was starved for 48 hours and given 1 $\mu g\ g^{-1}$ leptin (ip) twice daily (a gift from Dr M. Chises and Dr N. Levens, Novartis, Basel, Switzerland).

The mice were sacrificed and the weight of gastrointestinal tissues recorded. Areas of the small intestine and colon were fixed in Carnoy's fluid for 3 hours, and stored in 70% ethanol. Pieces of this tissue were prepared for micro-dissection by hydrolysis in 1 M hydrochloric acid for 8 minutes at 60 °C followed by Schiff's reagent. The number of mitotic figures in twenty crypts was scored.[8]

The area of 20 crypts and villi was calculated by crypt tracing using a drawing tube, a flatbed scanner and NIH Image.[8]

Results are presented as mean ± SEM (standard error of mean) and analysed by a one-way and two-way analysis of variance.

3 Results

The weight of the starved groups decreased by nearly 20% ($P < 0.001$). Starvation had no effect on the length of the small intestine, but the colon was shorter in the starved mice ($P < 0.05$). Leptin had no effect on either length.

The absolute weights of the small bowel, caecum and colon were all significantly lighter in the starved groups (Figure 1), but when these were expressed as a percentage of the (starved) body weight the stomachs were slightly heavier ($P = 0.065$), the small intestines were no different and the caecae and colons were (relatively) lighter. The relative weight of the caecum decreased by nearly 20% compared to the starved mice ($P = 0.043$). In the leptin-treated mice, and the colons were slightly lighter, but not significantly so.

There was a 40% decrease in the number of mitotic figures in the small intestine following starvation ($P < 0.001$) in all sites studied with no difference between sites (Figure 2). No effect of leptin treatment was observed.

Starvation caused a significant decrease in villus area ($P = 0.008$) in the proximal small intestine. There was a marked proximo-distal gradient in villus area ($P < 0.001$). In the distal small bowel there was no effect of starvation. No effect of leptin was seen in the proximal or mid small bowel villus areas, except in the distal region where there was an indication of a decrease ($P = 0.065$). Starvation caused a significant decrease in the area of the crypts in the small intestine ($P < 0.001$, Figure 3). No effect of leptin was noted.

In the colon the decrease in crypt areas following starvation was less pronounced (average of 13%) and not significant. There was a marked proximo-distal gradient in crypt area ($P < 0.001$).

The number of mitotic figures decreased by about 70% in the starved colon

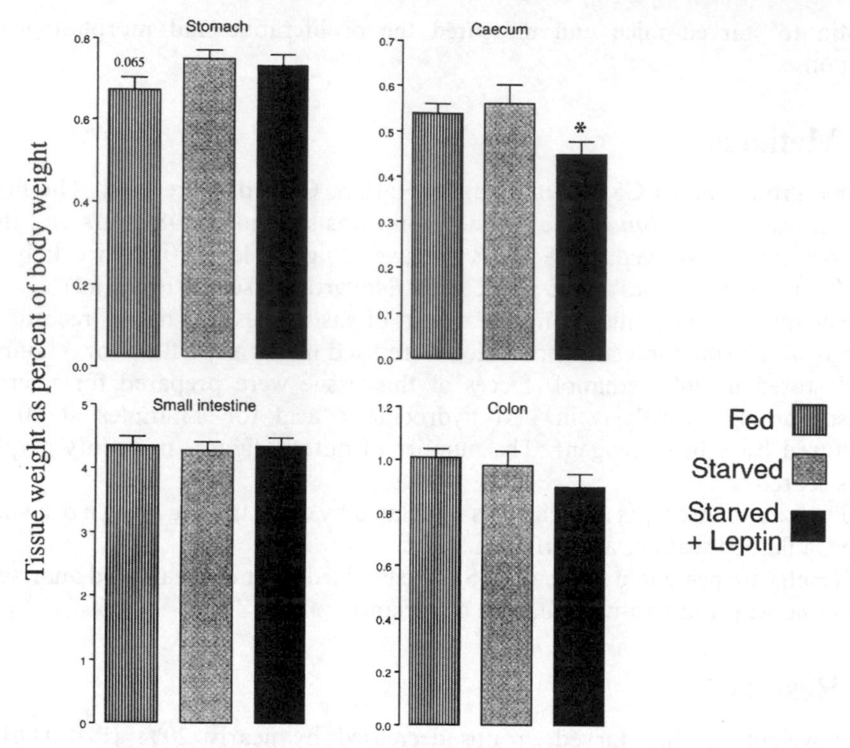

Figure 1 *The weights of the major regions of the gut expressed as a percentage of total body weight.* * $p < 0.05$ ** $p < 0.01$ *** $p < 0.001$ *(t-tests vs starved)*

(P < 0.001). A significant decrease in mitotic activity was seen in the distal colon of the leptin treated mice (P = 0.036) but no effect was seen in the proximal and mid colon.

Colonic crypt area decreased following leptin administration n in all the sites of the colon, significant at P < 0.001 by two-way analysis of variance.

4 Discussion

The results show the major importance of food intake for maintaining cell proliferation and preventing atrophy in the intestine.[9] These actions of food in the lumen may either be direct or indirect response, and can be attributed to 'luminal nutrition' or to the 'work of absorption' (intestinal workload).

The starved mice lost approximately 20% of their body weight, but several of the changes seen in total tissue weight were lost when these were corrected for the changes in body weight; thus starvation had little effect on relative tissue weight. This is surprising considering the high energy cost of maintaining the gut and indicates the importance of maintaining the capacity of the gut for the eventuality that food will become available soon. Some of this capacity may have been maintained by moderating cell migration and cell loss, which may be

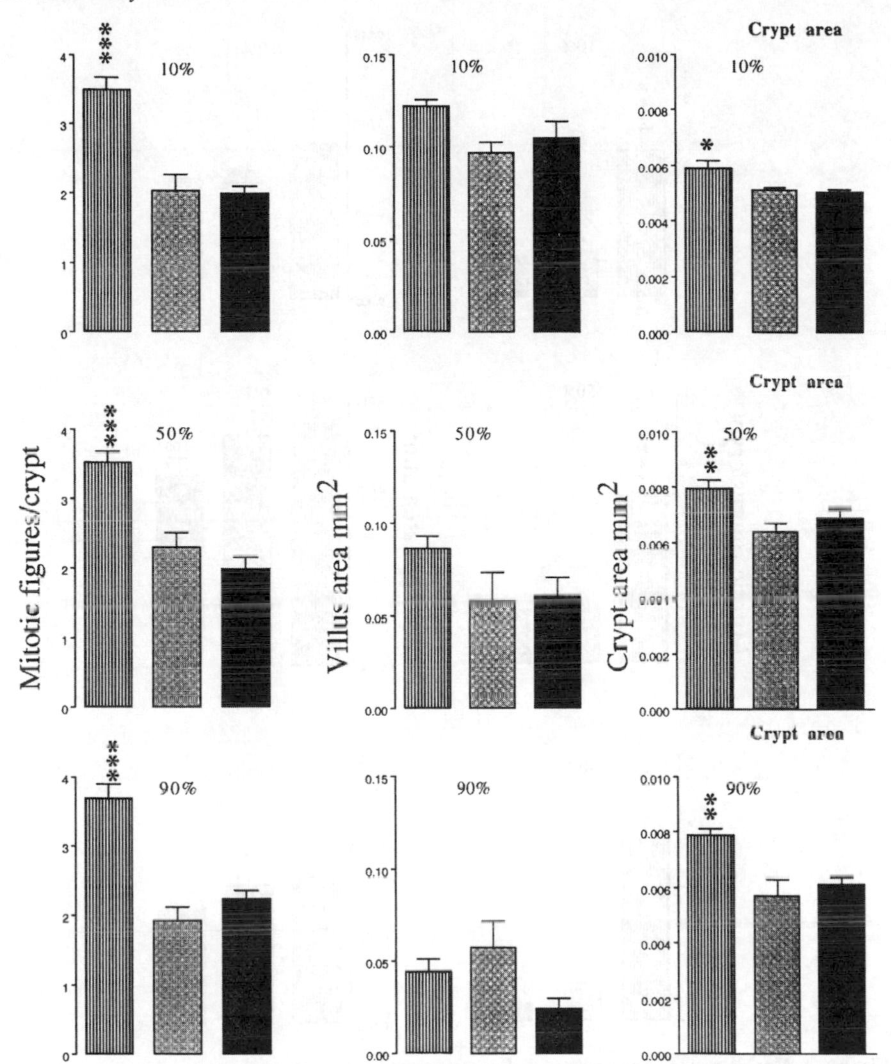

Figure 2 *Native mitoses and villus and crypt areas in the small intestine*

reflected in the relative differences between the decreases in mitotic activity and in compartment size. Starvation decreased mitotic activity by 40% and 70% whereas crypt area only decreased by 20% and 13% in the small bowel and colon respectively.

Leptin had no proliferative actions on the small intestine or colon, and in the colon appeared to be antiproliferative.

Leptin administration in fed animals should decrease food intake, and perhaps the antiproliferative actions seen in this study are in anticipation of the decreased need for digestive capacity.

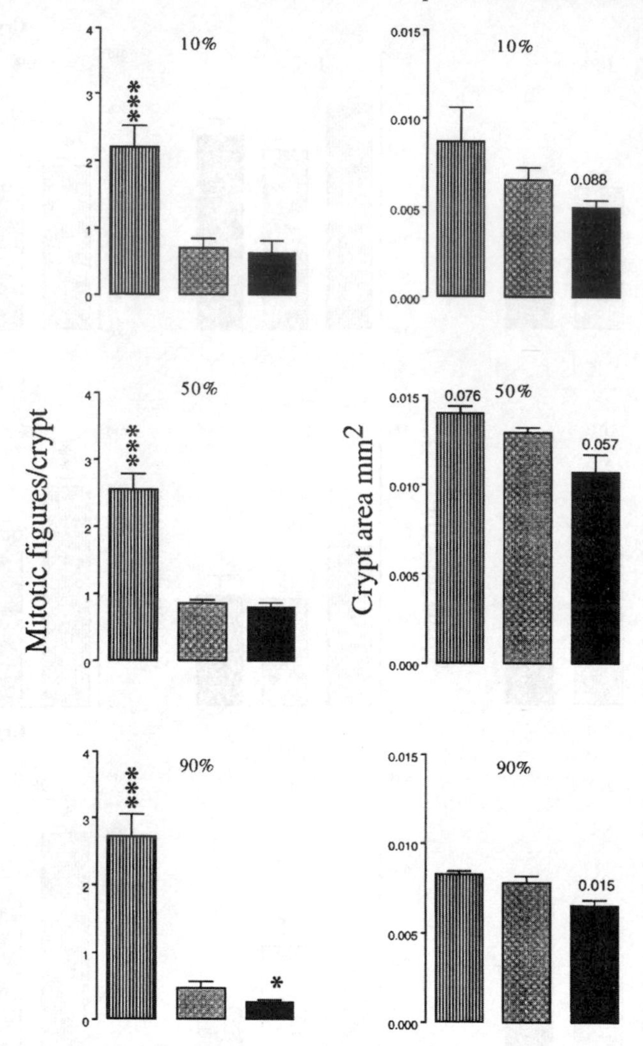

Figure 3 *Native mitoses and villus and crypt areas in the colon*

5 References

1 Friedman JM. Leptin, leptin receptors and the control of body weight. [Review] [46 refs]. *European Journal of Medical Research* 1997, **2**, 7–13.

2 Bray GA, York DA. Clinical review 90: Leptin and clinical medicine: a new piece in the puzzle of obesity. [Review] [77 refs]. *Journal of Clinical Endocrinology & Metabolism* 1997, **82**, 2771–6.

3 Morton NM, Emilsson V, Liu YL, Cawthorne MA. Leptin Action in Intestinal Cells. *J Biol Chem* 1998, **273**, 26194–26201.

4 Spicer LJ, Francisco CC. The adipose obese gene product, leptin: evidence of a direct inhibitory role in ovarian function. *Endocrinology* 1997, **138**, 3374–9.

5 Lord GM, Matarese G, Howard JK, Baker RJ, Bloom SR, Lechler RI. Leptin

modulates the T-cell immune-response and reverses starvation-induced immunosuppression. *Nature* 1998, **394**, 897–901.

6 Flier JS. lowered leptin slims immune response. *Nature Medicine* 1998, **4**, 1124–1125.

7 Takahashi Y, Okimura Y, Mizuno I, Iida K, Takahashi T, Kaji H, *et al.* Leptin induces mitogen-activated protein kinase-dependent proliferation of C3H10T1/2 cells. *Journal of Biological Chemistry* 1997, **272**, 12897–900.

8 Goodlad RA. Microdissection-based techniques for the determination of cell proliferation in gastrointestinal epithelium: application to animal and human studies. In: Celis JE, ed. *Cell biology: A laboratory handbook*. New York: Academic Press, 1994, 205–216.

9 Goodlad RA, Wright NA. The effects of starvation and refeeding on intestinal cell proliferation in the mouse. *Virchows Archives cell pathology* 1984, **45**, 63–73.

6.8

Inhibition of Hereditary Intestinal Carcinoma Development by Flavonoids

G. Jacobasch, S. Florian, H. Pforte, J. Hempel, K. Schmehl and D. Schmiedl

GERMAN INSTITUTE OF HUMAN NUTRITION, POTSDAM-REHBRÜCKE, GERMANY

1 Introduction

From epidemiological studies and *in vitro* experiments with different cell lines potential health effects for distinct flavonoids have been postulated.[1] Antioxidative, antiarteriosclerotic, anticarcinogenic, anti-inflammatory, immunoprotective and prebiotic properties are described.[2] Among the more than 4000 widespread plant flavonoids quercetin is the most common and effective aglycon. The aim of our experimental Min-mice studies was to examine the chemopreventive effects of quercetin on the intestinal carcinogenesis. Colorectal carcinomas are the third most common neoplasms in the world, initiated frequently by primary mutations of the apc-gene. Nearly all apc-gene mutations result in the synthesis of a shortened APC polypeptide chain accompanied by a loss of the β-catenin binding domain.[3] This domain is essential for the dimerization of β-catenin and APC, which is a prerequisite for phosphorylation and succeeding proteolytic degradation of β-catenin. Consequently the β-catenin concentration accumulates in both cytosol and nucleus, where it initiates a malignant transformation. Hereditary apc-gene defects cause the clinical manifestation of the familial adenomatous polyposis.[4] Polyposis appears mainly in the colon in humans, but in the small intestine in mice.[5] Answers to the following questions have been expected from results of our animal studies.

- Do both the aglycon quercetin and the glycoside rutin inhibit the programmed intestinal carcinogenesis in apc-gene defected Min-mice?
- Do flavonoid-degrading intestinal bacteria influence anticancerogenic effects?
- Does the end product of microbial quercetin degradation dihydroxyphenylacetic acid suppress the intestinal cancer development?

2 Methods

C57 BL/6J-Apc-Min-/+ supplied by Bomholtgard Breeding & Research Center (Bomgard, Denmark), were fed with a semisynthetic diet or supplemented with 14 μg pure quercetin g^{-1} or rutin. Furthermore two animal groups received the diet supplemented with 47 μg dried and pulverized buckwheat leaves as source of rutin or both, buckwheat rutin and a well fermentable resistant starch (100 mg g^{-1}). The period of the specific feeding regime was 60 days.

Histological investigations were accomplished by counting all polyps after fixation with 5% paraformaldehyde and staining with hamatoxylin. For immunochemistry 4 μm microtome sections were taken. Klenow DNA fragmentation detection assay (Calbiochem) and TUNEL *in situ* apoptosis kit (Boehringer-Mannheim) were performed as described in the user's guideline. Cell cycle proteins of the human colon adenocarcinoma were estimated by Western blot analysis. Cell lysates were subjected to sodium dodecyl sulfate electrophoresis on 14% polyacrylamide gels. The proteins were transferred on a polyvinylidene difluoride membrane by semi-dry blotting in a discontinuous buffer system and detected by incubation with the corresponding primary antibody followed by a peroxidase labeled secondary antibody.

3 Results and Discussion

Table 1 includes information about the study design, and the data illustrating the anticarcinogenic effectiveness of both quercetin and rutin. The intestinal cancer development was suppressed to the same extent by nearly 90%. The number of polyps and their size decrease significantly. Quercetin as well as pure quercetin-rutinoside can be absorbed very well in the stomach and in the small intestine. The glycoside does not require a microbial splitting off of the carbohydrate chain before. But absorption of rutin from dried buckwheat leaves takes place only with the assistance of microbial degradation. Due to this

Table 1 *Anticancerogenic potency of quercetin in dependence on the ability to absorb flavonoids in apc-gene defect Min-mice*

	Group				
	I	*II*	*III*	*IV*	*V*
Standard diet with	—	quercetin 14 μg g^{-1} diet	rutin 14 μg aglycon g^{-1} diet	rutin 47 μg g^{-1} diet + bwr	rutin 47 μg $^{-1}$ diet + bwr + RS
Number of polyps	72.0 ± 40.3	8.0 ± 4.2	8.0 ± 5.1	28.0 ± 8.1	58.0 ± 29.3

bwr = buckwheat rutin; RS = fermentable resistant starch.

neccessity the buckwheat material reaches the large intestine where it is degraded by *Eubacterium ramulus* to dihydroxyphenylacetic acid. With an increase of the microbial biomass by an intake of both, resistant starch and buckwheat material, the number of *Eubacterium ramulus* increases strongly.[6] Under this condition most of the rutin is metabolized microbially. When rutin is degraded more extensively by intestinal bacteria the anticarcinogenic effectiveness is lowered. Therefore we concluded that dihydroxyphenylacetic acid does not inhibit carcinogenesis. This postulate was proved directly in further Min-mice fed with a semisynthetic diet supplemented by dihydroxyphenylacetic acid. The results summarized in Table 1 give evidence that the anticarcinogenic activity can be only realized systemically. For this purpose an absorption of the flavonoid in its polyhydroxyphenylic structure including transport to the target cells via blood circulation is neccessary. Both quercetin and rutin can be absorbed well by the mucosa of the stomach as well as by the small intestine, and the glycosidic binding of rutin is split intracellularily by mucosal specific β-glucosidases.

Eubacteria ramulus growth is recorded only in the large intestine. In the oral part of the gastrointestine *Enterococcus casseliflavus* is established.[7] This microorganism splits only rutin, with the formation of the aglycon and the carbohydrate residue, but is not able to degrade the aglycon.

Rutin is more stable than quercetin at pH 7.0, a milieu condition occurring in the small intestine. Therefore it can be concluded that rutin absorption in the stomach offers the best precondition for a monitoring of the systemic anticancerogenic effect of quercetin.

Immunhistochemical investigations have been done to get insights into changes of the complex regulatory mechanisms of apoptosis, proliferation as well as time and space dependence of the cell cycle. The pattern of DNA fragmentation detected by using the Klenow assay in samples of histologically normal areas of colon demonstrates an increased repair activity in rutin treated animals. Nearly no difference has been found in the rate of apoptosis in colon between flavonoid-treated and untreated Min-mice by using the TUNEL *in situ* kit. Correspondingly also no changes in respect of bcl2 and p53 could be detected. Due to the fact that intestinal carcinogenesis is correlated with an increased expression of COX 2, the immunreactivity of COX 2 is decreased in tissues of the quercetin- or rutin-treated animals (Figure 1). Decisive for the anticancerogenic effect of quercetin is its antiproliferative potency. Key parameters for blocking tumor promotion seem to be changes in the expression of the cyclins A and D1 as well as in that of the cyclin dependent kinases cdc2 and cdk2, respectively. Whereas in presence of quercetin (10 μM) the concentrations of cyclin D1, cdk2 and cdc2 increase, the level of cyclin A decreases. These data give us the basis to the hypothesis that the enhanced cyclin D1 level inhibits DNA synthesis by an increased binding of PCNA, an essential subunit of the DNA polymerase. Additionally more cdk2 binds to cyclin D1 resulting in an inactive enzyme complex. Consequently the formation of the catalytically active cyclin A and cyclin E complexes decreases resulting in a blocking of the protein transcription necessary to initiate the S-phase.

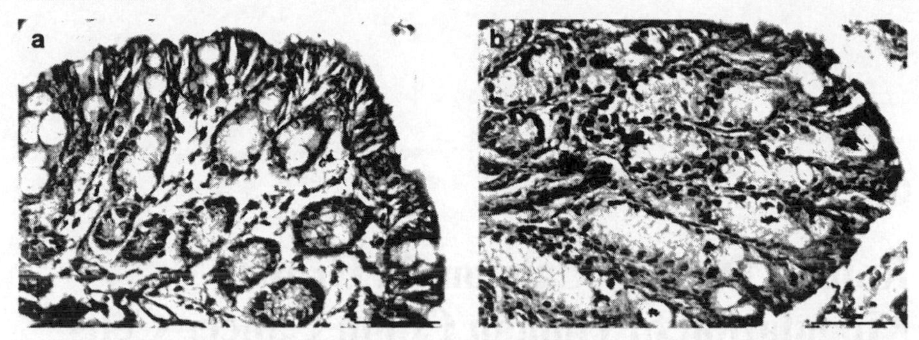

Figure 1 *Anti-COX-2 immunoreactivity. MIN-mouse, colon, 200-fold, POD/DAB. Positive structures show brown staining. (a) standard diet; (b) standard diet and quercetin*

4 References

1 M.G.L. Hertog, P.C.H. Hollman. Potential health effects of the dietary flavonol quercetin. *Eur. J. Clin. Nutr.*, **50**, 63–71 (1996).

2 J.V. Formica, W. Regelson. Review of the biology of quercetin and related bioflavonoids. *Fd. Chem. Toxic.*, **33**, 1061–1080 (1995).

3 G. Jacobasch, D. Schmiedl, K. Schmehl. Darmkrebsprävention durch resistente Stärke? *Ernährungs-Umschau*, **44**, 318–326 (1997).

4 J. Gordon, A. Thliveris, A. Samowitz, W. Carlson, M. Gelbert, L. Albertsen, H. Joslyn *et al.* Identification and characterization of the familial adenomatous polyposis coli gene. *Cell*, **66**, 589–600 (1991).

5 A.R. Moser, H.C. Pitot, W.F. Dove. A dominant mutation that predisposes to multiple intestinal neoplasia in the mouse. *Science*, **247**, 322–324 (1990).

6 R. Simmering, B. Kleeßen, M. Blaut. Quantification of the flavonoid-degrading bacterium Eubacterium ramulus in human fecal samples with a species-specific oligonucleotide hybridization probe. *Appl. Environ. Microbiol.*, **65**, 3705–3709 (1999).

7 H. Schneider, A. Schwiertz, M.D. Collins, M. Blaut. Anaerobic transformation of quercetin-3-glucoside by bacteria from the human intestinal tract. *Arch. Microbiol.*, **171**, 81–91 (1999).

6.9

Role of Dietary Calcium in Growth Modulation of Human Colon Cancer Cells

E. Kállay,[1] E. Bajna,[1] S. Kriwanek,[2] E. Bonner,[2] M. Peterlik[1] and H.S. Cross[1]

[1] INSTITUTE OF GENERAL AND EXPERIMENTAL PATHOLOGY, UNIVERSITY OF VIENNA MEDICAL SCHOOL, VIENNA, AUSTRIA
[2] HOSPITAL RUDOLFSSTIFTUNG, VIENNA, AUSTRIA

1 Introduction

There is ample evidence from epidemiological studies that the incidence of colorectal carcinoma in man is inversely related to calcium consumption.[1] In this respect it has been suggested that low calcium could reduce the extent of insoluble calcium salts formed from otherwise carcinogenic bile acids in the lumen of the intestine.[2] However, extracellular calcium ($[Ca^{2+}]_o$) is also a direct modulator of colonocyte proliferation. We were able to show that the proliferative potential of colon adenocarcinoma-derived Caco-2 cells is inversely related to $[Ca^{2+}]_o$ levels in the culture medium.[3] Recently we obtained evidence that Caco-2 cells are able to express the same G protein-coupled extracellular calcium-sensing receptor (CaR)[4] that had been cloned from parathyroid, kidney and other cells.[5]

2 Results and Discussion

The influence of calcium deficiency on the cell cycle was investigated by flow cytometry. Figure 1A illustrates the percentage of Caco-2 cells in the S phase of the cell cycle after treatment with decreasing concentrations of calcium. Growth for 24 hours in a 0.025 mM $[Ca^{2+}]_o$ medium significantly increased the number of cells in S phase over those grown in 1.8 mM $[Ca^{2+}]_o$ medium. A similar shift to the synthesis phase of the cell cycle was seen under calcium-free conditions.

To prove that growth modulation of Caco-2 cells by $[Ca^{2+}]_o$ requires binding of calcium to a specific receptor, the antimitotic concentration dependence was studied. Between 0.00 and 0.10 mM $[Ca^{2+}]_o$ Caco-2 cells exhibit high sensitivity to extracellular calcium with $[^3H]$thymidine incorporation into DNA being

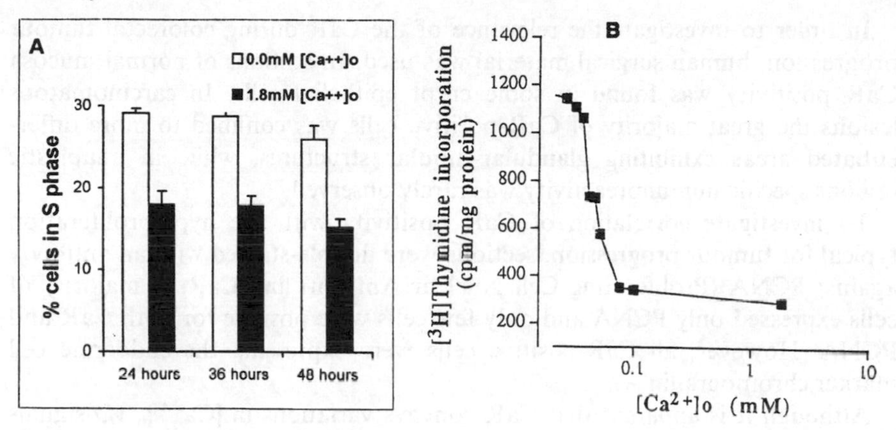

Figure 1 *Influence of $[Ca^{2+}]_o$ on proliferation of cOnfluent Caco-2 cells.* (A) *Flow cytometry.* (B) *$[^3H]$thymidine incorporation*

reduced by 70%. A further raise of $[Ca^{2+}]_o$ up to 1.8 mM had no additional suppressive effect on Caco-2 cell proliferation (Figure 1B).

The c-myc proto-oncogene transduces extracellular growth signals into entry of cells from the G_0 to G_1 cell cycle phase. While Caco-2 cells express high levels of c-myc mRNA when grown in calcium-free medium raising the calcium concentration to 1.8 mM results in reduction of c-myc mRNA levels by half within 2 hours (Figure 2A).

Additional evidence for the assumption that the anti-proliferative effect of increased $[Ca^{2+}]_o$ involves activation of an extracellular Ca^{2+}-sensing receptor comes from experiments in which a dose-dependent reduction of the DNA labelling index occurred when the well known receptor agonist Gd^{3+} was substituted for Ca^{2+} (Figure 2B).

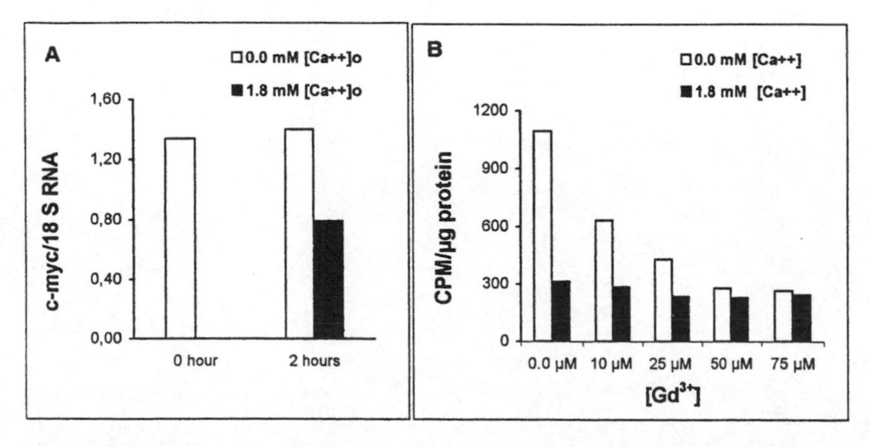

Figure 2 (A) *Density ratio of c-myc mRNA versus 18 S RNA determined by Northern blot.* (B) *Influence of CaR agonist Gd^{3+} on $[^3H]$thymidine incorporation*

In order to investigate the relevance of the CaR during colorectal tumour progression, human surgical material was used. In sections of normal mucosa CaR positivity was found in some crypt epithelial cells. In carcinomatous lesions the great majority of CaR-positive cells was confined to more differentiated areas exhibiting glandular-tubular structures, while in anaplastic regions specific immunoreactivity was rarely observed.

To investigate correlation of CaR positivity with the hyperproliferation typical for tumour progression, sections were double-stained with an antibody against PCNA (Proliferating Cell Nuclear Antigen) and CaR. A majority of cells expressed only PCNA and only few cells were positive for both CaR and PCNA. However, all CaR-positive cells were expressing the endocrine cell marker chromogranin A.

Although it is apparent that CaR conveys variations in $[Ca^{2+}]_o$ *via* signals along the PKC pathway into regulation of c-*myc* expression and, consequently, of cell growth, it is not clear yet whether CaR activity in the colon is directly coupled to PKC signaling. In parathyroid gland cells activation of the receptor inhibits the release of a cell-specific polypeptide. Intriguingly, the major proteolytic cleavage product of chromogranin A is pancreastatin, which has profound suppressive effects on growth of gastrointestinal cancer cells.[6]

3 References

1 C.F. Garland, F.C. Garland and E.D. Gorham, *Am. J. Clin. Nutr.*, 1991, **54**, 193S.
2 H.L. Newmark and M. Lipkin, *Cancer Res.*, 1992, **52**, 2067S.
3 H.S. Cross, M. Pavelka, J. Slavik and M. Peterlik, *J. Natl. Cancer Inst.*, 1992, **84**, 1355.
4 E. Kállay, O. Kifor, N. Chattopadhyay, *et al.*, *Biochem. Biophys. Res. Comm.*, 1997, **232**, 80.
5 E.M. Brown, G. Gamba, D. Riccardi, *et al.*, *Nature*, 1993, **636**, 575.
6 W.E. Fisher, P. Muscarella, L.G. Boros and W.J. Schirmer, *Int. J. Pancreatol.*, 1998, **24**, 169.

6.10

Human Colonic Crypt Calcium Signalling

Susanne Lindqvist, Paul Sharp, Ian Johnson and Mark Williams

UNIVERSITY OF EAST ANGLIA AND INSTITUTE OF FOOD RESEARCH, COLNEY, NORWICH NR4 7UA, UK

1 Introduction

There is great interest in the prevention of colorectal cancer as it is a leading cause of morbidity and mortality in developed countries. Colorectal carcinogenesis is now widely accepted to be a consequence of the accumulation of mutations in specific genes including those controlling cell division and apoptosis. Changes in the relative activities of these processes are also thought to pre-dispose the colonic mucosa to initiation along the adenoma-carcinoma sequence, but very little is known of the fundamental mechanisms that co-ordinate colonic epithelial cell production and cell loss. If we are to understand the way in which dietary factors influence the development of colorectal cancer it is imperative first to elucidate the cell to cell signalling pathways that determine normal cellular development in the colonic mucosa.

Our laboratory has developed a model that permits a study of human intestinal crypt cell signalling events in response to systemic and dietary factors. The major advantage of working with this system is that the crypts are of human origin and can be obtained from normal and diseased tissue. In addition, the crypt represents the differentiating-axis of the intestinal epithelium and the position of the stem cells at the base of the crypt, and all of the other native cell-to-cell contacts, are preserved (Figure 1). Ongoing research in our laboratory is concerned with understanding the role of calcium signalling in co-ordinating the physiology of the crypt and the way in which dietary components, such as luminal calcium, can modulate the spatiotemporal characteristics of colonic crypt calcium signals in health and disease. The present study demonstrates that, as is the case in the rodent,[1] the neurotransmitter acetylcholine initiates a calcium signal in the stem cell region at the crypt-base that propagates along the axis of isolated human colonic crypts. Furthermore, we identify the presence of the muscarinic receptor M3 subtype that is coupled to human colonic crypt calcium signals induced by acetylcholine and we implicate a role for intercellular communication in calcium signal propagation along the crypt-axis.

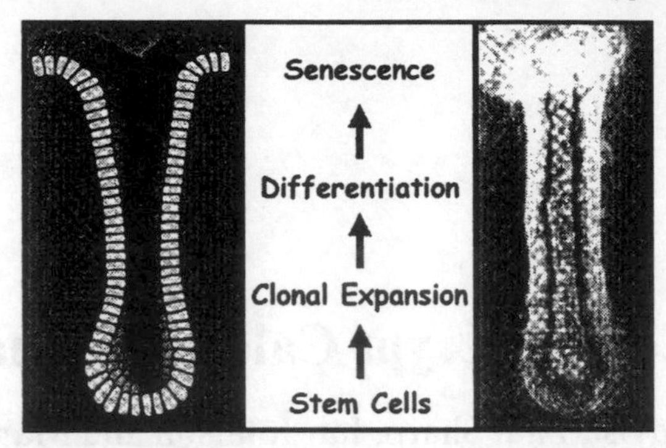

Figure 1 *Colonic crypt cell kinetics. The architecture and hierarchy of crypt cell kinetics along the differentiating pathway of the colonic crypt-axis is depicted in the left-hand and middle panels, respectively. A phase micrograph of an isolated human crypt is shown in the right-hand panel*

2 Materials and Methods

Biopsy samples were obtained from the distal colon at colonoscopy and the *ex vivo* tissue placed in a calcium-free, hepes buffered saline (HBS) for one hour. Vigorous shaking liberated isolated crypts that were affixed to collagen-coated coverslips. Crypts were loaded with the acetoxymethyl ester form of the calcium-sensitive dye Fura2 (1 μM) for 45 min, washed for 30 min in HBS, and placed on the stage of an inverted microscope coupled to a fluorescence imaging system (Photon Technology International). Crypts were continuously perifused with HBS and experimental solutions introduced by a two-way tap or *via* a perfusion micropipette attached to a microinjection system (Eppendorf 5242). In parallel experiments, crypts were fixed in 4% (w/v) paraformaldehyde, frozen in OCT and longitudinal 10 μm sections cut prior to labelling with a muscarinic M3 receptor antibody. Labelling was visualised using a FITC conjugated secondary antibody in conjunction with the above fluorescence imaging system.

3 Results and Discussion

Acetylcholine (10 μM) initiates a calcium signal in the stem cell region located at the base of isolated human colonic crypts (Figure 2). As the calcium signal increases in intensity at the crypt base it is also observed to propagate along the crypt-axis towards the surface epithelial cells. The kinetics of the acetylcholine-induced colonic crypt calcium signal are markedly biphasic. For all regions along the crypt-axis a dramatic increase in intracellular calcium is followed by a sustained elevated baseline that returns to the resting level upon removal of acetylcholine from the bathing medium. Classically, this behaviour is due to the initial release of calcium from intracellular stores followed by an influx of calcium from the bathing medium.[2]

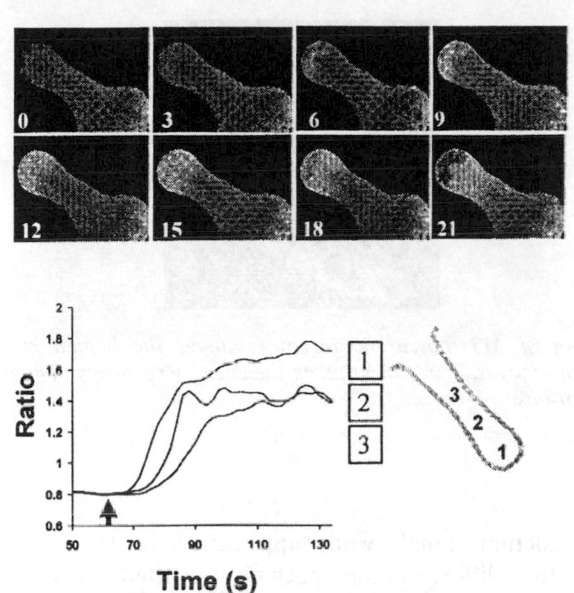

Figure 2 *Acetylcholine (10 μM)-induced intracellular calcium increase along the human crypt-axis.* (a) *Fluorescence grey scale ratio images (increase in brightness indicates an elevation of intracellular calcium) illustrates the spatial characteristics of the ACh induced calcium signal are (time in secs, to see* http://www.uea.ac.uk./~b289/ *for pseudo colour movie). The kinetics of the calcium signal along the crypt-axis are depicted in* (b). *The resting ratio corresponds to an intracellular calcium concentration of ~100 nM and the peak ratio to a level of ~800 nM*

In order to assess the relative sensitivity of regions located along the crypt-axis to acetylcholine the agonist was applied locally to the basolateral membranes of the crypt with the aid of a micro-pipette. The resultant calcium signals revealed a differential sensitivity gradient to acetycholine along the crypt-axis (Figure 3). Acetylcholine targeted to the crypt base

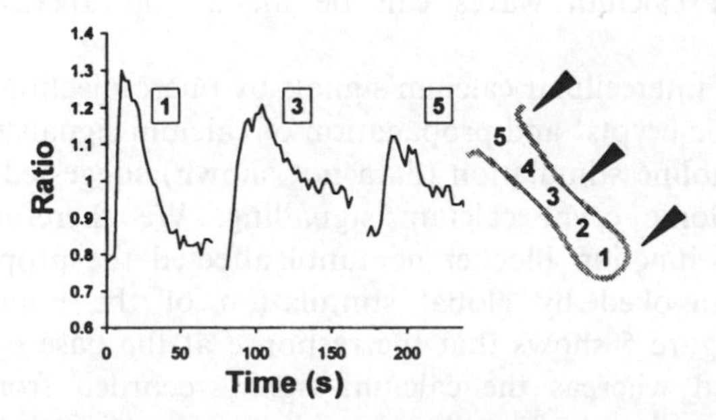

Figure 3 *Localised stimulation of the human colonic crypt-axis with acetylcholine (1 mM in the microperfusion pipette). The crypt-base exhibits the greatest response and sensitivity to the agonist*

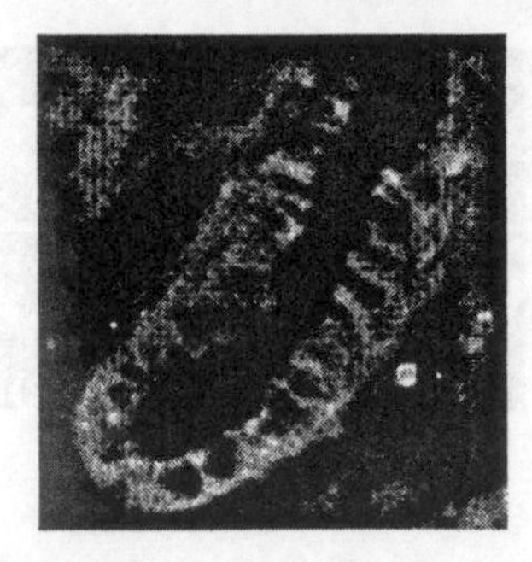

Figure 4 *Labelling of M3 muscarinic receptors along the human colonic crypt-axis. Receptor expression is evident along the entire structure and more pronounced at the crypt-base*

invoked large calcium signals while application of the agonist to regions located along the differentiation pathway elicited progressively smaller responses. As is the case in isolated rat colonic crypts[1] the pharmacology of the acetylcholine-induced colonic crypt calcium signal at the human crypt base suggests that the response is invoked by activation of the M3 muscarinic receptor subtype (data not shown). Consistent with this pharmacology data and the observed calcium signal sensitivity gradient to acetylcholine (Figure 3) is the pattern of M3 muscarinic receptor labelling shown in Figure 4. It is evident from these immunocytochemical studies that there is an M3 muscarinic receptor expresssion gradient along the crypt-axis with greatest labelling at the crypt-base. It is likely that the increased level of M3 receptor expression in the crypt base accounts for the increased sensitivity of the crypt base to acetylcholine and the consequent initation of the calcium signal in the stem cell region. Progressive stimulation of muscarinic receptors that are decreasingly expressed along the crypt axis would give rise to calcium signals with a longer latency and account for the appearance of the colonic crypt calcium wave, but, as is the case in vasopressin stimulated hepatocytes,[3] receptor-oriented calcium waves can be aided and abetted by ancillary mechanisms.

Stimulation of intercellular calcium signals by microinjection of IP_3 into the base of rat colonic crypts[1] and propagation of calcium signals from the site of targeted acetylcholine stimulation (data not shown) suggested a role for gap junctions in colonic crypt calcium signalling. We therefore investigated whether the gap junction blocker heptanol affected the propagation of the calcium signal invoked by global stimulation of the human crypt with acetylcholine. Figure 5 shows that the response at the base of the crypt was largely unaffected whereas the calcium signal recorded from a mid-crypt location was delayed and of smaller amplitude, suggesting that gap junctions play a key role in potentiating the acetylcholine-induced calcium signal along the crypt-axis.

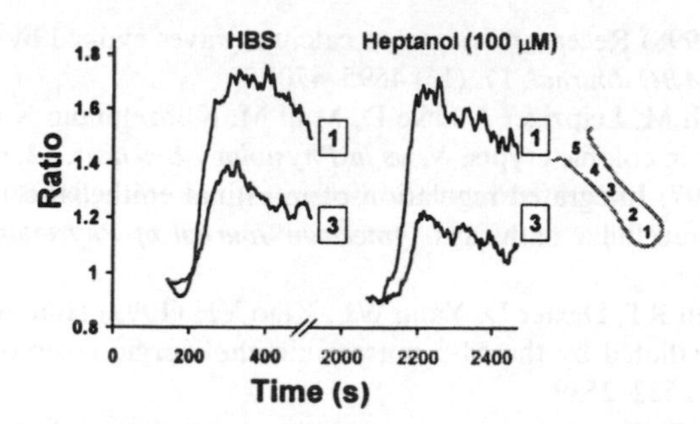

Figure 5 *The gap junction blocker heptanol delays the onset, and reduces the amplitude, of the acetylcholine-induced calcium signal in the mid-crypt region*

4 Summary and Conclusions

Acetylcholine induces a calcium signal in the stem cell region at the base of human colonic crypts that propagates along the crypt-axis. Colonic crypt calcium waves are minimally determined by the pattern of M3 mucarinic receptor expression along the crypt-axis and potentiated by intercellular communication through gap junctions. This study into human colonic crypt calcium signalling events has revealed a striking spatial correlation between acetylcholine-induced calcium signals and the physiology of the crypt. In the short term it is likely that acetylcholine-induced colonic calcium signalling plays a pivotal role in excitation-secretion coupling and serves to drive fluid along the lumen of the crypt towards the crypt-opening.[4,5] In the longer term, a perhaps less intuitive link to forge is that between the acetycholine-induced colonic crypt calcium wave and the coordination of colonic epithelial cell progression along the colonic crypt differentiation pathway. Interestingly, studies into the effects of acetylcholine on human colonic epithelial cell proliferation in our laboratory and others[6] support this notion (data not shown). Moreover, preliminary data in our laboratory indicates that the pattern of colonic crypt calcium signalling is aberrant in the mucosa of colorectal cancer patients. This suggests that there may be a link between colonic crypt calcium signalling, the colonic crypt differentiation pathway and colorectal carcinogenesis. Future studies will address the molecular basis of aberrant colonic crypt calcium signalling associated with colorectal carcinogenesis and investigate modulation by dietary factors such as calcium that have been implicated in the aetiology of colorectal cancer.

5 References

1 Lindqvist S, Sharp P, Johnson IT, Satoh Y and Williams MR (1998) Calcium signalling along the rat colonic crypt-axis. *Gastroenterology* **115**, 1131–1143.

2 Berridge MJ (1993) Inositol trisphosphate and calcium signalling. *Nature* **361**, 315–325.

3 Tordjmann T, Berthon B, Jacquemin E, Clair C, Stelly N, Guillon G, Claret M,

Combettes L (1998) Receptor orientated calcium waves evoked by vasopressin in rat hepatocytes. *EMBO Journal*, **17**, (16) 4695–4703.

4 Greger R, Bleich M, Leipziger J, Ecke D, Mall M, Kunzelmann K (1997) Regulation of ion transport in colonic crypts. *News in Physiology Sciences*, **12**, 62–66.

5 Barrett KE (1997) Integrated regulation of intestinal epithelial cell transport: Intercellular and intracellular pathways. *American Journal of Physiology*, **41** (4), C1069-C1076.

6 Frucht H, Jensen RT, Dexter D, Yang WL, Xiao YH (1999) Human colon cancer cell proliferation mediated by the M-3 muscarinic cholinergic receptor. *Clinical Cancer Research*, **5** (9), 2532–2539

6.11

Influence of Biologically Active Food Constituents on Apoptosis and Cell Cycle in Colorectal Epithelial Cells

Elizabeth K. Lund, Tracy K. Smith, Peter Latham, Rosemary Clarke and Ian T. Johnson

INSTITUTE OF FOOD RESEARCH, COLNEY, NORWICH NR4 7UA, UK

1 Introduction

Tissue architecture in the colonic mucosa is controlled by a balance between cell division and cell death. There are a wide range of dietary factors for which there is emerging evidence of effects on colonic mucosal integrity. Examples include butyrate, formed as a result of bacterial fermentation of fibre, and a range of polyphenolics found in fruit and vegetables. Two food constituents which we have recently observed to exert well defined effects in the rat colon are marine fish containing high concentrations of long chain polyunsaturated fatty acids (PUFA), and brassica vegetables such as Brussels sprouts which contain high concentrations of glucosinolates. Glucosinolate breakdown products cause the hot and bitter flavours of Brassica vegetables. The major glucosinolate in Brussels sprouts is sinigrin, and its breakdown product allyl isothiocyanate (AITC) has been shown to have biological activity. Marine fish oil contains two long chain PUFAs, C22:6, docosahexaenoic acid (DHA) and C20:5 eicosapentaenoic acid (EPA).

2 Experimental

2.1 Animal Studies

Wistar rats (male; 130 g) were housed in a purpose-built, air conditioned animal facility and received a semi-synthetic basal diet *ad lib* prior to random distribution into treatment groups. Typically, one group received the colon carcinogen dimethylhydrazine (DMH; sub cutaneous; 30 mg kg^{-1}) on two occasions separated by 5 days and the other received sham injections of saline. Both groups were then divided again after the second injection with DMH or saline into subgroups, to receive either no treatment or oral supplementation with fish oil or the glucosinolate sinigrin. Forty eight hours after the second injection the rats were killed and the colon was removed intact. Full thickness samples were removed from the mid and distal colon, fixed and stored in acetic acid-ethanol (25:75). Sub-samples were stained with Feulgen's reagent, Isolated crypts were prepared by microdissection, and the frequencies of mitotic and apoptotic crypt cells were determined by morphological criteria under light microscopy. In some experiments animals were maintained on basal diet for up to 18 weeks and numbers of ACF were determined in sheets of whole colon stained with Feulgen's reagent and examined under a dissecting microscope.

2.2 Cell Culture

HT29 human colorectal carcinoma cells were grown in Eagle's minimum essential medium supplemented with glutamine (2 mM), penicillin (50 units ml^{-1}), streptomycin (50 μg ml^{-1}) and fetal calf serum. Cultures were incubated in 5% CO_2 at 37 °C in a humidified incubator and split 1:3, using a mixture of 0.25% trypsin + 0.02% EDTA, every 7 days. The medium was changed in stock cultures every 4 days. Cells were plated out (10^5 cells per flask) and grown for 2 days prior to supplementation of the medium with EPA or allyl isothiocyanate (AITC; 1.2 μg ml^{-1} medium). Cells that had detached from the substrate were separated by removal and centrifugation of the culture medium. The pellet was then re-suspended in phosphate buffered saline. Cells that remained attached to the flask surface were harvested using trypsin (0.25%) and EDTA (0.02%), centrifuged and resuspended. The numbers of floating and adherent cells were determined using a haemocytometer. Cells obtained from duplicate incubations were fixed and stained for flow cytometry. In some experiments the cultures were treated with caspase inhibitors or antioxidants prior to determining numbers of floating and adherent cells.

3 Results and Discussion

Initial studies using rats treated with DMH have shown that, when given immediately after the carcinogen, both sinigrin[1] and EPA-enriched fish oil[2] increase the level of apoptosis near the crypt base. This is associated with a decrease in the numbers of ACF, which may indicate selective destruction of cells with DNA damage. This effect could result in reduced clonal survival of damaged cells, while allowing healthy cells to continue to proliferate. Although both EPA and AITC appear to act by increasing apoptosis, this apparent similarity may be misleading. While EPA increases apoptosis and suppresses mitosis in rat colon, even in the absence of DMH, sinigrin only exerts these effects after DMH treatment. We used the adenocarcinoma cell line HT29 to investigate in more detail how these dietary components can modulate cell death.

We have shown that EPA causes loss of cells from the plastic substrate in a dose- (5–15 μg ml^{-1}) and time- (24–96 h) dependent manner. The range of EPA concentrations that have a detectable effect on cell number are equivalent to those found in the plasma.[3] The loss of cells into the media may be part of a process of programmed cell death akin to the increased frequency of apoptotic cells seen in the rat model. Typical markers of apoptosis such as DNA ladders, labelling of double strand breaks with the TUNEL assay, sub-G0 DNA content and loss of membrane asymmetry with annexin V, were detected in the detached cells, but none of these classical indicators of apoptosis were found in the adherent cell population. The cell loss was however associated with a dose-dependent increase in caspase 3 expression, a critical enzyme in the induction of DNA cleavage during apoptosis (Figure 1). Conversely, cell loss was largely eliminated by treatment with the caspase 3 inhibitor Z-DEVD-FMK.

We were unable to detect any change in cell proliferation rate, measured as rate of incorporation of ^{3}H thymidine per cell, in response to EPA. It may therefore be that EPA reduces the rate of cell division of untransformed, but not transformed cells, and its critical function in controlling cancer risk is by apoptosis.

There are at least two plausible hypotheses as to how EPA may effect changes in cell function. The EPA in the cell may compete with arachidonic acid in the conversion to a range of eicosanoids by Cox-2 and thus alter cell physiology, or alternatively the presence of such a highly unsaturated fatty acid in the membrane of the cell may modify the redox conditions experienced by the cell. We have chosen to examine the latter hypothesis. Cells pre-incubated with the free radical trap, ebselen (40 μM), before addition of EPA to the media, did not lose contact with the substratum any faster than in control conditions. Addition

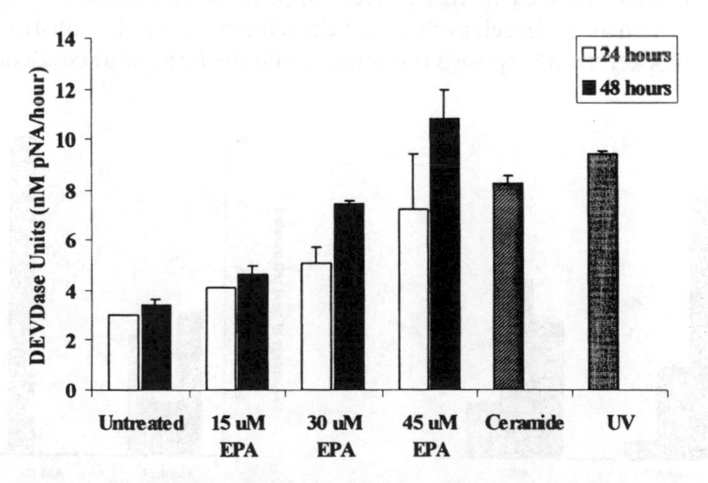

Figure 1 *Activity of DEVDase caspases (chiefly caspase-3) in HT29 cells, following treatment with increasing concentrations of EPA (15–45 μM), for 24 h or 48 h. Positive control cells were also treated with 25 μM C_2-ceramide (3 h), or 12000 μJ cm^{-2} UV (30 minutes). Data is expressed as means ± SEM for pNA liberated from a DEVDase consensus cleavage site per hour*

of buthionine sulfoximine (BSO, 100 μM), which acts to reduce glutathione concentration in the cells, caused no cell loss when given in the absence of EPA but when EPA was added to the medium adherent cell number was significantly reduced (5% of control number) compared to EPA alone (72%) at an EPA dose of 30 μM. These results suggest that EPA causes cell loss as part of a free radical mediated induction of apoptotic cell death.

Phenethyl isothiocyanate (PEITC) has been shown to induce apoptosis, CPP32 activity and DNA cleavage in HeLa cells,[4] and induction of NFκ B can be blocked by the addition of the antioxidant n-acetyl cystein (NAC).[5] These findings suggest that isothiocyanates may induce cell loss *via* a free-radical mediated induction of apoptosis. However, we have found that antioxidants do not protect HT29 cells subsequently exposed to AITC (20 μM). Neither can the cell loss be inhibited by the caspase 3 inhibitor Z-DEVD-FMK. We do however know that the reduction in cell number is at least partly explained by cell loss rather than a lower proliferation rate. Flow cytometric analysis reveals that AITC causes an accumulation of cells in S phase or G2 suggesting a block in the cell cycle during these stages possibly at the G2/M check point (Figure 2).

The high percentage of cells in G2/M in the floating population suggests that AITC causes not only a block in cell cycling but also an accompanying loss of cell adhesion to the substratum. Although histologically the nuclei appear apoptotic, with condensed chromatin and nuclear blebbing, these cells show no sign of apoptosis as detected using Annexin V or the presence of a sub-G0 peak.

In summary, both AITC and EPA can protect the colon against induction of ACF by increasing apoptotic cell death in colorectal crypts, but the cell signalling events involved in this process appear to be distinct. *In vitro*, the former causes an arrest in cell cycling and detachment from the substrate which cannot be blocked by a caspase 3 inhibitor, while the latter induces detachment

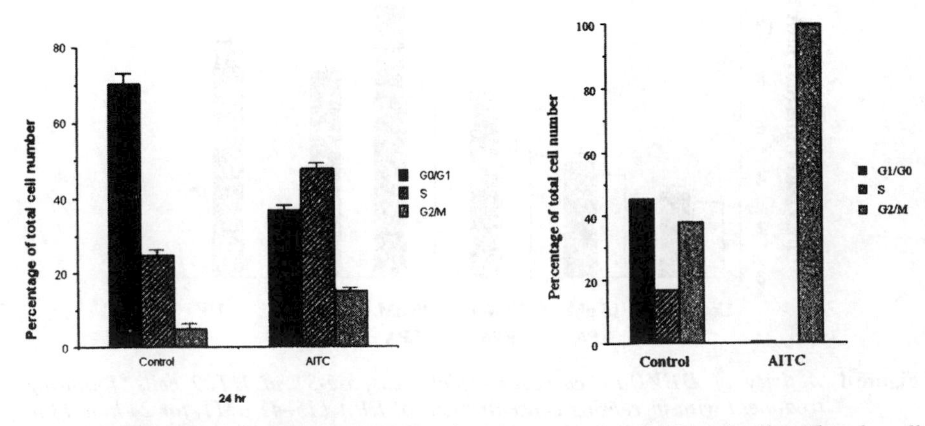

Figure 2 *The effect of AITC (15 μM) on the percentage distribution of cells within the cell cycle in adherent HT29 cells (left) and non adherent (floating) cells after 24 h exposure*

by a caspase 3 dependent pathway, although the cells only express other markers of apoptosis after leaving the substrate.

4 References

1 Smith, T. K., Lund, E. K. and Johnson, I. T. Inhibition of dimethylhydrazine-induced aberrant crypt foci and induction of apoptosis in rat colon following oral administration of the glucosinolate sinigrin. *Carcinogenesis* **19**, 267–73 (1998).
2 Latham, P., Lund, E. K. and Johnson, I. T. Dietary n-3 PUFA increases the apoptotic response to 1,2-dimethylhydrazine, reduces mitosis and suppresses the induction of carcinogenesis in the rat colon. *Carcinogenesis* **20**, 645–50 (1999).
3 Clarke, R. G., Lund, E. K., Latham, P., Pinder, A. C. and Johnson, I. T. Effect of eicosapentaenoic acid on the proliferation and incidence of apoptosis in the colorectal cell line HT29. *Lipids* **34**, 1287–95 (1999).
4 Yu, R., Mandlekar, S., Harvey, K. J., Ucker, D. S. and Kong, A. N. Chemopreventive isothiocyanates induce apoptosis and caspase-3-like protease activity. *Cancer Res* **58**, 402–8 (1998).
5 Chen, Y. R., Wang, W., Kong, A. N. and Tan, T. H. Molecular mechanisms of c-Jun N-terminal kinase-mediated apoptosis induced by anticarcinogenic isothiocyanates. *J Biol Chem* **273**, 1769–75 (1998).

6.12

Raw Brussels Sprouts Block Mitosis in Colorectal Cancer Cells (HT29) and Induce Apoptosis in Rat Colonic Mucosal Crypts *In vivo*

T.K. Smith, R. Clarke, J. Scott and I.T. Johnson

INSTITUTE OF FOOD RESEARCH, NORWICH RESEARCH PARK, COLNEY, NORWICH NR4 7UA, UK

1 Introduction

Epidemiological studies have shown a strong inverse relationship between the consumption of vegetables rich in glucosinolates and colorectal cancer risk.[1] Experimental evidence supports the hypothesis that secondary metabolites present in plant tissues may be responsible for this protective effect. Glucosinolates, and their breakdown products isothiocyanates, have been shown to block carcinogenesis by inducing Phase I and II enzymes, and to suppress tumour formation after treatment with model carcinogens.[2] Previously we have demonstrated that allyl isothiocyanate (AITC) selectively induces cell death in the HT29 cell line.[3] In a rat model, dietary sinigrin induces increased apoptosis in colonic crypt cells and suppresses aberrant crypt foci (ACF) after carcinogen treatment.[4] In the present study we demonstrate that juice obtained from fresh Brussels sprouts, which are rich in sinigrin,[5] blocks mitosis in colorectal cancer cells *in vitro* and induces apoptosis in colonic crypt cells *in vivo*.

2 Methods

2.1 Preparation and Analysis of Sprout Juice

Freshly harvested sprouts (variety 'Stephen') were passed through a commercial juicer; the resulting juice was filtered through a double layer of muslin and stored on iced before use. For *in vitro* studies the juice was subsequently centrifuged (× 5000 rpm) and passed through a 0.22 μm filter before further use. The sinigrin content of whole sprouts and sprout juice was assessed by HPLC after extraction with methanol:water and purification on A25.

2.2 *In Vitro* Study

HT29 human colorectal carcinoma cells were grown in Eagle's minimum essential medium (Sigma Chemicals, Poole, UK) supplemented with glutamine (2 mM), penicillin (50 units ml^{-1}), streptomycin (50 μg ml^{-1}) and fetal calf serum (10%; Imperial, Andover, UK). Cultures were incubated in 5% CO_2 at 37 °C in a humidified incubator and split 1:3, using a mixture of 0.25% trypsin + 0.02% EDTA, every 7 days. The medium was changed in stock cultures every 4 days. Cells were plated out (10^5 cells per flask) and grown for 2 days prior to treatment with sprout juice (60 μl per flask) or allyl isothiocyanate (1.2 μg ml^{-1} medium). Sprout juice was added to treatment flasks at 60 μl per flask, and control and treatment flasks were sealed and incubated in triplicate for a further 24 h. After incubation, cells that had detached from the substrate ('floaters') were separated by removal and centrifugation of the culture medium. The pellet was then re-suspended in phosphate buffered saline. Cells that remained attached to the flask surface ('adherents') were harvested using trypsin (0.25%) and EDTA (0.02%), centrifuged and resuspended. The numbers of 'floaters' and 'adherents' were determined using a haemocytometer. Cells obtained from duplicate incubations were fixed and stained for flow cytometry.

2.3 *In Vivo* Study

Forty male Wistar rats (130 g) housed in a purpose-built, air conditioned animal facility received a semi-synthetic basal diet *ad lib* for 9 days and were then randomly distributed into two treatment groups of 20. One group received DMH (sub cut; 30 mg kg^{-1}) the other received sham injections of saline, and the injections were repeated after a further 5 days. Both treatment groups were then divided again into subgroups of 10 animals to receive either no treatment or freshly prepared sprout juice (2 ml) 23 h and 45 h after the second injection with DMH or saline. After the second injection the rats were killed and the colon was removed intact. Full thickness samples (*ca.* 0.5 cm) were removed from the mid (*ca.* 55% of total length) and distal (*ca.* 95% of total length) colon, fixed and stored in acetic acid-ethanol (25:75). Sub-samples were stained with Feulgen's reagent. Isolated crypts were prepared by microdissection, and the frequency of mitotic and apoptotic crypt cells was determined by light microscopy using morphological criteria.

3 Results

Challenging HT29 cells with raw sprout juice *in vitro* caused a significant increase in the number of detached HT29 cells floating in the medium

Table 1 *Effects of raw sprout juice and isolated allyl isothiocyanate on the proportion of HT29 cells detached and floating in the incubation medium*

	Detached cells (percent total cell number)	
Treatment	*mean (n = 3)*	*sem*
Control	0.18	0.10
Sprout juice	3.18	0.61
AITC	3.71	0.53

(Table 1). Floating cells harvested after treatment with either raw sprout juice or AITC typically had condensed chromatin resembling apoptotic nuclei, but flow cytometric analysis indicated that they were arrested in the G2/M phase of the cell cycle.

Two days after treatment with the colorectal carcinogen DMH there were significantly more apoptotic cells in mid and distal colonic crypts of DMH treated rats compared with controls. Oral gavage with sprout juice had no significant effect on the level of apoptosis in control rats receiving no DMH but there were significantly more apoptotic cells in the basal regions of crypts from DMH-treated rats given sprout juice compared to those given no gavage (Figure 1).

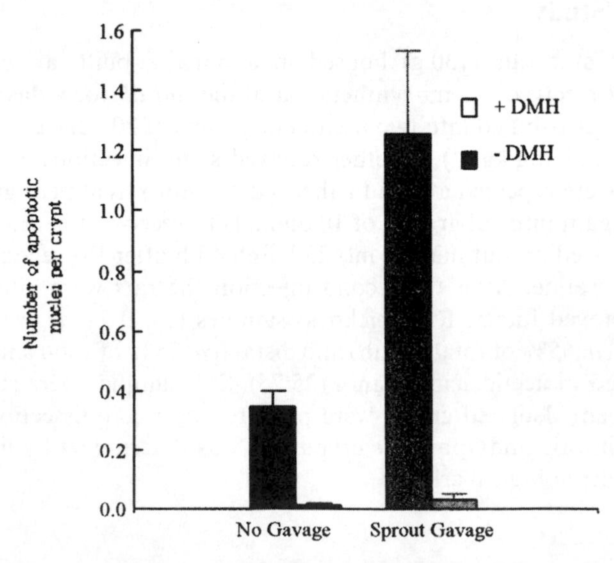

Figure 1 *Numbers of apoptotic cells, expressed as apoptotic nuclei per crypt, in the mid colon of rats given DMH or saline (controls) by sub-cutaneous injection, and subsequently given either no treatment, or Brussels sprout juice by gavage*

4 Discussion

Brussels sprouts contain a high level of sinigrin and other intact glucosinolates which are degraded to isothiocyanates by myrosinase when the plant tissue is physically disrupted. In the present study the juicing technique used to obtain raw juice caused a rapid disappearance of sinigrin. When cultured HT29 cells were exposed to the juice of raw sprouts an increased proportion became detached from the substratum and underwent arrest at the G2/M phase of the cell cycle.

Oral administration of raw sprout juice to rats *in vivo* caused increased apoptotic cell death in crypt cells of the mid and distal colon, but only in animals which had been treated with DMH. These results support previous observations from this laboratory demonstrating suppression of aberrant crypt foci in DMH-treated rats given isolated sinigrin or AITC by gavage, and they are consistent with other published studies on the effect of ITCs on the induction of apoptosis in cancer cells.[6]

The protective effects of Brussels sprouts *in vivo* may be due in part to the high levels of sinigrin present in the plant tissue. We infer that AITC or other degradation products derived from the enzymatic breakdown of sinigrin cause initiated cells to undergo arrest at the G2/M stage of the cell cycle. This may lead to an increased tendency for cells with DNA damage to undergo apoptosis. Enhanced deletion of the initiated cells could in turn lead to fewer pre-neoplastic and neoplastic lesions.

5 References

1 Steinmetz, K.A. and Potter, J.D. (1991). Vegetables, fruit, and cancer. I. Epidemiology. *Cancer Causes Control* **2**, 325–357.

2 Verhoeven, D.T., Goldbohm, R.A., van Poppel, G., Verhagen, H. and van den Brandt, P.A. (1996). Epidemiological studies on brassica vegetables and cancer risk. *Cancer Epidemiology, Biomarkers and Prevention* **5**, 733–748.

3 Musk, S.R. and Johnson, I.T. (1993). Allyl isothiocyanate is selectively toxic to transformed cells of the human colorectal tumour line HT29. *Carcinogenesis* **14**, 2079–2083.

4 Smith, T.K., Lund, E.K., Musk, S.R.R. and Johnson, I.T. (1998). Inhibition of DMH-induced aberrant crypt foci, and induction of apoptosis in rat colon, following oral administration of a naturally occuring glucosinolate. *Carcinogenesis* **19**, 267–273.

5 Fenwick, G.R., Heaney, R.K. and Mullin, W.J. (1983). Glucosinolates and their breakdown products in food and food plants. *Crit Rev Food Sci Nutr* **18**, 123–201.

6 Yu, R., Mandlekar, S., Harvey, K.J., Ucker, D. and Kong, A.N. (1998) Chemopreventive isothiocyanates induce apoptosis and caspase-3-like proteasa activity. *Cancer Research* **58**, 402–408.

Section 7

Anticarcinogenic Effects of Human Diets: Animal Models

7.1

Chemoprevention of Hepatocarcinogenesis in Rats by Diallyl Disulphide and Dipropyl Sulphide: Dose-Response Effect

R. Berges, A.M. Le Bon, M.F. Pinnert and M.H. Siess

UNITÉ DE TOXICOLOGIE NUTRITIONNELLE, INRA,
17, RUE SULLY, 21034 DIJON CEDEX, FRANCE

1 Introduction

It was shown that organosulfur compounds (OSCs) from garlic and onion can prevent carcinogenesis in many animal models and in different organs.[1,2] However, preventive effects are generally achieved by high OSC doses. Whether cells are exposed to such elevated concentrations in a physiological situation is doubtful. Thus dose-response studies are needed to assess low dose effects and to know the minimal dose producing an effect. In this study we determined the effects of 20, 200 and 2000 ppm of diallyl sulfide (DADS) and 200 ppm and 2000 ppm of dipropyl sulfide (DPS) on initiation of hepatocarcinogenesis by diethylnitrosamine (DEN) or aflatoxin B1 (AFB1).

2 Methods

2.1 Chemicals

DADS and DPS were obtained from Aldrich Chemical Co (L'Isle d'Abeau, France).

DADS DPS

2.2 Experimental Design

Five-week-old male SPF Wistar rats (Iffa Credo, L'Arbresles, France) were fed a semi-liquid purified diet. The initiation of carcinogenesis was induced by

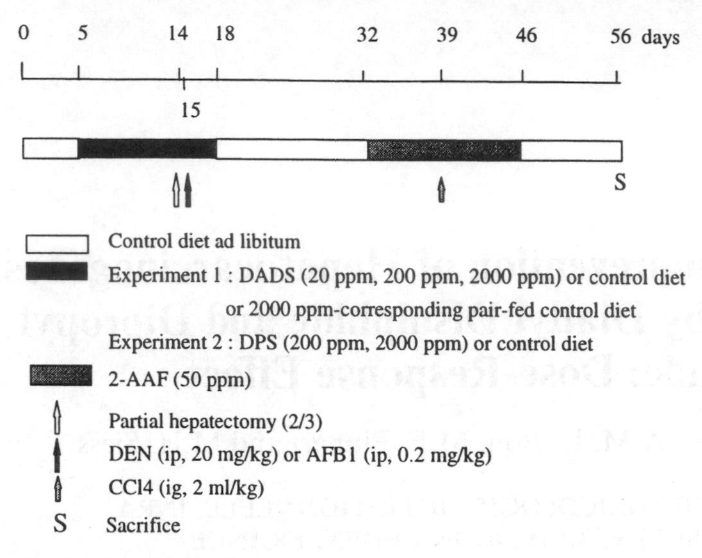

Figure 1 *Experimental design*

intraperitoneal injection of non necrogenic doses of DEN or AFB_1 24 h after a 2/3 partial hepatectomy.[3] DADS (Experiment 1) or DPS (Experiment 2) were added to the diet of rats during the phase of initiation. The rats were later submitted to a selection procedure by 2-acetylaminofluorene (2AAF) treatment and carbon tetrachloride (CCl_4) gavage and killed ten days after (Figure 1). The livers were removed and slices were frozen in isopentane at −150 °C.

2.3 Immunohistochemical Analysis and Quantification

Placental Glutathione-S Transferase (GST-P) positive foci were detected in frozen-cut liver sections by immunohistochemistry.[4] Morphometric analysis of the sections was performed with the aid of an image analysis software (Visilog, Noësis, Orsay, France).

3 Results

3.1 Experiment 1: Effect of Different Doses of DADS on the Initiation of Carcinogenesis in Rat Liver

Whatever the dose of DADS (20, 200 or 2000 ppm), no significant effect was observed on initiation of hepatocarcinogenesis by DEN (Table 1, Figure 2A). In contrast, when administered at 200 and 2000 ppm in the diet, DADS was very efficient in reducing the number and the size of preneoplastic foci initiated by AFB_1 (Table 1, Figure 2B).

Table 1 *Morphometric analysis of GST-P positive foci*

Carcinogen	*Dietary treatment*	*Number of rats*	*Number of foci* cm^{-2}	*% section area occupied by foci*
DEN	Control ad libitum	9	24.29 ± 3.77	0.19 ± 0.05
	20 ppm DADS	9	15.05 ± 2.60	0.10 ± 0.02
	200 ppm DADS	9	25.12 ± 3.74	0.17 ± 0.03
	Pair-fed control	9	26.90 ± 6.13	0.23 ± 0.05
	2000 ppm DADS	9	17.41 ± 3.50	0.11 ± 0.02
AFB_1	Control ad libitum	8	18.87 ± 4.66	1.39 ± 0.62
	20 ppm DADS	10	17.03 ± 2.61	0.80 ± 0.19
	200 ppm DADS	9	5.77 ± 2.55[a]	0.10 ± 0.05[a]
	Pair-fed control	9	10.45 ± 1.94	1.80 ± 0.78
	2000 ppm DADS	9	1.01 ± 0.44[a]	0.01 ± 0.004[a]

[a] Statistically different from the value of corresponding control group (Dunnett's test, $p \leqslant 0.05$).

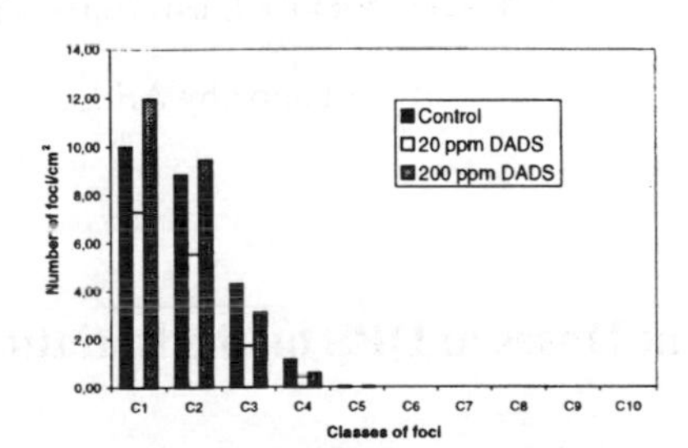

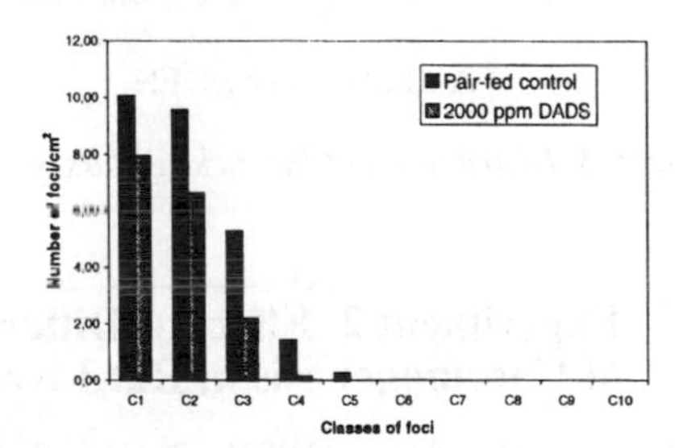

A : Initiation by DEN

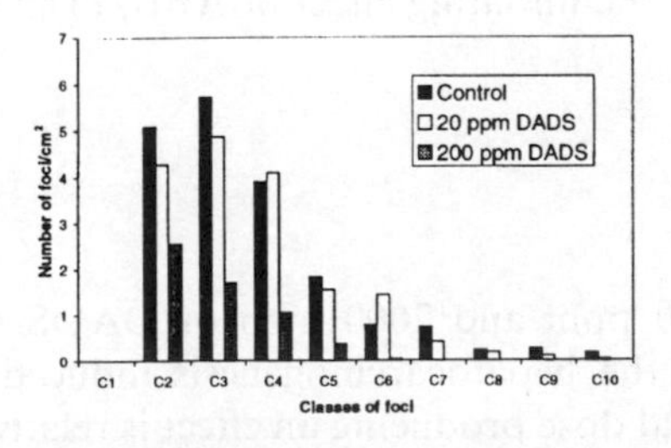

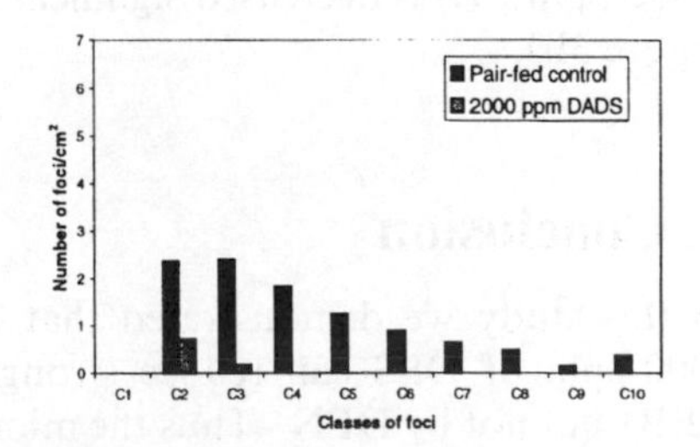

C1=0,25-0,50; C2=0,50-1; C3=1-2; C4=2-4; C5=4-8; C6=8-16; C7=16-32; C8=32-64; C9=64-128; C10>128 (10^{-2} mm^2)

B : Initiation by AFB_1

Figure 2 *Distribution of foci in size classes*

Table 2 *Morphometric analysis of GST-P positive foci*

Carcinogen	*Dietary treatment*	*Number of rats*	*Number of foci cm^{-2}*	*% section area occupied by foci*
DEN	Control ad libitum	10	9.19 ± 2.21	0.07 ± 0.02
	200 ppm DPS	10	12.81 ± 2.60	0.10 ± 0.02
	2000 ppm DPS	10	9.97 ± 2.82	0.07 ± 0.02
AFB_1	Control ad libitum	9	16.39 ± 3.15	0.96 ± 0.30
	200 ppm DPS	10	14.83 ± 3.89	1.18 ± 0.47
	2000 ppm DPS	10	2.80 ± 0.94^a	0.13 ± 0.07^a

[a] Statistically different from the value of corresponding control group (Dunnett's test, $p \leqslant 0.05$).

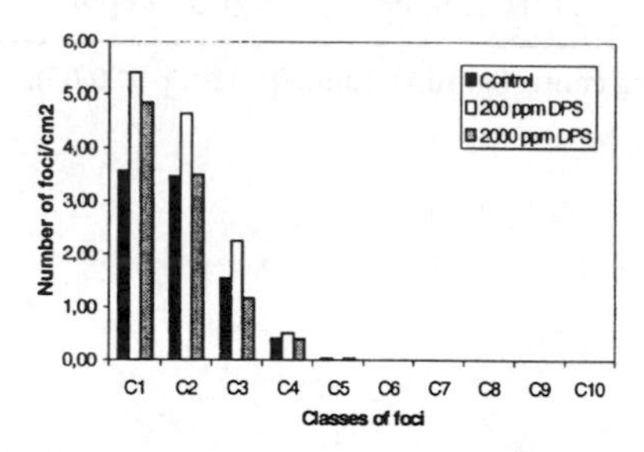

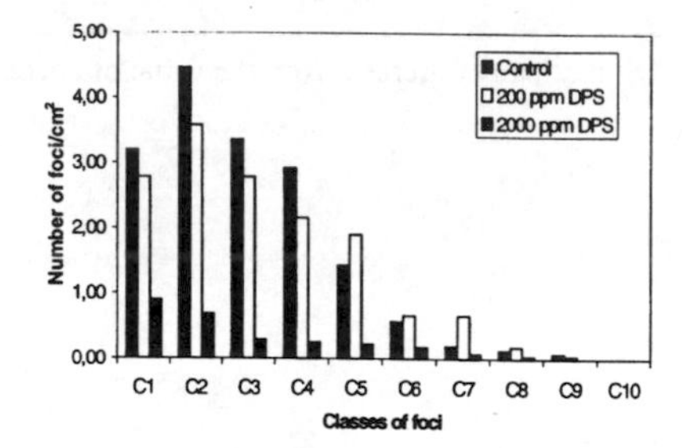

C1=0,25-0,50; C2=0,50-1; C3=1-2; C4=2-4; C5=4-8; C6=8-16; C7=16-32; C8=32-64; C9=64-128; C10>128 (10^{-2} mm^2)

A : Initiation by DEN

B : Initiation by AFB_1

Figure 3 *Distribution of foci in size classes*

3.2 Experiment 2: Effect of Different Doses of DPS on the Initiation of Carcinogenesis in Rat Liver

Whatever the dose, DPS had no influence on initiation of carcinogenesis by DEN (Table 2, Figure 3A). Conversely, when administered at the higher dose (2000 ppm), DPS decreased significantly the initiating effect of AFB_1 (Table 2, Figure 3B).

4 Conclusion

In this study we demonstrated that 200 ppm and 2000 ppm of DADS and 2000 ppm of DPS can reduce strongly the hepatocarcinogenesis induced by AFB1 but not by DEN . Thus the minimal dose producing an effect is relatively high. Previous results of our laboratory have shown that, at the same doses, DADS produced an induction of glutathione transferase.[5] Therefore the anti-carcinogenic effects of DADS could be related to its effect on enzymes involved in the detoxication of carcinogens.

5 References

1 V.L. Sparnins, G. Barany and L.W. Wattenberg, *Carcinogenesis*, 1999, **9**, 131.
2 D. Haber-Mignard, M. Suschetet, R. Bergès, P. Astorg and M.H. Siess, *Nutr. Cancer*, 1996, **25**, 61.
3 E. Cayama, H. Tsuda, D.S.R. Sarma and E. Farber, *Nature*, 1978, **275**, 60.
4 K. Sato, A. Kitahara, K. Satoh, T. Ishikawa, M. Tatematsu and N. Ito, *Gann*, **75**, 199.
5 D. Haber-Mignard, Thèse de Doctorat de l'Université René Descartes, Paris V, France, 1994.

7.2

Dietary Phytohaemagglutinin Reduces Growth of a Murine Non-Hodgkin Lymphoma

Ian F. Pryme,[1] Arpad Pusztai,[2] Susan Bardocz[2] and Stanley W.B. Ewen[3]

[1] DEPARTMENT OF BIOCHEMISTRY AND MOLECULAR BIOLOGY, UNIVERSITY OF BERGEN, ÅRSTADVEIEN 19, N-5009 BERGEN, NORWAY
[2] THE ROWETT RESEARCH INSTITUTE, GREENBURN ROAD, BUCKSBURN, ABERDEEN AB2 9SB, SCOTLAND, UK
[3] DEPARTMENT OF PATHOLOGY, UNIVERSITY OF ABERDEEN, UNIVERSITY MEDICAL BUILDINGS, FORESTERHILL, ABERDEEN AB9 2ZD, SCOTLAND, UK

1 Introduction

We have shown in a number of earlier studies that the bulk tumour mass (dry weight) of intraperitoneal (ascites) or subcutaneous (solid) Krebs II non-Hodgkin lymphoma (NHL) tumours in mice is less when phytohaemagglutinin (PHA), a lectin present in raw kidney bean (*Phaseolus vulgaris*), is incorporated into the diet.[1–5] We have also observed that the lectin reduces the rate of proliferation of a plasmacytoma tumour in Balb/c mice.[6] A three-fold increase in the survival time of injected mice has been observed.[7] The lectin, being resistant to proteolysis by digestive enzymes, retains its biological activity during passage through the gastrointestinal tract.[8,9] PHA causes a dose-dependent and fully reversible hyperplastic growth of the small intestine [for refs. see ref. 10]. Despite a loss in body weight due to the lipolytic effect of PHA,[11,12] there are apparently no long term detrimental effects on the animals. In addition to stimulating growth of the gut the lectin also results in an extensive absorption of amino acids and other nutrients from the intestinal lumen.[13–15]

The histological appearance of the murine tumours was similar to a human non-Hodgkin lymphoma. The predominant cell type was centroblastic lymphocytes with a variable number of centrocytic lymphocytes. A host lymphoid response to the transplanted tumour was absent and the tumour was locally invasive of surrounding structures, including skeletal muscle. The level of necrosis was somewhat higher in tumours from PHA-fed compared to control animals.[16]

The actual importance of the timing of feeding the animals the PHA-containing diet with respect to when the tumour cells were injected, was studied in a dietary shift experiment.[17] It was observed that Krebs II ascites tumours contained almost twice the number of cells when mice were pre-fed on lactalbumin (LA) control diet for three days, intraperitoneally (i.p.) injected with tumour cells and then kept on the same diet for a further eight days, compared to animals fed a diet containing PHA (7 mg g^{-1} diet) during the same period. The quantity of Krebs II cells recovered in ascites tumours was also significantly lower when the LA diet was replaced by one containing PHA on the same day as tumour cells were i.p. injected.

There is increased evidence which strongly suggests that diet and nutrients play a causative role in the development of cancer. These factors may be useful in the design of active methods of preventing the development of the disease or in the active management of tumour growth [for refs. see ref. 16]. Some data, for example, indicate that overnutrition significantly promotes the development of certain cancers, including those of the colon, pancreas, kidney, breast, ovary endometrium and prostate.[17] It is important, however, that nutrition should not be regarded as a sole means of cancer treatment/prevention but rather as a necessary component to be taken into consideration when planning anti-cancer strategy.[18]

Earlier observations indicated that the availability of polyamines from a common body pool is important for the development of non-Hodgkin lymphoma tumours in mice.[3,19–22] Experiments were thus designed to correlate tumour growth under conditions where diets were restricted in their nitrogen content (protein depletion), the intention thus being to severely reduce *de novo* polyamine synthesis by the developing tumour. Experiments were thus performed where mice were either fed diets of normal protein content or non-protein containing diets, in order to be able to study growth of a transplanted non-Hodgkin lymphoma tumour under conditions where animals were fed PHA to stimulate gut hyperplasia.

2 Materials and Methods

2.1 Experimental groups

All management and experimental procedures in this study using female NMRI mice were carried out in strict accordance with the requirements of UK Animals (Scientific Procedures) Act 1986 by staff licensed under this Act to carry out such procedures. The mice were caged in six groups (5 individuals per group) and initially fed a standard pellet diet with free access to water. Experiments were commenced by replacing the pellet diet with a semi-synthetic diet lacking protein: NPD (see ref. 4 for details of diets). Seven days later all mice were injected subcutaneously (s.c.) with 2×10^6 Krebs II cells[1] and diets were switched as appropriate. Group 1 was maintained on the same diet (NPD) while Groups 2 and 3 were fed NPD supplemented with 1.75 and 3.5 mg PHA/g diet respectively (for details of PHA-containing diets see ref. 4). Mice in Group 4 were switched to a normal diet containing 10% lactalbumin (LA) as the protein source, while Groups 5 and 6 were switched to LA diets containing either 1.75 or 3.5 mg PHA/g diet respectively. After 14 days on the respective diets all mice were sacrificed and tumours and organs dissected out and immediately frozen in liquid nitrogen. Tissues were later lyophilised for dry weight determination.

3 Results and Discussion

Figure 1 shows tumour growth, expressed as dry tumour mass related to body dry weight, in the experimental groups. In mice pre-fed on NPD for seven days and then maintained on the same diet for 14 days after s.c. injection of cells, tumours were smaller than those in mice which were fed the LA (control) diet. Prefeeding on NPD followed by a switch to LA on the day of injection resulted in tumours which were up to a factor of five times larger than those both pre- and post-fed on NPD. The dry mass of tumours which developed in mice

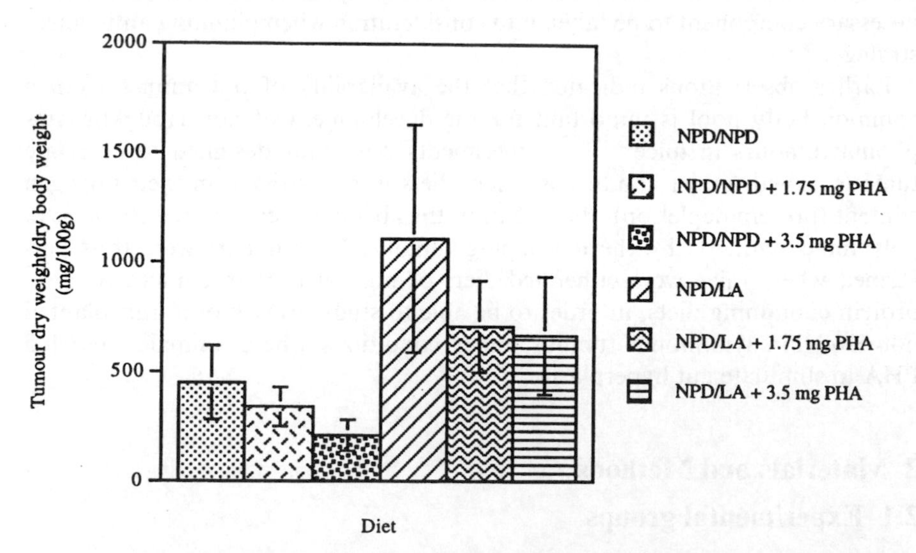

Figure 1 *NHL tumour growth expressed as dry tumour mass related to dry body weight. Eight groups of mice were fed the diets indicated in the figure (for details see Materials and Methods). The diet (7 days) before s.c. injection of tumour cells is indicated before the slashed line and the diet fed post-injection after. All mice were sacrificed after fourteen days. Tumours were excised and the dry weight determined following lyophilisation. The values were related to dry body weight and are expressed* $\pm$ *SD*

switched from NPD to NPD containing 1.75 or 3.5 mg PHA g/diet on the day of injection of tumour cells was markedly reduced compared to that in mice only fed NPD. It thus appeared that PHA, when included in the NPD, was able to effectuate a further reduction in tumour growth. The major increase in tumour mass caused by the switch from NPD to LA diet on the day of s.c. injection of Krebs II cells was reversed by the inclusion of increasing concentrations of PHA in the diet. The highest concentration of PHA (7.0 mg) reduced tumour mass to a level intermediate between those of feeding LA or NPD for the duration of the experiment. These results demonstrated that a switch from a non-protein containing diet (NPD) to one containing a normal level of protein (LA) on the day of tumour cell injection had a dramatic effect on dry tumour mass. Significantly, however, inclusion of PHA in the LA diet had a profound effect on reducing tumour size.

It appears, therefore, that PHA triggers an event, or a series of events, which cumulate in a considerable reduction in tumour growth. At present the actual biochemical nature of the manner in which PHA exerts its suppressive effect on tumour growth is not known. Earlier experiments have shown that there appears to be a correlation between the requirement of the NHL cells for extraneous dietary polyamines and the hyperplastic stimulation of gut growth which causes a sequestration of polyamines in the intestinal tissue.[3,19] The observations described here indicate that the feeding of a nitrogen-rich diet to mice fed a nitrogen-depleted diet (NPD) for seven days before injection of tumour cells favoured growth of the transplanted NHL tumour. Interestingly, addition of PHA to the LA diet appeared to effectively compete with the switch to the protein-rich diet. An increase in the incidence of necrosis in the tumour was observed. Since it has been established that PHA induces hyperplasia of the gut it is possible that this 'normal' growth competes with the developing tumour for important growth factors and nutrients, including polyamines. It is perceived that this may effectively starve the tumour, causing its decreased rate of proliferation.[21]

In conclusion the data clearly indicate that the excessive feeding of protein to tumour-bearing individuals should, if possible, be avoided. Since addition of the plant lectin phytohaemagglutinin to diets caused a marked reduction in tumour growth, the use of plant lectins may perhaps play an important role in the future development of novel ways of designing anti-cancer strategy. This is supported by recent exciting observations where a mistletoe lectin (ML-1), when added to the LA diet, was shown to produce a complete histological disappearance of transplanted murine NHL tumours in 60% of individuals in a test group.[23]

Acknowledgements

This work was supported by EU COST Actions 98 and 917. SB and AP at RRI received support from SOAEFD. Financial support from the Norwegian Cancer Society and Kaptein L.A. Hermansens og hustru I. Hermansens legat is gratefully acknowledged.

4 References

1 I.F. Pryme, S. Bardocz and A. Pusztai, *Cancer Lett.*, 1994, **76**, 133.
2 I.F. Pryme, A. Pusztai and S. Bardocz, *Int. J. Oncol.*, 1994, **5**, 1105.
3 S. Bardocz, G. Grant, T.J. Duguid, D.S. Brown, A. Pusztai and I.F. Pryme, *Int. J. Oncol.*, 1994, **5**, 1369.
4 I.F. Pryme, A. Pusztai and S. Bardocz, *J. Exp. Therap. Oncol.*, 1996, **1**, 171.
5 I.F. Pryme, A. Pusztai, G. Grant and S. Bardocz, *J. Exp. Therap. Oncol.*, 1996, **1**, 273.
6 I.F. Pryme, S. Bardocz, G. Grant, T.J. Duguid, D.S. Brown and A. Pusztai, 'Effects of Antinutrients on the Nutritional Value of Legume Diets', S. Bardocz, E. Gelencsér, A. Pusztai (Eds.), EC publications, 1996, COST 98 vol. 1, p. 34.
7 I.F. Pryme, A. Pusztai, G. Grant and S. Bardocz, *Cancer Lett.*, 1996, **103**, 151.
8 A. Pusztai, S.W.B. Ewen, G. Grant, W.J. Peumans, E.J.M. Van Damme and S. Bardocz, 'Lectin Reviews', D.C. Kilpatrick, E. Van Driessche and T.C. Bog-Hansen (Eds.), Sigma Library, St Louis, MO, 1991, Vol. 1, pp. 1–15.
9 A. Pusztai and S. Bardocz, *Glycosci. Glycotechnol.*, 1996, **8**, 149.
10 I.F. Pryme, A. Pusztai, S. Bardocz and S.W.B. Ewen, *Histol. Histopathol.*, 1998, **13**, 575.
11 S. Bardocz, G. Grant and A. Pusztai, *Brit. J. Nutr.*, 1996, **76**, 613.
12 S. Bardocz, G. Grant, S.W.B. Ewen, I.F. Pryme and A. Pusztai, 'Effects of Antinutrients on the Nutritional Value of Legume Diets', EC publications, 1998, COST 98, S. Bardocz, U. Pfüller and A. Pusztai (Eds.), vol. 5, p. 208.
13 A. Pusztai, G. Grant, L.M. Williams, D.S. Brown, S.W.B. Ewen and S. Bardocz, *Med. Sci. Res.*, 1989, **17**, 215.
14 A. Pusztai, 'Plant Lectins', Cambridge Univ. Press, Cambridge, 1991, pp. 1–251.
15 S. Bardocz, *Eur. J. Clin. Nutr.*, 1993, **47**, 683.
16 I.F. Pryme, A. Pusztai, S. Bardocz and S.W.B. Ewen, *Cancer Lett.*, 1999, **139**, 145.
17 E.L. Wynder, *Fed. Proc.*, 1976, **35**, 1309.
18 R.S. Rivlin, *J. Am. Coll. Nutr.*, 1982, **1**, 75.
19 S. Bardocz, G. Grant, T.J. Duguid, D.S. Brown, M. Sakhri, A. Pusztai, I.F. Pryme, D. Mayer and K. Wayß, *Med. Sci. Res.*, 1994, **22**, 101.
20 I.F. Pryme, S. Bardocz, G. Grant, T.J. Duguid, D.S. Brown and A. Pusztai, *Cancer Lett.*, 1995, **93**, 233.
21 S. Bardocz, G. Grant, T.J. Duguid, D.S. Brown, A. Pusztai and I.F. Pryme, *Cancer Lett.* 1997, **121**, 25.
22 I.F. Pryme, G. Grant, A. Pusztai and S. Bardocz, 'Polyamines in Health and Nutrition', S. Bardocz and A. White (Eds.), Kluwer Academic Publishers, Norwell, Mass., 1999, Chapter 22, p. 283.
23 S.W.B. Ewen, S. Bardocz, G. Grant, I.F. Pryme and A. Pusztai, 'Effects of Antinutrients on the Nutritional Value of Legume Diets', S. Bardocz, U. Pfüller and A. Pusztai (Eds.), EC publications, 1998, COST 98 vol. 5, p. 221.

7.3

Inhibitory Effects of Beer and Other Alcoholic Beverages on the Mutagenesis and DNA-Adduct Formations Induced by Several Carcinogens: Identification of One of the Antimutagenic Factors

Sakae Arimoto-Kobayashi, Sachiko Kimura, Chitose Sugiyama, Nanaho Harada, Miyuki Takeuchi, Miyuki Takemura and Hikoya Hayatsu

FACULTY OF PHARMACEUTICAL SCIENCES, OKAYAMA UNIVERSITY, TSUSHIMA, OKAYAMA 700-0082, JAPAN

Beer and other alcoholic beverages have been consumed at great amounts by humans. Heterocyclic amine (HA) mutagens are present in daily food as a result of cooking by heat. The HAs are currently suspected as human carcinogens. 2-Chloro-4-methylthiobutanoic acid (CMBA) was found from salt-nitrite treated Sanma hiraki fish as a potent mutagen.[1] We have explored the possibility that beer and other alcoholic beverages are antimutagenic against HAs and other carcinogens. In the *Salmonella* mutation assays, beer showed inhibitory effects against HAs (Trp-P-2(NHOH) and Glu-P-1(NHOH)) and several other agents. Those mutagens include CMBA, in addition to *N*-methyl-*N'*-nitro-*N*-nitrosoguanidine (MNNG) and benzo [*a*]pyrene-*trans*-7,8-dihydrodiol-9,10-epoxide (anti) (BPDE) that are directly mutagenic towards the bacteria (Figure 1). Japanese sake, red and white wines, and brandy were also effective. Since ethyl alcohol did not show these effects, other components of the beverages must be responsible for these inhibitions.

Non-volatile beer-components were administered orally to CDF1 mice together with Trp-P-2. Adducts in the liver DNA, detected by ^{32}P-postlabeling,[2] were significantly decreased by the administration of the beer components, as compared to those in control animals fed Trp-P-2 only.

We found that several phenolic compounds known to be present in beer were antimutagenic towards these mutagens in the *Salmonella* assays, but their

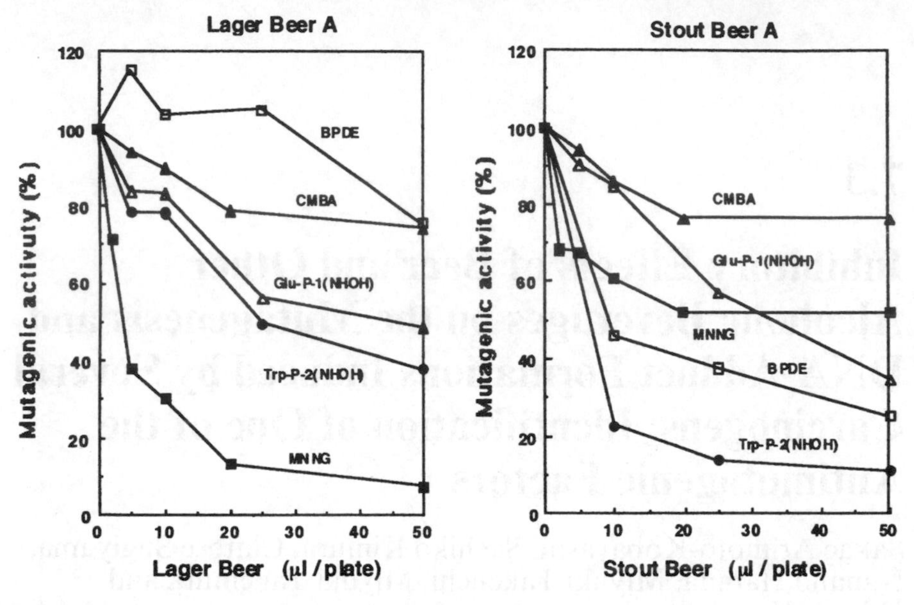

Figure 1 *Antimutagenic effects of beer samples on the mutagenicity*

activities were very small. We conclude that some component(s) of beer and of the other beverages are antimutagenic towards a series of mutagens. To isolate the antimutagenic component in a freeze-dried beer a sample was subjected to successive column chromatographies (Figure 2).

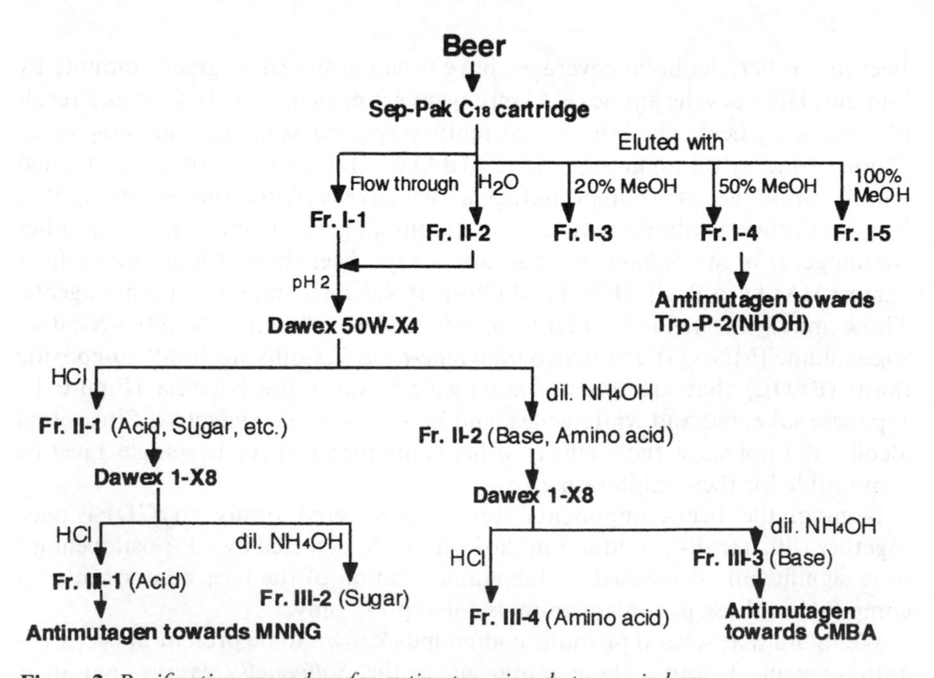

Figure 2 *Purification procedure for antimutagenic substances in beer*

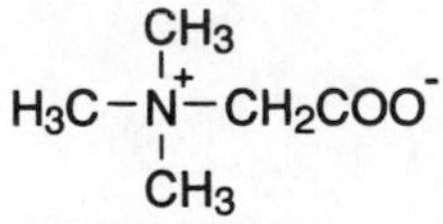

$$H_3C-\overset{CH_3}{\underset{CH_3}{\overset{|}{\underset{|}{N^+}}}}-CH_2COO^-$$

Figure 3 *Glycine betaine*

We found that the antimutagenic compounds towards CMBA, MNNG and Trp-P-2(NHOH) are not identical. To isolate the antimutagenic component toward CMBA, fraction III-3 was subjected to HPLC analysis. A single peak in an HPLC exhibited antimutagenic activity. The compound was isolated and characterized by ^{1}H- and ^{13}C-NMR and ESI-mass spectrometry. It was identified as glycine betaine (Figure 3).

Identification of the antimutagenic compounds towards other mutagens is under way.

References

1 W. Chen, J.H. Weisburger, E.S. Fiala, S.G. Carmella, D. Chen, T.F. Spratt and S.S. Hecht, Unexpected mutagen in fish, *Nature*, 1995, **374**, 599.

2 M. Ochiai, H. Nagaoka, K. Wakabayashi, Y. Tanaka, S.-B. Kim, A. Tada, H. Nukaya, T. Sugimura and M. Nagao, *Carcinogenesis*, 1993, **14**, 2165–2170.

3 S. Arimoto-Kobayashi, C. Sugiyama, N. Harada, M. Takeuchi, M. Takemura and H. Hayatsu, *J. Agric. Food Chem.*, 1999, **47**, 221–230.

4 S. Kimura, H. Hayatsu and S. Arimoto-Kobayashi, *Mutation Res.*, 1999, **439**, 267–276.

7.4

The Effect of Selected Complex Carbohydrates on Intestinal Tumour Risk in *Min* Mice

C.A. Higgins, J. Coaker, F. Armstrong and J.C. Mathers

HUMAN NUTRITION RESEARCH CENTRE, DEPT. BIOLOGICAL AND NUTRITIONAL SCIENCES, UNIVERSITY OF NEWCASTLE, NEWCASTLE UPON TYNE NE1 7RU, UK

1 Introduction

Diet is a major environmental determinant of colorectal cancer (CRC) with one recent epidemiological study finding a strong negative correlation (R = −0.76) between starch and CRC incidence between countries.[1] Starch which escapes digestion in the small bowel (resistant starch; RS) and nonstarch polysaccharides (NSP) are fermented in the large bowel resulting in the production of short chain fatty acids (SCFAs), including butyrate. At physiological concentrations (1–5 mM), butyrate supresses cell proliferation and induces differentiation and apoptosis of colon cancer cells.[2] Mutations in the *APC* (adenomatous polyposis coli) gene are an early event in the majority of human colorectal tumours. The *Min* (multiple intestinal neoplasia) mouse model, which has a nonsense mutation at codon 850 of the *Apc* gene, develops several bowel tumours spontaneously making it a valuable model for dietary studies of intestinal tumourigenesis.[3] This study aimed to determine the effects of selected complex carbohydrates on tumour multiplicity within the *Min* mouse and to ascertain whether increased crypt cell proliferation (CCP) or decreased apoptosis in histologically normal mucosa was a biomarker of tumour risk.

2 Materials and Methods

2.1 Experimental Design

At weaning, heterozygous *Min* mice were allocated at random to one of four experimental diets (40 mice/diet). The semipurified 'western' style experimental diets were high in animal fat, sucrose and animal proteins. The basal diet was modified by the addition (100 g kg^{-1}) of guar gum (GG), a highly fermentable NSP, raw potato starch (RPS) or Hylon VII (HYL), both forms of RS in place of cellulose in the basal diet. Outcome measurements were made at 60 and 90 days after commencing feeding of the experimental diets. Animals were injected intraperitoneally with vincristine sulphate (1 mg kg^{-1} body weight) 2 hours prior to sampling to arrest cells in metaphase in preparation for measurement of CCP.[4] Tumour number, size and location were recorded. All tumour material was processed to 10% neutral buffered formalin for routine histological diagnosis. Apparently normal intestinal mucosa for CCP and apoptosis measurement was collected at five sites throughout the intestine and fixed overnight in 10% neutral buffered formalin. Apoptosis was detected in hemi-crypts by morphological criteria[5] on H&E stained tissue sections.

2.2 Statistical Analyses

For tumour multiplicity data, results for both cohorts (60 and 90 days on test diets) were pooled and analysed using a general linear model with cohort and test diet as fixed effect factors. For CCP and apoptosis measurements, only data from the 60 day cohort are presented.

3 Results

3.1 Tumour Multiplicity

Total intestinal tumour multiplicity was almost doubled ($p < 0.01$) by GG feeding and was higher but not significantly so in the HYL animals (Table 1).

3.2 Crypt Cell Proliferation (CCP) and Apoptotic Index

CCP decreased from the proximal duodenum to the terminal colon with all treatments. No significant differences between diets on CCP were observed.

Table 1 *Effect of selected polysaccharides on intestinal tumour multiplicity in Min mice (means of 60 and 90 days dietary intervention) (*significant at $p < 0.05$)*

	Diet					*Probability of effects*		
	Basal	*+GG*	*+RPS*	*+HYL*	*SEM*	*GG*	*RPS*	*HYL*
				Tumour multiplicity				
Small Intestine	12.3	22.0	13.2	19.2	2.42	0.01*	0.98	0.09
Colon	0.53	0.65	0.35	0.38	0.13	0.85	0.44	0.57
Total tumours	12.8	22.7	13.6	19.5	2.47	0.01*	0.99	0.11

Only a small proportion (maximum mean = 1.37% for any treatment) of gut epithelial cells were identified as apoptotic and there tended to be a higher proportion of apoptotic cells in the small intestine than the large bowel. The apoptotic index was generally lower in GG fed animals than those fed RS but no statistically significant treatment effects were observed.

4 Discussion

We have demonstrated that the type of dietary polysaccharides fed to *Min* mice can alter tumour multiplicity markedly. GG and Hylon VII increased tumour multiplicity in the small intestine but the significance of this for humans remains uncertain since most human tumours are in the colon. In this study we did not detect any significant difference in CCP between diets at any intestinal site despite the differences observed in tumour multiplicity. This confirms that as little as 50% of the normal cell complement of Apc protein may be adequate to ensure normal proliferation.[6] There was no evidence that CCP or apoptotic index in normal mucosa was a reliable biomarker of cancer risk.

Acknowledgement

This study was funded by MAFF contract AN0317.

5 References

1 A. Cassidy, S.A. Bingham and J.H. Cummings, *Br. J. Canc.*, 1994, **69**, 937.
2 A. Hague, A.J. Butt and C. Paraskeva, *Proc. Nutr. Soc.*, 1996, **55**, 937.
3 A.R. Shoemaker, A.R. Moser and W.F. Dove, *Canc. Res.*, 1995, **55**, 4479.
4 J.C. Mathers, J. Kennard and O.F.W. James, *Br. J. Nutr.*, 1993, **70**, 567.
5 J.F.R. Kerr, A.H. Wyllie and A.R. Currie, *Br. J. Nutr.*, 1972, **26**, 239.
6 J.C. Mathers, M. Kooshkghazi, M.J. Coaker, S. Williamson, A. Kartheuser, J. Burn and R. Fodde, *Proc. Nutr. Soc.*, 1998, **57**, 43A.

7.5

8-Methoxypsoralen Compromises the Development of Ovarian Large Antral Follicles in Wistar Rats

Moussa M. Diawara,[1]* Kathryn J. Chavez,[1] David E. Williams[2] and Patricia B. Hoyer[3]

[1] DEPARTMENT OF BIOLOGY, UNIVERSITY OF SOUTHERN COLORADO, PUEBLO, CO 81001, USA
[2] DEPARTMENT OF ENVIRONMENTAL AND MOLECULAR TOXICOLOGY, OREGON STATE UNIVERSITY, CORVALLIS, OR 97331, USA
[3] DEPARTMENT OF PHYSIOLOGY, UNIVERSITY OF ARIZONA, TUCSON, AR 85721, USA

Abstract

The psoralens are naturally occurring metabolites found in many crop plants. Bergapten (5-methoxypsoralen) and xanthotoxin (8-methoxypsoralen) are widely used in skin photochemotherapy. Research has shown that dietary bergapten (5-methoxypsoralen) and xanthotoxin (8-methoxypsoralen) reduced birthrate in female rats when males and females were exposed to these chemicals. In addition, the two compounds, administered to female rats (mated to undosed males), reduced levels of circulating 17β-estradiol, and the number of implantation sites, pups, and corpora lutea, suggesting a direct effect on ovarian follicles. The present study was initiated to determine whether effects of xanthotoxin can be observed in ovarian follicles. Female Wistar rats were dosed daily (30 days) with vehicle control or xanthotoxin (180 mg kg^{-1}, p.o.). Ovaries were collected and prepared for histological evaluation. Antral follicles (diameter >600 μm) were found to be larger ($p < 0.05$) in ovaries from xanthotoxin-dosed compared with control rats. Moreover, the granulosa cell layer thickness, and granulosa cell layer:follicle diameter ratio were reduced ($p < 0.05$) in dosed animals. This provides visual evidence that the granulosa cell population in large antral follicles is affected by xanthotoxin, and is consistent with its observed reduction in circulating 17β-estradiol levels.

* Corresponding author: moussa@uscolo.edu

1 Introduction

The psoralens occur naturally as secondary metabolites in various plant species including many grocery fruits and vegetables.[1–3] The concentrations of psoralens found in produce have been reported to be hazardous to human and animal health.[2,3] In addition to the potential agricultural exposure, oral administration of synthetic forms of 5-methoxypsoralen (bergapten) and 8-methoxypsoralen (xanthotoxin or methoxsalen) in combination with UVA irradiation, a procedure referred to as PUVA, remains the treatment of choice for skin disorders such as psoriasis and vitiligo.[4,5] The psoralens are also used in the treatment of skin conditions such as skin depigmentation (leprosy, vitiligo, and leucoderma), mycosis fungoides, polymorphous dermatitis, and eczema.[2,5,6] The medicinal use of the psoralens has been, however, associated with an increased incidence of skin cancer in patients.[3,7,8] The psoralens proved to be both mutagenic and carcinogenic.[9–14] The United States National Toxicology Program (NTP/NIEHS) tested 8-methoxypsoralen (xanthotoxin or methoxsalen) for subchronic toxicity in rats using oral administration.[15] The study found that the high doses of 8-methoxypsoralen (200–400 mg kg^{-1} body) altered histology of liver, testes, and adrenals, reduced body weight gain, increased liver somatic index, and even caused death.

A number of recent reports documented other potential risks associated with PUVA. Complications of burns received in a tanning saloon by a patient under PUVA resulted in the death of the patient.[1] PUVA has been implicated in the increased incidence of malignant melanoma in several patients, resulting in the death of some.[8] Gunnarskog *et al.*[16] reported a 'marked increase in low-birth weight infants' to women exposed to PUVA prior to pregnancy. Animal studies by Diawara and colleagues[17] showed that oral administration of 8-methoxypsoralen and/or 5-methoxypsoralen (at 0–180 mg kg^{-1} body) to male and female rats also significantly reduced animal birthrate. Subsequent studies demonstrated that the two compounds administered to females only (mated to undosed males) reduced the numbers of implantation sites, fetuses and corpora lutea in pregnant females compared with the control group.[18] The two compounds also reduced levels of circulating blood 17β-estradiol in non-pregnant females in a dose-dependent manner. However, female cyclicity was not significantly disrupted.[18] Taken together, these findings are consistent with a reduction in number of ovulations in psoralens-dosed females. We initiated the current investigations to determine whether direct effects of xanthotoxin can be observed in ovarian follicles.

2 Materials and Methods

Ten non-pregnant female Wistar rats (from Charles River Laboratory; Charles River; WI/BR; Wilmington, MA) were dosed daily (30 days) with vehicle control or xanthotoxin (180 mg kg^{-1}, p.o.). The drug was dissolved in acetone, adsorbed onto alphacel, and then incorporated (after evaporation of acetone) in a powdered AIN 76-A (ICN) diet/agar gel mixture.[18] The control diet was prepared in a similar manner without xanthotoxin. The doses administered were based on previous research.[15,18] The rats were individually housed on each dietary treatment in standard sized cages (under UVA radiation for 45 min–1 h/day) and had free access to a known amount of diet mixture in non-spill stainless steel containers, and to deionized water. At the end of the 30 day dosing period, females were sacrificed. For each animal, the ovaries were removed, fixed and serially sectioned for histological examination at the light microscope level.[19] The number and size of follicles with diameter between 360–600 μm and follicles with diameter >600 μm were recorded by examination of all sections in each ovary from each animal (n = 10 animals/group). The thickness of granulosa cells layer around the follicles was also estimated using a calibrated ocular.

3 Results and Discussion

Antral follicles with diameter >600 μm were significantly ($p < 0.05$) larger in ovaries from xanthotoxin-dosed females as compared with control groups ($3.776 \times 10^5 \pm 0.292 \times 10^5$ μm^2 control; $5.872 \times 10^5 \pm 0.408 \times 10^5$ μm^2 xanthotoxin). In addition, the thickness of the granulosa cell layer (81.600 ± 5.820 μm control; 47.300 ± 5.060 μm xanthotoxin), and granulosa cell layer:follicle diameter ratio (0.003 ± 0.00024 control; 0.001 ± 0.00017 xanthotoxin) were both significantly ($p < 0.05$) reduced in the dosed females. Figures 1 and 2 illustrate the size of a large antral ovarian follicle and thickness of the granulosa cell layer in a control female and in a xanthotoxin-dosed female. However, the number of large antral follicles per ovary was not significantly affected (4.67 $\pm$

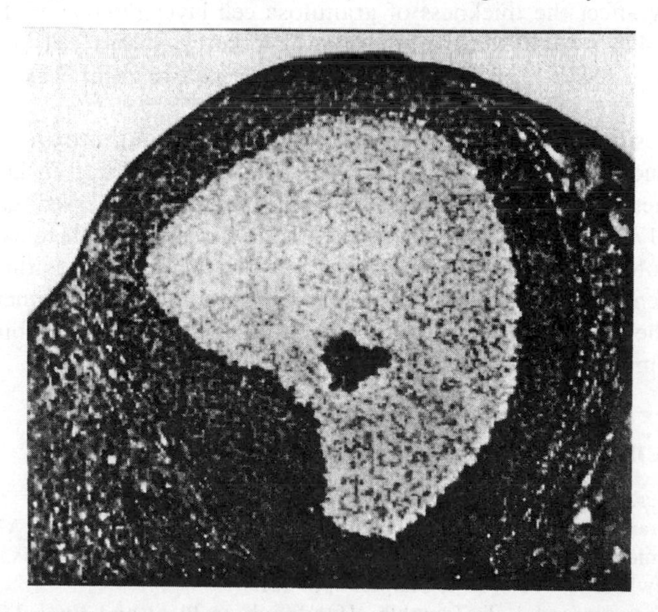

Figure 1 *Size of a large ovarian follicle and thickness of granulosa cell layer in a control female Wistar rat.* × *40*

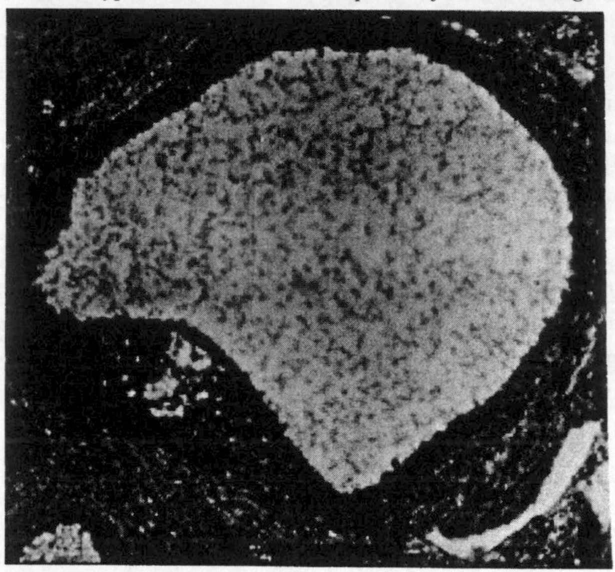

Figure 2 *Size of a large ovarian follicle and thickness of granulosa cell layer in a xanthotoxin-dosed female Wistar rat.* $\times 40$

1.76 control; 5.00 $\pm$ 2.08 xanthotoxin). In addition, xanthotoxin did not affect the number (5.333 $\pm$ 3.390 control; 5.333 $\pm$ 1.764 xanthotoxin) or the size ($1.576 \times 10^5 \pm 0.132 \times 10^5\ \mu m^2$ control; $1.290 \times 10^5 \pm 0.106 \times 10^5\ \mu m^2$ xanthotoxin) of small antral follicles (diameter $< 600\ \mu m$). Dosing did not also significantly affect the thickness of granulosa cell layer (76.044 $\pm$ 10.992 μm control; 82.428 $\pm$ 7.404 μm xanthotoxin) or the granulosa cell layer:follicle diameter ratio (0.006 $\pm$ 0.0010 control; 0.008 $\pm$ 0.0010 xanthotoxin) in small follicles.

In conclusion, the study provides evidence that xanthotoxin affects the thickness and/or distribution of the granulosa cell population in large antral follicles. These findings are consistent with the previously observed reduction in circulating 17β-estradiol levels in treated females. It is reasonable to hypothesize that xanthotoxin exposure compromises synthesis of 17β-estradiol in preovulatory females by a direct or indirect effect on granulosa cell function. This results in the reduced ovulations and the subsequent decrease in birthrate of pregnant females that we previously reported.[18]

4 References

1 R. C. Beier, *Rev. Environ. Contam. Toxicol.*, 1990, **113**, 47.

2 M. R. Berenbaum, 'Herbivores Their Interactions with Secondary Plant Metabolites, The Chemical Participants', G. A. Rosenthal and M. R. Berenbaum, Eds., Academic Press, New York, 1991, Vol. I, Chapter 6, p. 221.

3 M. M. Diawara and J. T. Trumble, 'Handbook on Plant and Fungal Toxicants', J. P. F. D'Mello, Ed., CRC Press, Inc., Boca Raton, 1997, Chapter 12, p. 175.

4 R. Brickl, J. Schmid and F. W. Koss, *Photodermatology*, 1984, **1**, 174.
5 E. J. Van Scott, *J. A. M. A.*, 1975, **235**, 197.
6 M. A. Pathak and T. B. Fitzpatrick, *Photochem. Photobiol. B*, 1992, **14**, 3.
7 R. S. Stern, L. A. Thibodeau, R. A. Kleinerman, J. A. Parrish and T. B. Fitzpatrick, *N. Engl. J. Med.*, 1979, **300**, 809.
8 R. S. Stern, K. T. Nichols and L. H. Vakeva, *N. Engl. J. Med.*, 1997, **336**, 1041.
9 M. J. Ashwood-Smith, G. A. Poulon, M. Barker and M. Mildenberger, *Nature*, 1980, **285**, 407.
10 R. Roelandts, *Arch Dermatol.*, 1984, **120**, 662.
11 A. R. Young, *J. Photochem. Photobiol. B–Biol.*, 1990, **6**, 237.
12 C. Bauluz, J. M. Paramio and R. de-Vidania, *Cell. Mol. Biol.*, 1991, **37**, 481.
13 M. Takasugi, A. Guendouz, M. Chassignol, J. L. Degout, J. Lhomme, N. T. Thuong and C. Helene, *Biochemistry*, 1991, **88**, 5602.
14 S. C. Yang, J. G. Lin, C. C. Chion, L. Y. Chen and J. L. Yang, *Carcinogenesis*, 1994, **15**, 201.
15 J. K. Dunnick, W. E. Davis Jr., T. A. Jorgenson, V. J. Rosen and E. E. McConnell, *National Cancer Institute Monographs*, 1984, **66**, 91.
16 J. G. Gunnarskog, A. J. Bengt Kallen, B. G. Lindelof and B. Sigurgeirsson, *Arch. Dermatol.*, 1993, **129**, 320.
17 M. M. Diawara, T. G. Allison, P. Kulkosky, S. McCrory, L. A. Martinez and D. E. Williams, *J. Natural Toxins*, 1997, **6**, 183.
18 M. M. Diawara, K. J. Chavez, P. B. Hoyer, D. E. Williams, J. Dorsch, P. Kulkosky and M. R. Franklin, *J. Biochem. Mol. Toxicol.*, 1999, **13**, 195.
19 J. A. Flaws, J. K. Doerr, I. G. Sipes and P. B. Hoyer, *Reprod. Toxicol.*, 1994, **8**, 509.

7.6

Effect of Ascorbigen on 7,12-Dimethylbenzanthracene-induced Mammary Tumor in Female Sprague-Dawley Rats

Xiaokang Ge,[1] Gad Rennert[2] and Shmuel Yannai[1]*

[1] DEPARTMENT OF FOOD ENGINEERING AND BIOTECHNOLOGY, TECHNION – ISRAEL INSTITUTE OF TECHNOLOGY, HAIFA 32000, ISRAEL
[2] DEPARTMENT OF COMMUNITY MEDICINE AND EPIDEMIOLOGY, NATIONAL CANCER CONTROL CENTER, THE LADY DAVIS CARMEL MEDICAL CENTER, HAIFA, ISRAEL

1 Abstract

Epidemiological studies have suggested that consumption of cruciferous vegetables may significantly decrease the risk of various cancers in humans. The anticarcinogenic activity of crucifers may result from the presence of active compounds in the vegetables. Ascorbigen is one of the indole derivatives formed from cruciferous vegetables. In this study, the effect of ascorbigen on the formation of 7,12-dimethyl[*a*]benzanthracene (DMBA)-induced mammary tumor was studied in female Sprague-Dawley rats. Rats were divided into four groups which received: (1) a single intubation of 15 mg DMBA without ascorbigen; (2) 15 mg DMBA and two daily intubations of 16 mg ascorbigen; (3) 15 mg DMBA and two intubations of 32.5 mg ascorbigen day^{-1}; (4) 15 mg DMBA and two intubations of 65 mg day^{-1} of ascorbigen. The first tumor appeared in rats treated only with DMBA was found 30 days later; while in all ascorbigen-treated groups it appeared after 40 days. After 90 days, 63% of rats receiving only DMBA showed mammary tumors. Rats receiving ascorbigen at dose levels of 32.5 mg day^{-1} and 65 mg day^{-1} had 33% and 39% incidence of mammary tumors, respectively, while rats receiving 16 mg day^{-1} of ascorbigen did not show decreased tumor incidence. The results of this study suggests that ascorbigen may be a promising chemopreventive agent in DMBA-induced mammary carcinogenesis in rats.

* Correspondence author: Fax: 00972-4-8320742

2 Introduction

It has been shown that frequent consumption of vegetables decreases the risk for human cancer.[1] Identifying dietary photochemicals which have the ability to inhibit carcinogenesis has been a recent focus. Ascorbigen is a natural indole derivative from cruciferous vegetables such as cabbage, cauliflower, Brussels sprouts, broccoli. It is formed in damaged plants and in food processing after myrosinase is liberated. High-pressure liquid chromatography analysis found that the concentration of this compound in stomach juice is much higher than that of indole-3-carbinol, another major indole derivative formed in cruciferous vegetables, when the vegetable is consumed. Recent studies have suggested that ascorbigen has the ability to induce the hepatic microsomal cytochrome P450 enzymes and elevate the activity of C2 hydroxylation of estradiol.[2] The increase of the C2 hydroxylation of estrogens has been considered to have the potential to decrease the incidence of mammary tumors in animals.[3] However, the anti-carcinogenic activity of this compound has not been studied yet. The purpose of this study was to investigate whether ascorbigen was effective in the inhibition of the formation of DMBA-induced mammary tumor in rats.

3 Materials and Methods

3.1 Chemicals

Ascorbigen was synthesized according to the method of Kiss and Neukom.[4] 7,12-Dimethyl[*a*]anthracene (DMBA) and indole-3-carbinol were purchased from Sigma.

3.2 Animals

Virgin female Sprague-Dawley rats were used in this study. The animals were housed in polycarbonate cages (two to five rats/cage), and were kept in a room lighted 12 h day^{-1} and maintained at a temperature of 22 ± 1 °C. Rodent diet (Koffolk 1949, purchased from Koffolk, Inc., Tel Aviv, Israel) and tap water were given *ad libitum*. Ascorbigen and the carcinogen, DMBA, were dissolved in dimethyl sulfoxide (DMSO), either alone or mixed together according to the different experimental protocols described below. The animals used here were approved by the 'Technion Committee for Care and Use of Laboratory Animals'.

3.3 Treatment of Animals

Female Sprague-Dawley rats, at 48 days of age were divided into four groups randomized by weight.

- Group 1. At 48 days of age, animals were fed by oesophageal intubation with 0.5 ml DMSO, once a day, for two days. On the third day (age, 50 days), animals were fed by one intubation of 15 mg DMBA in 0.5 ml DMSO.

- Group 2. At 48 days of age, animals were fed by intubation with 16 mg ascorbigen in 0.5 ml DMSO, once a day, for two days. On the third day (age, 50 days), animals were fed by intubation with both 16 mg ascorbigen and 15 mg DMBA in 0.5 ml DMSO.
- Group 3. At 48 days of age, animals were fed by intubation with 32.5 mg ascorbigen in 0.5 ml DMSO, once a day, for two days. On the third day (age, 50 days), animals were fed by intubation with both 32.5 mg ascorbigen and 15 mg DMBA in 0.5 ml DMSO.
- Group 4. At 48 days of age, animals were fed by intubation with 65 mg ascorbigen in 0.5 ml DMSO, once a day, for two days. On the third day (age, 50 days), animals were fed by intubation with both 65 mg ascorbigen and 15 mg DMBA in 0.5 ml DMSO.

The animals were weighed every one or two weeks. Rats were palpated for mammary tumors every one or two weeks starting 6 weeks after the day of administration of DMBA. The experiments were finished 90 days after the day of administration of DMBA. At the end of experiment, all the animals were sacrificed by CO_2 asphyxiation.

4 Results

The development of palpable mammary tumors in rats treated with DMBA only or DMBA with different doses of ascorbigen is shown in Figure 1 and Table 1. At 60 days after the day of given DMBA, tumor incidence was decreased in all ascorbigen treated groups. Table 1 shows that at 90 days after the day of given DMBA, 63% of the rats receiving DMBA only (group 1) showed mammary tumors. Rats receiving dose levels of 32.5 mg day^{-1} (group 3) and 65 mg day^{-1} (group 4) had a decreased incidence of mammary tumors,

Table 1 *Effect of ascorbigen on DMBA-initiated mammary tumors*

Group	*Number of rats*	*Carcinogen*[a]	*Treatment*[b]	*Mammary tumor incidence (%)*[c]	*Average no./rat*
1	27	DMBA	Vehicle	63	1.32
2	20	DMBA	Ascorbigen, 16 mg day^{-1}	60	1.20
3	18	DMBA	Ascorbigen, 32.5 mg day^{-1}	33[d]	0.53[e]
4	18	DMBA	Ascorbigen, 65 mg day^{-1}	39[d]	0.72

[a] DMBA (15 mg/rat) was fed by intubation to female Sprague-Dawley rats at 50 days of age.
[b] Ascorbigen was given by intubation when rats were at 48 days of age for continuous 3 days; vehicle, 0.5 ml DMSO (dimethyl sulfoxide).
[c] Tumor incidence (%) at 90 days after the day of administration of DMBA.
[d] Different from group 1. X^2-test, $p < 0.05$.
[e] Different from group 1. t-test, $p < 0.05$.

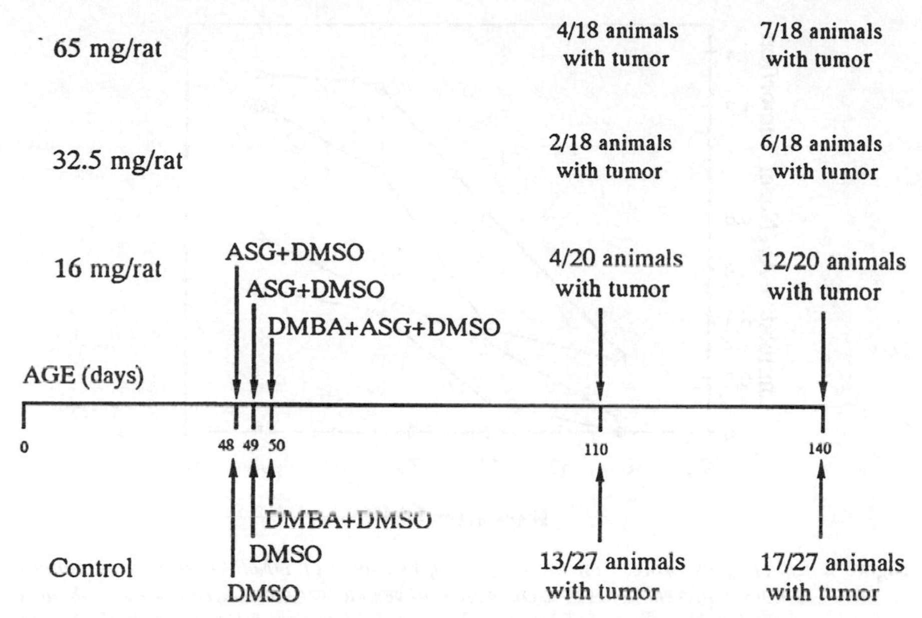

Figure 1 *Scheme for the study of the effect of ascorbigen on the formation of DMBA-initiated mammary tumors. AGE, the age of the anaimals. DMSO, dimethyl sulfoxide. DMBA, dimethylbenz[a]anthracene. ASG, ascorbigen*

while rats receiving 16 mg day^{-1} dose level (group 2) did not show the decreased incidence (Table 1). Table 1 also shows that at the dose of 65 mg day^{-1} the tumor incidence has not been further decreased compared to that occurred in 32.5 mg day^{-1}. Figure 2 shows that the latency period of the mammary tumors occurring in rats that received 32.5 and 65 mg kg^{-1} ascorbigen was also increased. The appearance of first tumor in the group receiving DMBA only was 30 days after the day of administration of DMBA, while in all the ascorbigen-treated groups the first tumor was found after 40 days.

5 Discussions

In this study, ascorbigen has been investigated for its capacity to suppress carcinogen dimethylbenzanthracene (DMBA)-induced mammary tumor formation. The experiment described here demonstrates that ascorbigen has an activity against DMBA-initiated mammary tumor formation in rats. The carcinogen-initiated rat mammary tumor model has been extensively used to investigate the development of rat mammary tumors using diverse carcinogens and chemicals which modulate tumor formation (reviewed by Welsch[5]). Ascorbigen inhibited DMBA-induced tumor incidence at doses 32.5 mg/rat and 65 mg/rat. In this study, ascorbigen was given two days before DMBA was given. That means it was given before and during the tumor initiation stage. Compounds which decrease enzyme activity responsible for carcinogen activa-

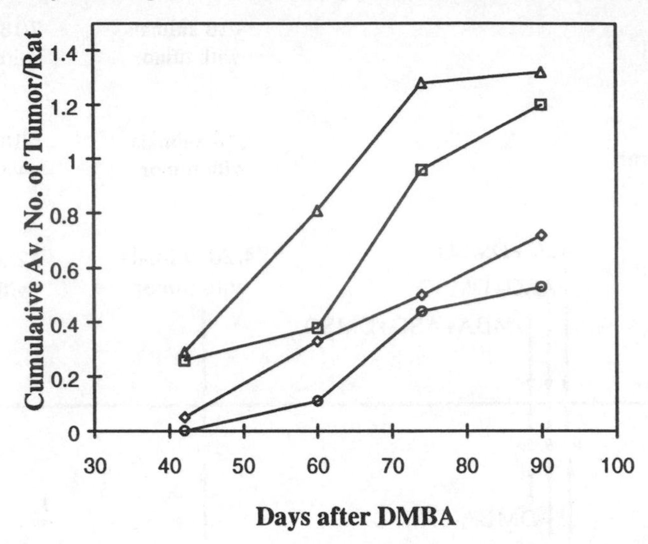

Figure 2 *Effect of ascorbigen on the time of appearance of DMBA-initiated mammary tumors. Different doses of ascorbigen and vehicle were intubated to rats at 48 days of age for three days. At 50 days of age, all the rats were intubated a single dose of 15 mg/rat of DMBA. The groups were: △ Group 1, DMBA; □ Group 2, DMBA + ascorbigen 16 mg day^{-1} for 3 days; ○ Group 3, DMBA + ascorbigen 32.5 mg day^{-1} for 3 days; ◇ Group 4, DMBA + ascorbigen 65 mg day^{-1} for 3 days*

tion, increase enzyme activity responsible for carcinogen detoxification, and scavenge amount of reactive ultimate carcinogen can lead to the inhibition of tumor initiation.

Ascorbigen is a dietary ingredient and present in cruciferous vegetables. Mukhanov and colleagues have found that some ascorbigen derivatives inhibited the growth of transplanted adenocarcinoma of large intestine in athymic BALB/c mice. The stable and unstable transformation products of ascorbigen could determine the biological properties of the original compound. In gastric juice ascorbigen can be transformed to 5,11-dihydroindole[3,2-*b*]carbazole, a compound with an affinity for the aromatic hydrocarbon.[7] It has recently been reported that when 2000 ppm ascorbigen was fed to female Sprague-Dawley rats, the activity of 2-hydroxylation of estradiol (cytochrome P450 1A1 dependent) was induced in liver microsomes.[2] It has been reported that increased 2-hydroxylation of estradiol may be a result of the decreased mammary tumor incidence.[3] The alteration in cytochrome P450 1A1 activity by ascorbigen may be responsible for its detoxification of carcinogens. Moreover, ascorbigen may also have an ability to sequester the reactive carcinogen. Our *in vitro* study has shown that ascorbigen may also act as antioxidant.[8]

In conclusion, this study shows that ascorbigen may be an anticarcinogenic agent in rats. Although the exact mechanisms by which ascorbigen inhibited the DMBA-initiated mammary tumor formation are not completely understood, the beneficial effects of dietary ascorbigen in cancer prevention are being established.

6 References

1 Freudenheim JL, Marshall JR, Vena JE, Laughlin R, Brasure JR, Swanson MK, Nemoto T and Graham S. 1996. Premenopausal breast cancer risk and intake of vegetable, fruit and related nutrients. *J Natl Cancer Inst* **88**, 340–348.

2 Sepkovic DW, Bradlow HL, Michnovicz J, Murtezani S, Levy I and Osborne MP. 1994. Catechol estrogen production in rat microsomes after treatment with indole-3-carbinol, ascorbigen, or β-naphthaflavone: A comparision of stable isotope dilution gas chromatography-mass spectrometry and radiometric methods. *Steroids* **59**, 318–323.

3 Bradlow HL, Michnovicz JJ, Telang T and Osborne M. 1991. Effects of dietary indole-3-carbinol on estradiol metabolism and spontaneous mammary tumors in mice. *Carcinogenesis* **12**, 171–1574.

4 Kiss G and Neukom H. 1966. Uber die Struktur des Ascorbigens. *Helv Chim Acta* **49**, 989–992.

5 Welsch CW. 1985. Host factors affecting the growth of carcinogen-induced rat mammary carcinogens: a review and tribute to Charles Brenton Huggins. *Cancer Res* **45**, 3415–3443.

6 Mukanov VI. 1984. Study of ascorbigen and its derivatives. *Bioorg. Khim* **10**, 544–559.

7 Preobrazhenskaya MN, Bukhman VM, Korolev AM and Efimov SA. 1993. Ascorbigen and other indole-derived compounds from Brassica vegetables and their analogs as anticarcinogenic and immunomodulating agents. *Pharm Therapeutics* **60**, 301–313.

8 Ge X, Karmansky I and Yannai S. 1999. Indole derivatives, 3,3′-diindolymethane and ascorbigen, inhibit oxidation of human low-density lipoproteins *in vitro*. *Pharmacy Pharmacol Commun* in press.

7.7

Inhibitory Effects of Curcumin on Mouse Stomach Neoplasia and Human Carcinoma Cell Line

Bingqing Chen, Xia Li, Jiaren Liu, Shuran Wang, Weijia Feng, Guifang Lu and Xiaohui Han

PUBLIC HEALTH COLLEGE, HARBIN MEDICAL UNIVERSITY, HARBIN 150001 P.R. CHINA

1 Introduction

Curcumin (diferuloylmethane), the naturally occurring yellow pigment in turmeric and curry, is isolated from the rhizomes of the plant *Curcuma longa* Linn. It is a phenolic compound, used widely as a spice and a yellow colouring agent in food, possesses potent antioxidant,[1–4] anti-inflammatory[5–7] and anti-carcinogenic properties.[8–10] Recently, curcumin was demonstrated to have an anti-carcinogenic effect.[11–13] In this study, we have shown that curcumin can inhibit forestomach carcinogenesis induced by B(a)P and micronuclei formation induced by cyclophosphamide (CP) in mice, and also inhibit the growth of human gastric carcinoma cell (SGC-7901).

2 Materials and Methods

The model of neoplasia induced by B(a)P in mice: Female Swiss mice were divided into five groups of 25 mice. Positive control group was administered B(a)P (1 mg dissolved in 0.1 ml salad oil per mouse) by gavage twice a week for four weeks. Negative control groups were given salad oil (0.1 ml) only and fed 1.6% curcumin respectively. Test groups were given B(a)P + 0.4% curcumin; B(a)P + 1.6% curcumin in diet. The experimental period was 20 weeks from the first time of carcinogen administration.

Micronucleus test of bone marrow cell in mice: Curcumin was dissolved in salad oil. Mice were divided into five groups of 10. Three test groups were given curcumin 60, 120 and 240 mg kg^{-1} bw, respectively. The positive control group was administered carcinogen CP (50 mg kg^{-1} bw). Apart from the negative control group (salad oil 4 ml kg^{-1} bw), all animals were administered CP (50 mg kg^{-1} bw) 6 h before being sacrificed. Micronuclei of bone marrow cell were observed under microscope. The effect of curcumin at concentrations of 12.5, 25 and 50 μM on cell growth was measured in human gastric carcinoma cell line.

3 Results and Discussion

The effect of the addition curcumin in diet was to inhibit and decrease the incidence of forestomach neoplasia in mice. Table 1 shows that with increasing curcumin in the diet, the numbers of tumors and gross tumor incidence were decreased. Supplements of 0.4% and 1.6% curcumin in diet can inhibit forestomach neoplasia induced by B(a)P in mice. Moreover, curcumin can reduce micronuclei formation of bone marrow cell induced by CP in mice by 43.7%, 56.9%, and 63.9% respectively (Table 2).

Curcumin protects against human malignancies and the cellular changes in human gastric carcinoma cell line. The inhibitory effect of curcumin on human gastric carcinoma cell line was found to be significant at 25 μM group

Table 1 *Inhibitory effect of curcumin on B(a)P-induced neoplasia tumor in mice*

Treatment	*No. of animals at risk*	*No. of animals with tumor*	*Total tumor*	*No. of tumors/ mice (X ± SD)*	*Gross tumor incidence (%)*
Oil alone	9	0	0		0
B(a)P alone	15	15	78	5.28 ± 3.6	100
B(a)P + 0.4% curcumin	10	7	25	3.57 ± 1.98	70*
B(a)P + 1.6% curcumin	14	7	9	1.28 ± 0.66	50**
Curcumin (1.6%) alone	10	0	0		0

Compared with B(a)P alone group, [a]P < 0.05, [b]P < 0.01.

Table 2 *Inhibition of curcumin on micronuclei formation induced by CP in mice*

	No of animals	*Micronuclei (1/1000, X ± SD)*	*Inhibition (%)*
Oil alone	10	4.8 ± 1.3	
Curcumin (mg kg^{-1}) + CP (50 mg kg^{-1})			
60	10	17.5 ± 3.5[a]	43.7
120	10	13.4 ± 3.1[a]	56.9
240	10	11.2 ± 1.9[a]	63.9
CP alone (50mg/kg)	10	31.1 ± 2.0	

[a]Compared with CP alone, P < 0.01.

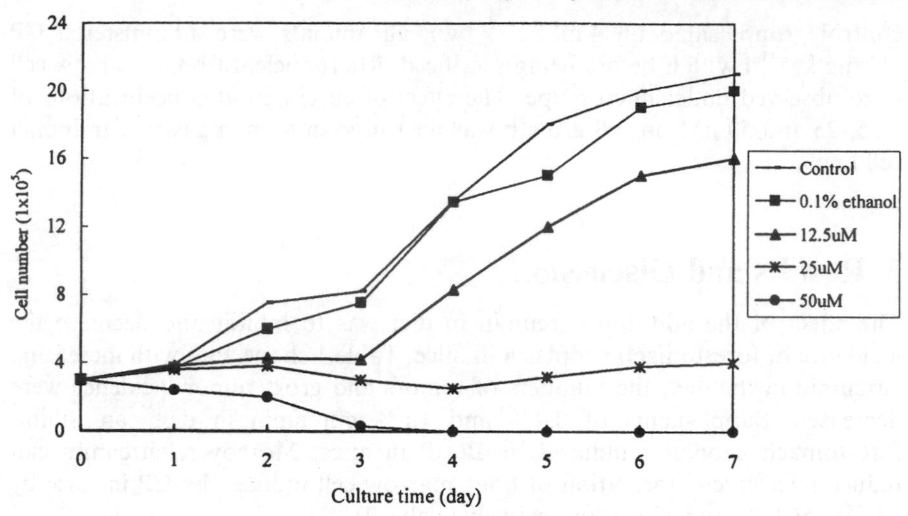

Figure 1 *Effect of curcumin on the growth of SGC-7901 cells*

(Figure 1). This result suggests that curcumin can inhibit the growth and proliferation of tumor cells.

4 Conclusion

Curcumin in the diet fed to female Swiss mice can inhibit B(a)P-induced forestomach neoplasia and inhibit micronucleus formation of bone marrow cell in mice. Curcumin also inhibits the growth of human gastric carcinoma cell *in vitro*. These results indicate that curcumin may be a potent agent in chemoprevention against cancer.

5 References

1 R. Parshad, K. K. Sanford, F. M. Price, V. E. Steele, R. E. Tarone, G. J. Kelloff and C. W. Boone, *Anticancer Res.*, 1998, **18**, 3263–3266.

2 T. Nagano, Y. Oyama, N. Kajita, L. Chikahisa, M. Nakata, E. Okazaki and T. Masuda, *Jpn J. Pharmacol.*, 1997, **75**, 363–70.

3 D. V. Rajakumar and M. N. Rao, *Free Rad. Res.*, 1995, **22**, 309–17.

4 K. I. Priyadarsini, *Free Rad. Biol. Med.*, 1997, **23**, 838–843.

5 N. Venkatesan and G. Chandrakasan, *Mol. Cell Biochem.*, 1995,**142**, 79–87.

6 Y. X. Xu, K. R. Pindolia, N. Janakiraman, C. J. Noth, R. A. Chapman and S. C. Gautam, *Exp. Hematol.*, 1997, **25**, 413–22.

7 G. Chandrakasan,N. Venkatesan and V. Punithavathi, *Life Sci.*, 1997, **61**, 51–58.

8 C. V. Rao, A. Rivenson, B. Simi and B. S. Reddy, *Cancer Res.*, 1995, **55**, 259–66.

9 G. D. Stoner and H. J. Mukhtar, *Cell Biochem. Suppl.*, 1995, **22**, 169–80.

10 A. J. Ruby, G. Kuttan, K.D. Babu, K. N. Rajasekharan and R. Kuttan, *Cancer Lett.*, 1995, **94**, 79–83.

11 S. S. Deshpande, A. D. Ingle and G. B. Maru, *Cancer Lett.*, 1997, **118**, 79–85.
12 K. Krishnaswamy, V. K. Goud, B. Sesikeran, M. A. Mukundan and T. P. Krishna, *Nutr. Cancer*, 1998, **30**, 163–166.
13 A. Gescher, U. Pastorino, S. M. Plummer and M. M. Manson, *Br. J. Clin. Pharmacol.*, 1998, **45**, 1–12.

7.8

Effect of Fermented Milk on Colorectal Cancer Biomarker in Rats Induced by a Chronic Exposure to Several Heterocyclic Amines

Emmanuelle Tavan,[1,*] Chantal Cayuela,[2] Jean-Michel Antoine[2] and Pierrette Cassand[1]

[1] LABORATOIRE DE TOXICOLOGIE ALIMENTAIRE, UNIVERSITÉ BORDEAUX 1, AVENUE DES FACULTÉS, 33405 TALENCE CEDEX, FRANCE
[2] DANONE, CENTRE INTERNATIONAL DE RECHERCHE DANIEL CARASSO, 15 AVENUE GALILÉE, 92350 LE PLESSIS-ROBINSON, FRANCE

1 Abstract

Beneficial aspects of fermented dairy products in human nutrition have been underlined by many authors. Particularly, fermented milks could play a role in prevention of cancers such as colon cancer which is highly related to alimentation. We have performed a study to evaluate inhibition by fermented milk of colorectal carcinogenicity induced by heterocyclic amines (HA), mutagenic compounds found in cooked food and suspected to be implicated in human colorectal cancer. Rats were given during seven weeks 252 mg kg^{-1} per os of either 2-amino-3-methylimidazo[4,5-*f*]quinoline (IQ) or a mixture of three HA: IQ, 2-amino-3,4-dimethylimidazo[4,5-*f*]quinoline (MeIQ) and 2-amino-1-methyl-6-phenylimidazo[4,5-*b*]pyridine (PhIP). Different groups of rats were used: controls and those eating diets supplemented with milk fermented by a *Bifidobacterium* lactic acid bacteria strain. The results showed a higher induction of microadenomas in rats induced with the three HA mixture than with IQ alone. An inhibition by the fermented milk of the preneoplasic lesions multiplicity was observed when the rats were induced by the HA mixture only.

* Corresponding author: e.tavan@istab.u-bordeaux.fr

2 Introduction

Colon carcinogenesis is a complex multistep process where preneoplasic phenotypic lesions in epithelial cells represents an intermediate stage, leading finally to tumours.[1] Rodents treated with colon carcinogens develop some lesions, aberrant crypt foci (ACF), which are visualised with a microscope at low magnification after colon staining.[2] ACF have also been described in the colon of humans with high risk of colon cancer, and have therefore been suggested to be preneoplasic lesions and so good markers of colon carcinogenesis.[3]

The development of colon cancer is strongly affected by dietary habits and food may contain various carcinogenic compounds but could be also a source of protective nutriments.[4] In the class of bad compounds, new highly mutagenic, heterocyclic amines (HA) have been found to be produced in food upon cooking meat and fish.[5] Chronic intake of HA by the human population may play a role in the etiology of human colorectal cancer; many of these extremely genotoxic compounds have been shown to induce tumours in many different organs of rodents or monkeys in long-term animal studies.[6] In order to define carcinogenicity mechanisms, *in vivo* rodent models are commonly used, more particularly to define precisely the influence of diet on the carcinogenic potential of heterocyclic amines. On the other hand, lactic acid bacteria (LAB) and the final fermented milk products are well known for their 'probiotic' effects on human health, and their action in cancer prevention has also been investigated.[7] Particularly, Reddy and Rivenson[8] have shown colon tumour inhibition with *Bifidobacterium longum* in rats induced by a high-fat diet supplemented with 2-amino-3-methylimidazo[4,5-*f*]quinoline (IQ) and the LAB.

The goal of this study was therefore to investigate the colon carcinogenesis inhibition potential of a milk fermented with a LAB, *Bifidobacterium* sp., in rats induced either with IQ or with a mixture of several HA: IQ, 2-amino-3,4-dimethylimidazo[4,5-*f*]quinoline (MeIQ) and 2-amino-1-methyl-6-phenylimidazo[4,5-*b*]pyridine (PhIP).

2 Materials and Methods

Fisher F344 male rats were given 252 mg kg^{-1} per os of either 2-amino-3-methylimidazo[4,5-*f*]quinoline (IQ) or a mixture of three HA: IQ, 2-amino-3,4-dimethylimidazo[4,5-*f*]quinoline (MeIQ) and 2-amino-1-methyl-6-phenylimidazo[4,5-*b*]pyridine (PhIP). The carcinogens were administered by seven gavages of 34 mg kg^{-1} and fourteen gavages of 1 mg kg^{-1} during seven weeks, in order to have a good reproducibility in the carcinogenesis induction but also to simulate low and chronic HA ingestion with sometimes a high intake, like in human diet. Different groups of rats were constituted for each carcinogens treatment: control rats which were eating standard diet and rats eating diets supplemented with milk fermented by *Bifidobacterium* strain.

The fermented milk supplemented diet was performed as follows: *Bifidobacterium* strain was inoculated (3% v/v) in reconstituted non-fat milk supplemen-

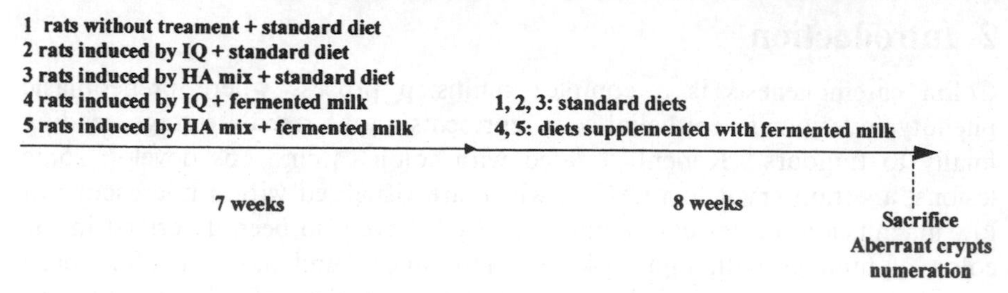

Figure 1 *Carcinogenesis protocol*

ted with 1% yeast extract and 0.03% cysteine and incubated 6 hours at 40 °C. This was repeated three times to obtain the inoculum. Finally, pasteurised liquid non-fat milk (Finesse, Monoprix, France) was heated at 95 °C for 10 minutes and then cooled to fermentation temperatures, before being inoculated by the inoculum (3% v/v) and incubated for 6.5 hours. The number of bacteria at the end of fermentation was 10^8 per ml of milk. The fermented milk was added at 30% w/w to the standard ground diet, daily.

Rats receiving no treatment, neither carcinogens nor fermented milk, were also used; however, they were administered the HA solvent during the seven induction weeks (Figure 1).

At the end of the seven induction weeks, all the rats were kept on their respective diets during eight more weeks and then sacrificed. The colon was fixed in phosphate buffered formalin (10%), stained with methylene blue 1% and scored for aberrant crypt foci following Bird's procedure[2] with a light microscope at magnification 40.

Statistical Analysis

The level of significance was tested using one-way analysis of variance (ANOVA) followed by HSD-Tuckey test, the minimum level of significance accepted being $p < 0.05$.

4 Results and Discussion

The results are presented in Figure 2. It was verified that the rats without carcinogen treatment had no preneoplasic lesions. We observed that rats induced by the mix of HA presented significantly more microadenomas than rats treated with IQ alone. The rats which were eating the diet supplemented with fermented milk had a significant inhibition (45%) of preneoplasic lesions multiplicity induced by the three HA mixture. These results suggest that fermented milk could have a preventive action, as aberrant crypt foci multiplicity is linked to the tumour induction risk.[2] However, this effect depends on the carcinogen employed; perhaps this action could show particular efficacy towards PhIP, which is a more specific colon carcinogen. These *in vivo* results are in accordance with antimutagenesis assays performed in our laboratory with

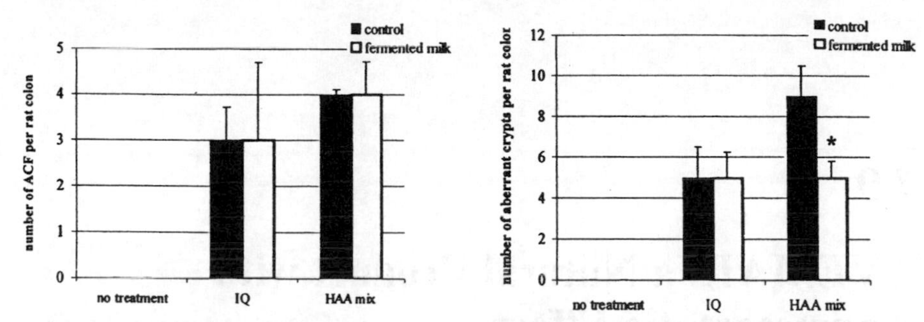

Figure 2 *Number of aberrant crypts and aberrant crypt foci per rat colon following the different treatments*

the *Bifidobacterium* strain, which has been tested in the Ames test alone in MRS and after milk fermentation. Complementary studies and results have been carried out in our laboratory during this *in vivo* experiment, in particular HA excretion and DNA lesions in epithelial colon cells.

In conclusion, fermented milk may have a preventive effect towards colorectal cancer induced by food genotoxics. It seems more interesting to test a mixture of different HAs, since the effect is increased and better matches the reality of human diet; however, our HA induced-colon carcinogenesis rat model should be improved to increase the fermented milk effect sensitivity.

Acknowledgement

This investigation was supported by a grant from Danone, Centre International de Recherche Daniel Carasso, Le Plessis-Robinson, France.

5 References

1 D.W. Day, *Scand. I. Gastroenterol.*, 1984, **19**, 99.
2 R.P. Bird, *Cancer Lett.*, 1987, **37**, 147.
3 L. Roncucci, D. Stamp, A. Medline and W.R. Bruce, *Hum. Pathol.*, 1991, **22**, 287.
4 W.C. Willet, *Science*, 1994, **264**, 532.
5 T. Sugimura, M. Nagao, T. Kawachi, M. Honda, T. Yahagi, Y. Seino, S. Sato, N. Matsukura, T. Matsushima, A. Shirai, M. Sawamura and H. Matsumoto, 'Origins of Human Cancer', Eds. H.H. Hiatt, J.D.Watson and J.A. Winsten, Cold Spring Harbor Laboratory, New York, 1977, p. 1561.
6 D.W. Layton, K.T. Bogen, M.G. Knize, F.T. Hatch, V.M. Johnson and J.S. Felton, *Carcinogenesis*, 1995, **1**, 39.
7 R. Fuller, *Gut*, 1991, **32**, 439.
8 B.S. Reddy and A. Rivenson, *Cancer Res.*, 1993, **53**, 3914.

7.9

AVEMAR: a Natural Product with Antimetastatic Effect

R. Tömösközi-Farkas,[1,3] E. Ráső,[2] K. Lapis,[2] S. Paku[2] and M. Hidvégi[3]

[1] CENTRAL FOOD RESEARCH INSTITUTE, DEPARTMENT OF ENZYMOLOGY, BUDAPEST, HERMAN O. ÚT 15, H-1022, HUNGARY
[2] SEMMELWEIS UNIVERSITY OF MEDICINE, 1ST INSTITUTE OF PATHOLOGY AND EXPERIMENTAL CANCER RESEARCH, BUDAPEST, ÜLLOI ÚT 26, H-1085, HUNGARY
[3] BIROMEDICINA LTD., BUDAPEST, PUSKIN U. 4, H-1088, HUNGARY

1 Introduction

In the last considerable part of his activities, Albert Szent-Györgyi dealt with the biological effect of quinones.[1] Using the theory of Szent-Györgyi, an orally applicable fermentation product of wheat germ containing 0.04% methoxy-substituted p-benzoquinone has been invented under the trade name of AVEMAR. This product has shown several biological effects. Oral administration of AVEMAR enhances blastic transformation of splenic lymphocytes and shortens the survival time of skin grafts in a co-isogenic mouse skin transplantation model.[2] A highly significant antimetastatic effect of AVEMAR has been observed in three metastasis models.[3]

The biological effect of these quinones is related to their property of taking part in redox cycles in their free radical form. In wheat germ, 2-methoxy-p-benzoquinone (2-MBQ) and 2,6-dimethoxy-p-benzoquinone (2,6-DMBQ) appear in the form of glucosides. During the fermentation of wheat germ with yeast (*Saccharomyces cerevisiae*), the quinones are released by the glucosidase enzyme of the yeast fungus. The original perception of Szent-Györgyi was that, by means of the biological activity of these related quinones, the fermented wheat germ may possess immune-stimulatory effect.[4,5]

In the following, a review is given of the the cytostatics-combined study results concerning the product AVEMAR.

2 Materials and Methods

2.1 Preparation of AVEMAR

Seventy kilograms of *Saccharomyces cerevisiae*, obtained from a local grain distillers company, was suspended in a 3 m^3 isothermic (30 ± 1 °C) fermenter containing 2 m^3 tap water. After mixing, 210 kg of freshly ground wheat germ, obtained from a local grain milling company, was added to the yeast suspension. The mixture was then fermented, filtered, concentrated and spray dried. The resulted pulver was homogenized and kept in sealed containers.[6]

2.2 Animals

In all experiments inbred mice from our Institute were used. The animals were 8–10 weeks old and weighed 20–22 g. They were kept in plastic cages (5 per cage) and were fed with rodent pellets (Charles River Hungary) and tap water *ad libitum*. The room temperature was 20–22 °C, relative humidity 50 ± 5%.

2.3 Treatments

MSC treatment was started 24 hours after tumor implantation. MSC was dissolved in water and administered by means of a gastric tube. The daily dose was 3 g kg^{-1} body weight p.o. administered in 0.1 ml water. Control animals received tap water (0.1 ml) also via gastric tube daily. The daily dose was 3 g kg^{-1} body weight p.o. The experiments were completed 14 (3LL-HH), 21 (B16), 20 (C38) and 51 (HCR-25) days after tumor inoculation, by means of exsanguination during anaesthesia.

2.4 Tumor Models

In the experiments the following transplantable tumor lines were used, grown on mice or rats, B16 mouse melanoma and C38 mouse colorectal tumor. *B16 melanoma* line was maintained in C57Bl/6 mice by serial intramuscular transplantation (5×10^5 cells, hind leg muscle). The C38 mouse colorectal carcinoma was maintained in C57Bl/6 mice by subcutaneous transplantation. The B16 melanoma was used as muscle-lung metastasis model, while the C38 mouse colorectal carcinoma cell line was applied as a spleen-liver metastasis model.

3 Results and Discussion

Mice bearing the C38 colorectal carcinoma implanted into the spleen were treated with 5-FU administered i.p. 3 times in a dosage of 1 mg kg^{-1}, while the mice inoculated with the B16 melanoma received DTIC treatment daily (60 mg kg^{-1} i.p.) Synchronously the animals treated with antineoplastic agents also received MSC daily (3 g kg^{-1}).

In these experiments the body mass of the animals was checked with particular case every three days. The size of the B16 melanoma growing in the muscle of hind leg was also measured every three days with a caliper. At the time of evaluation the animals were narcotised with diethyl ether and bled out. The mass (weight) of the primary tumor and of the organs with metastases was registered with an accuracy of 0.1 g. The number of the liver and lung metastases was counted and their diameters were also measured accurately with a scale built into a stereo microscope. The results are shown in Figures 1 and 2.

Figure 1 shows that in the case of combined (MSC + DTIC) treatment the number of lung metastases of B16 melanoma practically decreased to zero, and this effect is significant. The results show that in the therapeutic composition the metastasis inhibitory fermented plant extract (AVEMAR, MSC) having a metastasis inhibitory effect also alone exerted a more than additive effect, that is it synergically has the metastasis inhibitory effect of DTIC used in clinical practice to decrease metastasis in protocols for treatment of patients with metastasis.

Figure 2 shows that a 20 day treatment with the therapeutic composition decreased the number of liver metastases synergically. This effect is significant. The mass of the diseased spleen also decreased significantly as a consequence of the treatment. Althuogh the therapeutic effects were considerable the usual toxic side effects of cytostatica, *e.g.* decrease of body mass were not observed.

The product MSC has been put on the market as a non toxic food supplement. Its immune-reconstructive effect recommends it for all conditions where the immune status has been injured. It is an adequate supplement to the

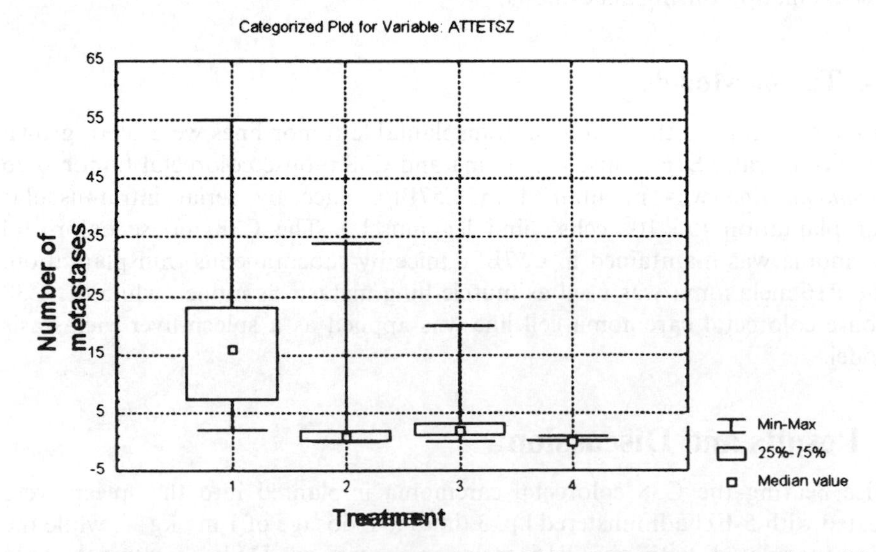

Figure 1 *Effect of the therapeutic composition (MSC + DTIC) on the number of lung metastases of B16 melanoma inoculated into the muscle of the hind leg (1 control, 2 DTIC, 3 AVEMAR, 4 AVEMAR + DTIC)*

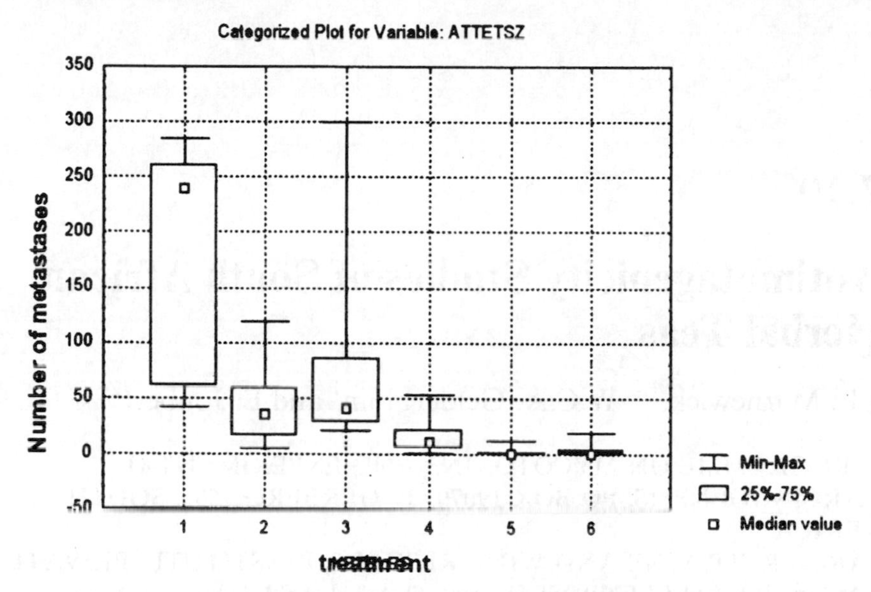

Figure 2 *The effect of the therapeutic composition (MSC + 5FU) on the liver metastases of C38 colorectal tumor injected into the spleen and on the mass of the spleen bearing the implanted 'primary' or parent tumor (1 control, 2–3 AVEMAR, 4 FU, 5–6 AVEMAR + 5-FU)*

pharmaceutical treatment of malignant tumors. Based on our *in vivo* and *in vitro* experimental data, the product can be recommended in a daily dosage of 5–15 g.

4 References

1 Szent-Györgyi, A.: Academic Press, New York, London, 1972, **71**.
2 Hidvégi, M., Ráso, E., Tömösközi-Farkas, R., *et al.*: *Immunopharmacology*, 1999, **41**, 183–186.
3 Hidvégi, M., Ráso, E., Tömösközi-Farkas, R., *et al.*: *Anticancer Res.*, 1998, **18**, 2353–2358.
4 Pethig, R., Gascoyne, P.R.C., McLaughlin, J.A. *et al.*: *Proc. Natl. Acad. Sci. USA*, 1984, **81**, 2088–2091.
5 Pethig, R., Gascoyne, P.R.C., McLaughlin, J.A. *et al.*: *Proc. Natl. Acad. Sci. USA*, 1985, **82**, 1439–1442.
6 Tömösközi-Farkas, R., Hidvégi, M. and Lásztity R.: *Acta Biol. Hung.*, 1998, **49**(1), 79–87.

7.10

Antimutagenicity Studies of South African Herbal Teas

J.L. Marnewick,[1,*] W.C.A. Gelderblom[1] and E. Joubert[2]

[1] PROGRAMME ON MYCOTOXINS AND EXPERIMENTAL CARCINOGENESIS, PO BOX 19070, TYGERBERG 7505, SOUTH AFRICA
[2] ARC-FRUIT, VINE AND WINE RESEARCH INSTITUTE, PRIVATE BAG X5013, STELLENBOSCH 7599, SOUTH AFRICA

The leaves and stems of *Aspalathus linearis* (Rooibos) and *Cyclopia intermedia* (Honeybush) are used to brew traditional South African herbal teas. Both Rooibos and Honeybush tea are popular as health beverages as they contain no harmful stimulants like caffeine and only trace amounts of tannins. The polyphenolic compounds of Rooibos and Honeybush tea have been found to be very different from those found in black and green teas, but they also exhibit strong antioxidant activity. However, very little is known about the antimutagenic and anticarcinogenic properties of these two herbal teas.

Aqueous extracts of fermented and green (unfermented) Rooibos and Honeybush tea exhibited no direct mutagenic activity in the *Salmonella typhimurium* mutagenicity assay. Fermented Honeybush tea exhibited a lower protective effect than the green tea against 2-acetylaminofluorene (2-AAF)- and aflatoxin B_1 (AFB_1)-induced mutagenesis. A similar effect was obtained with Rooibos tea when using AFB_1 as a mutagen. However, fermented Rooibos tea exhibited a similar protective effect against 2-AAF-induced mutagenesis than the green tea. A far less inhibitory effect was noticed against the direct acting mutagens, methyl methanesulfonate (MMS), cumene hydroperoxide and hydrogen peroxide (H_2O_2) using tester strain TA102, designed to detect oxidative mutagens and carcinogens. In some cases the tea extracts even enhanced the mutagenicity. This lack of protective effects was confirmed when utilizing the double-layered technique. Other investigators also showed a low protective effect against mutagenicity by direct-acting mutagens, when using extracts of green and black tea (*Camellia sinensis*).

The extracts of green Rooibos and Honeybush tea at 10% (w/v) displayed the highest antimutagenicity (>90% inhibition) against the metabolic activated mutagens (2-AAF, and AFB_1). The lower inhibition rates of their fermented counterparts, could be due to the lower soluble solid content of the fermented tea preparations and/or due to the fermentation process during which the polyphenols are likely to be oxidised. Gravimetric analyses of the aqueous extracts indicated that the green tea preparations had a higher soluble solid content as well as a higher percentage total polyphenols than their fermented counterparts. The methanol extract of green Rooibos tea also showed a protective effect against these mutagens, indicating that the anti-mutagenic activity is extracted in this organic phase.

Mechanistically the inhibitory effect detected could be due to either

(i) an interaction between the different components of the tea extracts and the enzyme systems catalyzing the metabolic activation of the various promutagens or
(ii) a direct interaction between the tea constituents (presumably the polyphenolic compounds) and the promutagens or the reactive intermediates.

Differences seem to exist in the degree of protection against mutagenesis of a specific carcinogen, suggesting that different mechanisms, as mentioned above, could be involved. Studies are in progress to identify the compounds in Rooibos and Honeybush tea that protects against mutagenesis by diverse mutagens and carcinogens and to clarify some of these mechanisms involved in the antimutagenic effect of the teas.

7.11

Vegetables as Nutraceuticals – Falcarinol in Carrots and Other Root Crops

Kirsten Brandt and Lars P. Christensen

DANISH INSTITUTE OF AGRICULTURAL SCIENCES,
DEPARTMENT OF FRUIT, VEGETABLE AND FOOD SCIENCE,
PO BOX 102, DK-5792 AARSLEV, DENMARK

1 Introduction

Many compounds with effects on human physiology and disease have been identified through studies of toxic or allergenic plants, or of plants used in traditional medicine. Some of these compounds occur in smaller concentrations in food plants. We propose that the available knowledge about pharmacological or harmful effects is used systematically to guide the search for anti-cancer components of food.

1.1 Plant Secondary Metabolites

Plants contain a great number of different secondary metabolites, many of which have some kind of biological activity. Food plants have been selected for nutrient value and taste, so their contents of toxic and bitter compounds are lower than average. Still thousands of different compounds, belonging to many different classes, are found in one or the other plant food. Only a fraction of these have been tested for the ability to prevent or retard the development of cancer.[1] However, due to the low average calorie intake in the modern sedentary society, it is increasingly important to identify the most healthy food components, to ensure as adequate nourishment as possible.

1.2 Screening for Beneficial Plant Components

Due to the great variation in biological activities it is important to select those structures for study, where there is some prior indication of useful properties. Until now, the most popular property in this respect has been antioxidative capacity, covering several classes of compounds, followed by fibres, also a heterogenous group with some common properties. This approach has been led by hypotheses about the mode of action of the compounds in question, guided by epidemiological studies. However, it is very difficult to use epidemiological studies to test for effects of almost ubiquitous types of compounds, so the results tend to be inconclusive.[2–4] To identify cancer-preventing properties of foods by thoroughly and systematically testing every group of plant components is unrealistic.

While many types of compounds probably affect various stages of cancerogenesis, several criteria can be used to select interesting structures for further study:

A. Chemically reactive functional groups in the molecule
B. Known effect on the immune system, *e.g.* allergenicity
C. Known effect on other organisms, *e.g.* microbes
D. Occurrence in plants with known medicinal properties
E. Occurrence in foods with a significant intake
F. Absence of toxicity (at relevant concentrations)
G. Large differences in intake among normal diets

The first six criteria regard the possible effects of a compound, the last whether the discovery of an effective one will easily lead to improvement of the average diet and thus population health. Several types of compounds fulfil these criteria; among those onion oils and glucosinolates have already received some attention.

2 Polyacetylenes

The present work concentrates on the polyacetylene falcarinol and related structures (Figure 1), as examples of a type of compound that deserves further study. Thus the data presented are selected to illustrate how this type of compound fulfils the criteria mentioned above, primarily from literature from other fields than food science.

Many scientists have investigated the biological activities of polyacetylenes, but always from other viewpoints than general health prevention, primarily plant protection, pharmacology, etnobotany, plant stress physiology and toxicology.[5] But despite regular occurrence in food plants, no studies have addressed the possibility that normal levels of polyacetylenes in food may be beneficial to healthy people.

The polyacetylenes described here are relatively unstable (Figure 2), so degradation during cooking should be considered important when assessing the potential chemopreventive effect.

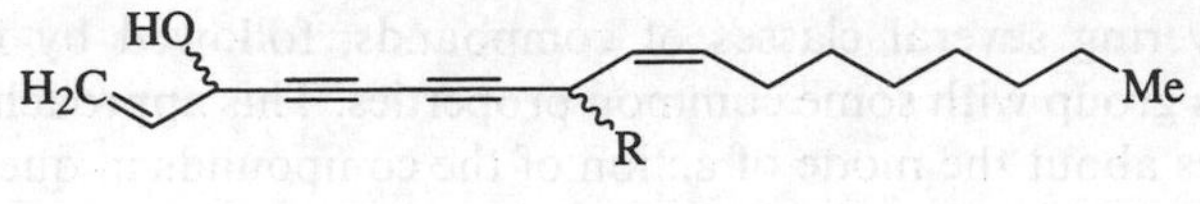

1 R = H, Falcarinol

2 R = OH, Falcarindiol

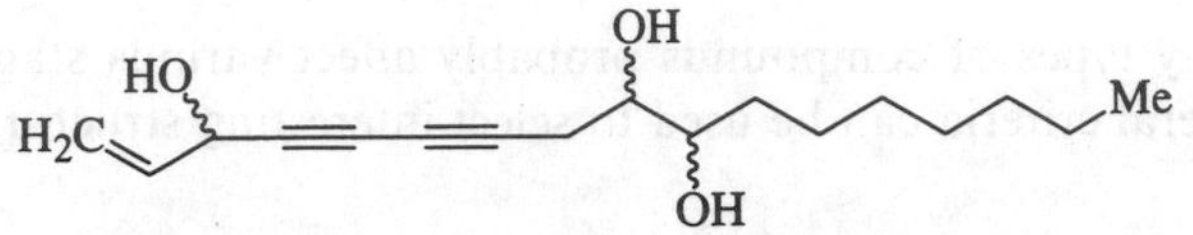

3 Panaxytriol

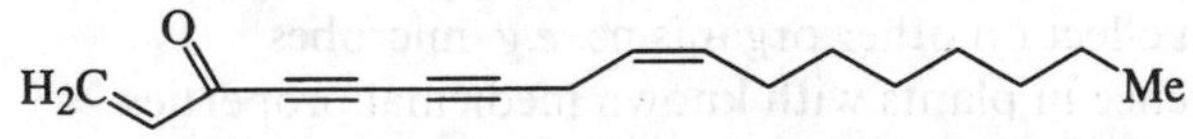

4 Falcarinone

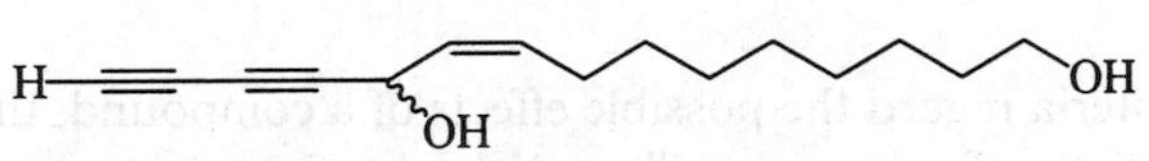

5 *cis*-Pentadeca-6-ene-1,3-diyne-5,15-diol

Figure 1 *Structures of selected polyacetylenes from vegetables and medicinal plants*

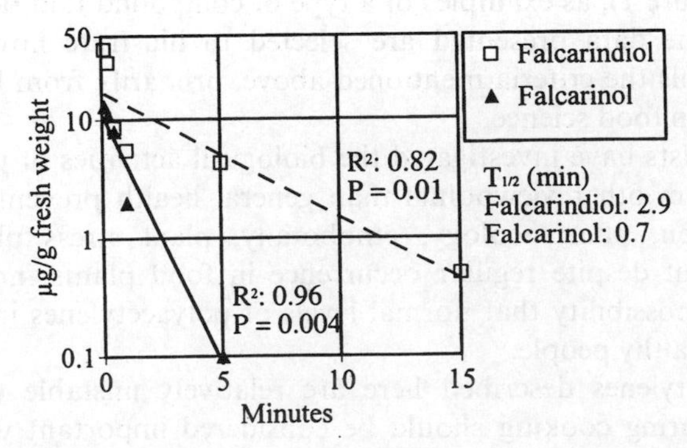

Figure 2 *Degradation of polyacetylenes after placing 1 cm carrot slices in boiling water*

Table 1 *Studies of biological activities of falcarinol and related compounds*

Plant and compound	*Effect*	*Effective concentration*	*Reference*
Studies of cancer-related pharmacological properties			
Panax ginseng, **1**	Inhibition of MK-1 tumour cell growth	0.07 $\mu g\ ml^{-1}$ in cell culture	6
1 and **3**	Cytotoxicity against MK-1 tumour cells	0.03 and 0.17 $\mu g\ ml^{-1}$ in cell culture, respectively	5
3	Pharmacokinetics in rats after injection or oral administration (1 $mg\ kg^{-1}$)	1 μg or 50 $ng\ ml^{-1}$ in serum, respectively	7
3	Potentiation of cytotoxicity of mitomycin C against MK-1 tumour cells	3 $ng\ ml^{-1}$ in cell culture	8
Peucedanum praeruptorum **2**	Suppression of mutagenicity of furylfuramide and Trp-P-1 in *Salmonella typhimurium*	20 $\mu g\ ml^{-1}$	9
Studies of phytoprotective properties in vegetable crops			
Carrot (*Daucus carota*), **1**	Inhibition of spore germination of *Botrytis cinerea*	10 $\mu g\ ml^{-1}$ in water	5
1	Accumulation after infection with *Botrytis cinerea*	Increase from 52 to 300 $\mu g\ ml^{-1}$ in fresh root	5
1	Stimulation of oviposition of carrot fly (*Psila rosea*)	50 μg on artificial leaf	10
1 and **2**	Inhibition of hyphal growth of *Mycocentrospora acerina*	30 and 40 $\mu g\ ml^{-1}$, respectively, in agar	5, 11
2	Resistance to infection by *Mycocentrospora acerina*	80 $\mu g\ g^{-1}$ in peel of fresh root	12
Tomato (*Lycopersicon esculentum*) **1** and **2**	Accumulation after infection with *Cladosporium fulvum*	Increase from 0 to 28 and 22 $\mu g\ ml^{-1}$, respectively, in fresh leaf	13
2	Accumulation after infection with *Cladosporium fulvum*	Increase from 0 to 10 $\mu g\ ml^{-1}$ in green fruits and leaves	5
Eggplant (*Solanum melongena*) **2** and **5**	Inhibition of growth of *Monilinia fructicola*. Accumulation after elicitation with RNase A	1.5 $\mu g\ ml^{-1}$ of **6** in culture medium. Increase from 0 to 0.3 and 3 $\mu g\ ml^{-1}$, respectively, in cell culture	14
Studies of toxicity and allergenicity			
Daucus carota, **1**	Toxicity in mice after injection	100 $mg\ kg^{-1}$	5
Crithmum maritimum, **1** and **2**	Toxicity against *Artemia salina*	0.7 and 3.3 $\mu g\ ml^{-1}$ in seawater, respectively	5
Schefflera arboricola, **1**	Allergic contact dermatitis (patch test)	2 $mg\ ml^{-1}$ in vegetable oil	15
Hedera helix, **1**	Allergic contact dermatitis (patch test)	300 $\mu g\ ml^{-1}$ in white petrolatum	5

Table 2 *Studies of occurrence of falcarinol and related compounds in edible parts of food plants*

Plant material and compound	*Concentration*	*Reference*
Carrot (*Daucus carota*), **1**	52 $\mu g\ g^{-1}$ in fresh roots	5
1 and **2**	40 and 80 $\mu g\ g^{-1}$ in fresh roots, respectively	16
1, **2** and 12 minor polyacetylenes	A total of 40 $\mu g\ g^{-1}$ in fresh roots, increasing to 60 $\mu g\ g^{-1}$ after storage	17
2	2–93 $\mu g\ g^{-1}$ in fresh roots, highest in peel. Located in oily droplets	5, 18
Parsley root (*Petroselinum crispum* spp. *tuberosum*) **1** and **2**	30 and 10 $\mu g\ g^{-1}$ in fresh roots, respectively	19
	15 and 10 $\mu g\ g^{-1}$ in fresh roots, respectively	20
Celery (*Apium graveolens*), **2** and **4**	10–100 $\mu g\ g^{-1}$ in fresh roots	20

3 Conclusion

Falcarinol and falcarindiol are likely to be responsible for part of the beneficial effect of carrot eating. This should be investigated further. If a direct link can be established with prevention of cancer or other diseases, the recommendation should be to eat raw carrots, preferably with peel, rather than cooked carrots. Today the recommendation, if any, is to cook the carrots thoroughly in order to increase bioavailability of β-carotene.

4 References

1 I.T. Johnson, G. Williamson and S.R.R. Musk, *Nutr. Res. Rev.*, 1994, **7**, 175–204.

2 R. Collins 'Natural Antioxidants and Anticarcinogens in Nutrition, Health and Disease', Eds. J.T. Kumpulainen and J.T. Salonen, Royal Society of Chemistry, Cambridge, 1999, 417.

3 V. Breinholt, 'Natural Antioxidants and Anticarcinogens in Nutrition, Health and Disease', Eds. J.T. Kumpulainen and J.T. Salonen, Royal Society of Chemistry, Cambridge, 1999, 93.

4 C.S. Fuchs, E.L. Giovannucci, G.A. Colditz, D.J. Hunter, M.J. Stampfer, B. Rosner, F.E. Speizer and W.C. Willet, *New England J. Med.*, 1999, **340**, 169–176.

5 L. P. Christensen, *Recent Res. Devel. Phytochem.*, 1998, **2**, 227, and references cited therein.

6 T. Saita, M. Katano, H. Matsunaga, H. Yamamoto, H. Fujito and M. Mori, *Chem. Pharm. Bull.*, 1993, **41**, 549–552.

7 T. Saita, H. Matsunaga, H. Yamamoto, F. Nagumo, H. Fujito, M. Mori and M. Katano, *Biol. Pharm. Bull.*, 1994, **17**, 798–802.

8 H. Matsunaga, M. Katano, T. Saita, H. Yamamoto and M. Mori, *Cancer Chemother. Pharm.*, 1994, **33**, 291–297.

9 M. Miyazawa, H. Shimamura, R.C. Bhuva, S.I. Nakamura and H. Kameoka, *J. Agric. Food Chem.*, 1996, **44**, 3444–3448.
10 Maki, J. Kitajima, F. Abe, G. Stewart and M.F. Rya, *J. Chem. Ecol.*, 1989, **15**, 1883–1897.
11 B. Garrod and B.G. Lewis, *Trans. Br. Mycol. Soc.*, 1982, **78**(3), 533–536.
12 K. Olsson and R. Svensson, *J. Appl. Genetics*, 1997, **38**, 219–223.
13 U. Batista and V.J. Higgins, *Can. J. Bot.*, 1991, **69**, 822.
14 S. Imoto and Y. Ohta, *Plant Physiol.*, 1988, **86**, 176–181.
15 L. Hansen and P.M. Boll, *Phytochemistry*, 1986, **25**, 529–530.
16 S.G. Yates and R.E. England, *J. Agric. Food Chem.*, 1982, **30**, 317–320.
17 E.D. Lund and J.H. Bruemmer, *J. Sci. Food Agric.*, 1991, **54**, 287–294.
18 B. Garrod and B.G. Lewis, *Trans. Br. Mycol. Soc.*, 1979, **72**(3), 515–517.
19 S. Nitz, M.H. Spraul and F. Drawert, *J. Agric. Chem.*, 1990, **38**, 1445–1447.
20 F. Bohlmann, *Chem. Ber.*, 1967, **100**, 3454.

Section 8

Anticarcinogenic Effects of Human Diets: Studies in Man

8.1

Food and Cancer Prevention: Human Intervention Studies

John C. Mathers

HUMAN NUTRITION RESEARCH CENTRE, DEPARTMENT OF BIOLOGICAL AND NUTRITIONAL SCIENCES, UNIVERSITY OF NEWCASTLE, NEWCASTLE UPON TYNE NE1 7RU, UK

1 Strategies for Cancer Chemoprevention in Humans

That variation in diet makes a major contribution to the several fold differences in cancer incidence between communities and over time[1] is no longer in doubt.[2] Approximately one third of cancer cases is potentially avoidable by changes in diet.[3] However, careful analysis of the evidence from observational epidemiological studies allows only very limited conclusions to be drawn about the enhancing or inhibiting effects of particular eating patterns, foods or food components on cancer risk.[2] Such evidence may be adequate for general public health advice[4,5] but the effectiveness of the latter remains to be quantified. The most compelling evidence about the benefits or risks of dietary exposure to particular foods will derive from well-designed human intervention trials.

2 Definitive Intervention Studies

Cancer chemoprevention is the use of specific chemical compounds to prevent, inhibit or reverse carcinogenesis.[6] Food components ingested as part of a normal diet are excluded from the accepted definition of chemoprevention.[7] As a field of study, cancer chemoprevention is in its infancy and there have been only nine definitive trials completed and published to date.[8] In their analysis, Lippman *et al.* (1998) considered that definitive chemoprevention trials had the following elements in their study design:

- Primary endpoint of cancer incidence
- Two-sided hypothesis testing
- Randomisation with placebo control versus interventions, and
- Large scale ($n \geq 1000$).

The nine definitive trials included 15 primary endpoints of cancers at specific sites (lung, skin, oesophagus-gastric cardia or breast) or total cancer. Most used nutrients or food derivatives (β-carotene, retinol, selenium, vitamin C, molybdenum, α-tocopherol, multivitamins and minerals, riboflavin, zinc or niacin) as the intervention agent – the only exception was the recently reported tamoxifen trial.[9] Of the 15 interventions, 9 produced no significant effect, 2 resulted in statistically significant negative effects (both used β-carotene) and only two demonstrated significant protection (retinol against skin squamous cell carcinoma[10] and tamoxifen against breast cancer.[9] Even for the trials with positive outcomes for the primary endpoints, there were some potential, or actual, adverse effects. With tamoxifen treatment for breast cancer prevention, there was a significant increase in the risk of endometrial cancer and greater rates of stroke, pulmonary embolism and deep vein thrombosis especially in the older (50+ years) women.[9] Daily supplementation with 25 000 IU retinol (which lowered skin cancer rates by about a quarter) was associated with a 3% increase in plasma total cholesterol and a 1% reduction in HDL cholesterol,[11] which could increase coronary artery disease risk by about 6%. Implementation of such interventions will need to proceed with caution to balance potential benefits in respect of the disease of primary interest against greater risks of other common diseases.

2.1 Lessons from the β-Carotene Supplementation Trials

On 18 January 1996, the Beta-carotene and Retinol Efficacy Trial (CARET) 'was stopped 21 months early because of clear evidence of no benefit and substantial evidence of possible harm'.[12] What had gone wrong? The hypothesis that dietary β-carotene reduces human cancer is at least 20 years old[13] and is supported by a wealth of data from epidemiological, animal and cell culture studies. Persons consuming diets rich in fruits and vegetables, especially carotenoids, and those with higher serum concentrations of β-carotene, have lower risk of cancer, particularly lung cancer.[14] There was a widely-accepted mode of action, *i.e.* the potent singlet oxygen quenching capacity of β-carotene and, as a chemoprevention agent, it held considerable promise since it was natural (food-derived), cheap, readily-available and safe.[8] Unfortunately, the risk of lung cancer among those who took pharmacological doses of β-carotene (20–30 mg d^{-1}) was significantly increased in both the Finnish study of male smokers,[15] and in the American CARET study where subjects were at increased risk because of asbestos exposure (men) or smoking (men and women).[12] Lack of evidence of benefit in terms of neoplasia or cardiovascular disease of supplementation of healthy men with 25 mg β-carotene d^{-1} for 12 years in the Physicians' Health Study[16] added to the disenchantment with this agent. Recent work with ferrets (a good model for β-carotene metabolism in humans) has confirmed the adverse effects of a combination of β-carotene supplementation and exposure to cigarette smoke on lung neoplasia and provided a possible mechanistic explanation.[17] Omenn[14] concluded that 'No one should be encouraged to take beta-carotene supplements' and the Department of Health[2] went further in recommending 'the avoidance of β-carotene supplements as a means of protecting against cancer'.

The adverse effects of high dose β-carotene supplements in some people at enhanced risk of lung cancer and their failure to produce any health benefit in others has underlined the need for well-designed trials which not only test efficacy but also seek evidence of potential harm from even the most apparently innocuous chemoprevention agents. There are wider legislative/regulatory implications for the availability and marketing of food supplements which may be consumed in amounts considerably greater than those found in usual diets. However, these concerns should not prevent the search for more effective and safer chemoprevention agents. The disappointment over β-carotene should not be allowed to:

- damage public health campaigns designed to increase intakes of vegetables and fruits or
- prevent investigation and exploitation of the great potential for reducing the burden of cancer by dietary means.

3 Life-style Modification

Just as prevention and cessation of smoking and avoidance of known carcinogens are the best measures for prevention of lung cancer,[14] so life-style changes in respect of diet may offer the greatest potential for primary prevention of cancers at sites such as colon/rectum, breast, prostate and stomach. I am unaware that any definitive life-style modification study with cancer as its primary endpoint has, as yet, reported. The Polyp Prevention Trial (PTT) is one of the most interesting attempts to induce substantial changes in diet with recurrence of adenomatous polyps of the large bowel (the best available surrogate for colorectal cancer (CRC)) as the outcome measure.[8] 2079 participants with one or more polyps recently resected have been randomised to a control arm or to a comprehensive 'diet plan' in a multi-centre, randomised, controlled trial. The PTT's multi-component eating plan has explicit consumption targets for dietary fat, dietary fibre and vegetables and fruits and employs the philosophy that a multi-component (as distinct from a single component) dietary intervention is more likely to capture the beneficial biological interactions among foods and food constituents.[18] The key measurement period for subjects in the PTT are the 3 years between a colonoscopy one year after entering the trial and the exit colonoscopy after 4 years. Analysis of baseline demographic, behavioural, nutritional and clinical characteristics showed that the Intervention and Control groups in the PTT were well matched.[18,19] Randomisation began in 1991 and follow-up was due to be completed in 1998 but, at the time of writing, no results are yet available.

The costs of such large scale intervention trials are very substantial. Because of this, it is probable that rather few large scale primary prevention of neoplasia trials based on dietary modification will be carried out so the evidence base for effective life-style modification is likely to be very limited. The challenge is to design cost effective intervention studies which will yield conclusive evidence of efficacy (and absence of risk) in well-defined population groups which can form the basis for public health strategies. Particular problems for such studies include:

- Impossibility of using double-blind designs
- Eliciting compliance with the intervention over a number of years, and
- Measuring the extent of compliance.

4 The Capp Studies – Use of Genetically-predisposed Individuals as a Novel Paradigm

4.1 The CAPP1 Study

At its most fundamental, cancer is a genetic disease due to mutations in, or altered expression of, tumour suppressor genes and proto-oncogenes in stem cells that result in loss of the normal controls on proliferation, migration and apoptosis. For CRC, the gatekeeper gene is the tumour suppressor gene *APC* which is disabled through mutation of both alleles in at least 75% of colorectal tumours.[20] Individuals who inherit a germline mutation in one *APC* allele acquire sporadic mutations in the second *APC* allele which results in the development of numerous adenomatous polyps in their colon and rectum (familial adenomatous polyposis, FAP). In the absence of total colectomy, which is the only proven treatment, such individuals are on a fast track to CRC. Since the molecular and pathological bases of CRC in both FAP and in the sporadic disease are very similar, we reasoned that subjects with FAP could provide a good model for chemoprevention of CRC studies. Since it is probable that all CRC is the result of interactions between environmental exposure (including diet) and inheritance, studies with FAP patients provide a paradigm for genetically targeted interventions. In the CAPP1 Study, young FAP gene carriers who retain their colons are being randomised to a double-blind, placebo controlled trial.[21] The intervention agents are aspirin (600 mg d^{-1}) and resistant starch (RS; provided as 30 g d^{-1} of a 1:1 mix of raw potato starch and Hylon VII) which are administered in a 2×2 factorial design. The primary outcome measure is the clinical appearance of the rectum that is recorded by video-endoscopy before and 1 year after initiation of treatment. Such recordings are scored by experienced observers blinded to knowledge of the treatment applied and whether it is a 'before' or 'after' endoscopy. In addition, multiple rectal mucosal biopsies are taken for assessment of crypt cell proliferation, a potential surrogate endpoint in CRC chemoprevention studies.

4.2 The CAPP2 Study

The human DNA mismatch repair gene family, including *hMSH2, hMSH6, hMLH1, hPMS1 and hPMS2*, encode a group of proteins that co-operate to detect, excise and repair single and double base pair error during DNA replication. Germ-line mutations in one of these genes (most commonly hMSH2 and hMLH1) are responsible for the syndrome hereditary non-polyposis colorectal cancer (HNPCC) which is characterised by early onset CRC (median age 44 years) with a proximal preponderance and multiple synchronous and metachronous CRCs. In addition, HNPCC patients have excess extra-colonic cancers at sites including the endometrium, ovary, breast, stomach and upper urinary tract.[22] A molecular diagnostic feature of HNPCC tumours is a high level of microsatellite instability (MSI).[23] Failure to detect and repair replication errors in key regulatory genes in persons with HNPCC is believed to be responsible for their greatly enhanced risk of cancer.

The CAPP2 Study aims to recruit 1200 HNPCC gene carriers with an intact colon (or limited resection) to a randomised, double-blind, controlled trial.[24] Using a 2 × 2 factorial design, each subject will receive aspirin (or placebo) and RS (or placebo) for a 2 year period with the primary endpoints being the number, size and histological stage of colorectal carcinomas. The aspirin dose is 600 mg d^{-1} (2 enteric coated tablets per day) with one week's supply provided in a blister pack. The RS is provided as 30 g d^{-1} high amylose maize starch in two sachets that is expected to deliver about 13 g RS to the colon.

5 Design Issues in Cancer Chemoprevention Studies

5.1 Study Subjects

Many of the studies to date have recruited middle-aged or older subjects – often those at increased risk as a result of behavioural or medical history reasons – because the incidence of most cancers increases steeply in later life leading to more informative events. If an intervention agent worked by preventing very early steps in tumorigenesis, there could be theoretical benefits in initiating treatment at a younger age. However, these putative benefits should be weighed against the longer time-scale of the study which would bring with it problems of compliance, potential loss of quality of life and possibly greater risks of adverse effects through long term exposure to the agent.

5.2 Monitoring Compliance

With some intervention agents, assessment of compliance may be made objectively by measurement of the agent (or a derivative) in a body fluid, *e.g.* assay from serum β-carotene in the Finnish β-carotene trial.[15] For others, it may be possible to count tablet use. In the CAPP Studies, the RS sachets contain a small amount of *para*-amino benzoic acid, which is readily absorbed and quantitatively excreted in urine[25] that can be used as a compliance marker. Where the intervention is a lifestyle change, particularly if it is aimed at changing dietary behaviour, monitoring the extent of compliance is a significant challenge. Assessment tools which rely on participant recording/recalling of dietary intake may be compromised by the nature of the intervention leading to biased estimates of food intake.

5.3 Outcome Measures

Definitive intervention studies require malignant disease as the primary end-point. Such studies are likely to be relatively long term, require large numbers of subjects and be expensive but provide the most reliable evidence of efficacy. However, given the long list of potential chemoprevention agents,[26,27] the availability of reliable surrogate endpoints would be of considerable help. In contrast with cardiovascular disease where there are several well-established intermediate endpoints, *e.g.* raised blood lipids concentrations, raised blood pressure and abnormal blood coagulation, there are few usable surrogate endpoints for cancer chemoprevention studies. For CRC, the appearance, size and histology of colorectal adenomas are the most reliable precursors of malignant disease.[28] There has been considerable use of aberrant crypt foci (ACF) as a pre-cancerous marker in rodent studies but until recently[29] little evidence that ACF were precursors of adenomas and cancer in humans.

Since the cancer involves loss of control of cell kinetics there has been considerable interest in the utility of perturbations in colonic epithelial cell proliferation as a predictor of CRC.[31] There is a considerable body of evidence in support of the hypothesis that the rate of proliferation and/or the distribution of cycling cells within the crypt is altered in cancer and in those at high risk of CRC.[30,32,33] Indices of intestinal cell proliferation are altered readily by dietary and pharmacologically means but it remains uncertain whether these changes are of pathological significance.[34] Analysis of epithelial proliferation measurements (using *in vitro* bromodeoxyuridine (BrdUrd) labelling of rectal biopsies from 223 screenees, 132 of whom had adenomas removed > 3 years previously) showed that age and lifestyle factors (including tobacco smoking and intakes of 'fibre' and calcium) could have significant effects which should be taken into account when undertaking intervention studies.[35] Preliminary data from the PTT has been used to test the potential utility of several cell proliferation assays as surrogate endpoints in a large multi-centre intervention trial with adenoma recurrence as its primary outcome measure.[36] Proliferating cells nuclear antigen (PCNA) labelling index appeared to have little utility, BrdUrd data were inconclusive but PCNA proliferative height showed some promise as a marker in CRC intervention studies.[36] In both CAPP Studies, we are collecting rectal mucosal biopsies to assess the usefulness of proliferation indices in adenoma prevention trials. Although crypt cell proliferation is deregulated in FAP,[33] it appears similar in HNPCC gene carriers and in control subjects.[37] It remains to be seen whether the intervention agents alter cell proliferation in these at risk subject groups and the extent to which this predicts appearance of adenomas.

5.4 Ethical Issues

Nyrén has provided a thoughtful account of many of the ethical considerations arising from cancer chemoprevention trials.[38] These include cost-benefit considerations for society and risk/benefit ratios and quality of life issues for participants in such trials. Targeting high-risk populations, especially those from cancer families (as in the CAPP Studies), may yield direct benefits for the participant (if he/she receives an effective agent) or for the patient's family (which is likely to include others at risk). However, there is unlikely to be much direct benefit for healthy persons recruited to cancer chemoprevention trials. Indeed several hundred years of chemoprevention treatment may be necessary to gain one life year for otherwise healthy people who begin treatment in middle age.[39] There may be adverse psychological and social sequelae leading to loss of quality of life from participation in chemoprevention trials as a result of increased anxiety about cancer risks conveyed via the study protocol.[38] This is unlikely to be an issue for those with inherited predispositions to cancer – indeed participation in a trial may enhance quality of life by raising hopes of amelioration of the condition for the individual and his/her family. Lifestyle changes including adoption of healthier eating behaviours may require substantial effort from the participant but this cost may be offset by benefits in terms of an increased sense of well-being and reduced risks of other common diseases including cardiovascular disease and diabetes.

6 Conclusions

Prevention of cancer by pharmaceutical or behaviour modification methods is in its infancy. However, the evidence from epidemiological and limited experimental studies provides considerable hope that much human cancer is potentially avoidable. It is unlikely that observational studies will provide sufficiently strong evidence of benefits, risks and costs to allow the development of effective prevention strategies. There will be no substitute for well-designed human intervention trials that will require major resource inputs and last several years. Evidence of lack of efficacy, or indeed of increased risk, should be viewed as just as valuable an outcome from such trials as demonstration of a protective effect.

Acknowledgements

The CAPP Studies are supported by the EU BIOMED Programme, the Imperial Cancer Research Fund and the Ministry of Agriculture, Fisheries and Food (ANO317).

8 References

1 R. Doll and R. Peto. *J. Natl. Cancer Inst.*, 1981, **66**, 1191.
2 Department of Health *Nutritional Aspects of the Development of Cancer*. Report on Health and Social Subjects 48. London: The Stationery Office (1998).
3 W.C. Willett. *Environ. Health Pers.* 1995, **130** (Suppl 8): 165.
4 M.J. Hill, J. Faivre and A. Giacosa. *Prevention and early detection of colorectal cancer* London: WB Saunders Company Ltd, 1996, p. 144.
5 J.H. Cummings and S.A. Bingham. *B.M.J.*, 1998, **317**, 1636.
6 G.I. Kelloff, C.W. Boone, C.C. Sigmann and P. Greenwald. *Prevention and early*

detection of colorectal cancer, 1996 [G.P. Young, P. Rozen and B. Levin, editors]. London: WB Saunders Company Ltd, p. 115.
7 V.G. Vogel and B. Levin. *Cancer Bull.*, 1995, **47**, 473.
8 S.M. Lippman, J.J. Lee and A.L. Sabichi. *J. Natl. Cancer Inst.*, 1988, **90**, 1514.
9 B. Fisher, J.P. Constantino, L. Wickerham, C.K. Redmond, M. Kavanah, W.M. Cronin, V. Vogel, A. Ribidoux, N. Dimitrov, J. Atkins, M. Daly, S. Wieand, E. Tan-Chiu, L. Ford and N. Wolmark. *J. Natl. Cancer Inst.*, 1998, **90**, 1371.
10 T.E. Moon, N. Levine, B. Cartmel, J.L. Bangert, S. Rodney, Y.M. Dong, Q. Peng and D.S. Alberts. *Cancer Epid. Biomarkers Prev.* 1997, **6**, 949.
11 B. Cartmel, T.E. Moon and N. Levine. *Amer. J. Clin. Nutr.*, 1999, **69**, 937.
12 G.S. Omenn, G.E. Goodman, M.D. Thornquist, J. Balmes, R. Cullen, A. Glass, J.P. Keogh, F.L. Meyskens Jr, B. Valanis, J.H. Williams Jr, S. Barnhart and S. Hammar. *N. Eng. J. Med.*, 1996, **334**, 1150.
13 R. Peto, R. Doll, J.D. Buckley and M.B. Sporn. *Nature*, 1981, **290**, 291.
14 G.S. Omenn. *Ann. Rev. Pub. Health,* 1998, **19**, 73.
15 The Alpha-Tocopherol, Beta-Carotene Cancer Prevention Study Group, *N. Engl. J. Med.*, 1994, **330**, 1029.
16 Ch. Hennekens, J.E. Buring, J.E. Manson, M. Stampfer, B. Rosner, N.R. Cook, C. Belanger, F. LaMotte, J.M. Gaziano, P.M. Ridker, W. Willett and R. Peto. *N. Engl. J. Med.*, 1996, **334**, 1145.
17 X-D. Wang, C. Liu, R.T. Bronson, D.E. Smith, N.I. Krinsky and R.M. Russel. *J. Natl. Cancer Inst.*, 1999, **91**, 60.
18 A. Schatzkin, E. Lanza, L.S. Freedman, J. Tangrea, M.R. Cooper, J.R. Marshall, P.A. Murphy, J.V. Selby, M. Shike, R.R. Schade, R.W. Burt, J.W. Kikendall and J. Cahill for the PTT Study Group. *Cancer Epid. Biomarkers Prev.*, 1996, **5**, 375.
19 E. Lanza, A. Schatzkin, R. Ballard-Barbash, D.C. Clifford, E. Paskett, D. Hayes, E. Boté, B. Caan, M. Shike, J. Weissfeld, M. Slattery, D. Moteski and C. Daston. *Cancer Epid. Biomarkers Prev.*, 1996, **5**, 385.
20 M.A. Reale and E.R. Fearon. *Prevention and early detection of colorectal cancer*, 1996 [G.P. Young, P. Rozen and B. Levin, editors]. London: WB Saunders Company Ltd, p. 63.
21 J. Burn, P.D. Chapman, J. Mathers, L. Bertario, D.T. Bishop, S. Bülow, J. Cummings, R. Phillips and H. Vasen. *Eur. J. Cancer*, 1995, **31A**, 1385.
22 H.T. Lynch and T. Smyrk. *Cancer*, 1996, **78**, 1149.
23 J. Rüschoff, S. Wallinger, W. Dietmaier, T. Bocker, G. Brockhoff, F. Hofstädter and R. Fishel. *Proc. Nat. Acad. Sci. USA*, 1998, **95**, 11301.
24 J. Burn, P.D. Chapman, D.T. Bishop, and J.C. Mathers. *Proc. Nutr. Soc.*, 1998, **57**, 183.
25 S.A. Bingham, H. Vorster, J.C. Jerling, E. Magee, A. Mulligan, S.A. Runswick and J.H. Cummings. *Brit. J. Nutr.*, 1997, **78**, 41.
26 D.K. Singh and S.M. Lippman. *Oncol.*, 1998, **12**, 1643.
27 D.K. Singh and S.M. Lippman. *Oncol.*, 1998, **12**, 1787.
28 S.R. Hamilton. *Prevention and early detection of colorectal cancer*, 1996 [G.P. Young, P. Rozen and B. Levin, editors]. London: WB Saunders Company Ltd, p. 3.
29 T. Takayama, S. Katsui, Y. Takahashi, M. Ohi, S. Nojiri, S. Sakamaki, J. Kato, K. Kogawa, H. Miyaki and Y. Niitsu. *N. Engl. J. Med.*, 1998, **339**, 1277.
30 M. Lipkin. *Cancer*, 1974, **34**, 878.
31 M.J. Wargovich. *Prevention and early detection of colorectal cancer*, 1996 [G.P. Young, P. Rozen and B. Levin, editors]. London: WB Saunders Company Ltd, p. 89.

32 O.T. Terpstra, M. van Blankenstein, J. Dees and G.A.M. Eilers. *Gastroenterology*, 1987, **92**, 704.
33 S.J. Mills, N.A. Shepherd, P.A. Hall, A. Hastings, J.C. Mathers and A. Gunn. *Gut*, 1995, **36**, 391.
34 J.C. Mathers. *Proc. Nutr. Soc.*,1998, **57**, 219.
35 P. Rozen, F. Lubin, N. Papo and G. Zajicek. *Cancer,* 1998, **83**, 1319.
36 L.M. McShane, M. Kulldorff, M.J. Wargovich, C. Woods, M. Purewal, L.S. Freedman, D.K. Corle, R.W. Burke, D.J. Mateski, M. Lawson, E. Lanza, B. O'Brien, W. Jr. Lake, J. Moler and A. Schatzin. *Cancer Epid. Biomarkers Prev.*, 1998, **7**, 605.
37 S.E. Green, P. Chapman, J. Burn, A.D. Burt, M. Bennett, D.R. Appleton, J.S. Varma and J.C. Mathers. *Gut*, 1998, **43**, 85.
38 O. Nyrén. *Acta Oncol.*, 1998, **37**, 235.
39 M. Hakama. Chemoprevention of cancer. *Acta Oncol.*, 1998, **37**, 227.

8.2

Effect of Increased Fruit and Vegetables Consumption on Markers for Disease Risk: A Diet Controlled Human Intervention Trial

W.M.R. Broekmans,[1,2] W.A.A. Klöpping-Ketelaars,[1]
H. Verhagen,[1] H. van den Berg,[1] F.J. Kok[2] and G. van Poppel[1]

[1] TNO VOEDING, PO BOX 360, 3700 AJ, ZEIST, THE NETHERLANDS
[2] DEPARTMENT OF HUMAN NUTRITION AND EPIDEMIOLOGY, WAGENINGEN UNIVERSITY, THE NETHERLANDS

1 Introduction

Observational epidemiological studies have shown that a high consumption of fruit and vegetables is associated with a decreased risk of human cancer[1,2] at a number of common sites and a decreased risk of cardiovascular disease.[3,4] However, the association between health and vegetable consumption may also be due to the fact that high vegetable consumers have a more healthier dietary intake (less energy, less fat) or an otherwise healthier life-style. We therefore evaluated whether an increased intake of a varied mix of fruits and vegetables with an energy- and fat-controlled diet has beneficial physiological effects in human volunteers.

2 Methods

We performed a randomized, diet controlled, single blind, parallel, intervention trial. 23 apparently healthy volunteers (40–60 years) received a diet low in fruits and vegetables (100 grams/day; control group) during four weeks and 24 volunteers received a diet high in fruits and vegetables (500 grams/day; fruits and vegetables group) and drank 200 ml of fruit juice per day. Body weight was measured and controlled by adjusting dietary energy level if weight deviated 1.5 kg compared to the reference body weight at day 1. Blood was collected at day 1 and 29 and the following parameters were measured (Comet assay and micro nucleus assay in lymphocytes, α-class GST, total, HDL, LDL cholesterol, triglycerides in serum, homocysteine, vitamin B12, folate in plasma, fibrinogen, t-PA antigen, t-PA activity, F1 + F2, D-dimers, vitamin C, α-tocopherol, carotenoids in plasma and total antioxidant capacity, antibody response to hepatitis B vaccination and C-reactive protein).

A total of the weekly fruits and vegetables diet and weekly control diet was mixed and analyzed for energy, macronutrients, vitamins, carotenoids, folic acid, fibre, flavonoids and glucosinolates.

3 Results

The total energy content of the two diets was similar. The energy percentages from the macronutrients were near recommended energy percentages in both diets. The amount of vitamin C, carotenoids, glucosinolates and flavonoids were higher in the fruits and vegetables diet by respectively approximately 160, 360, 530 and 400 percent. The results of plasma carotenoids and vitamin C are presented in Figure 1.

After four weeks of intervention the changes of plasma concentrations of lutein, β-cryptoxanthin, α-carotene and vitamin C appeared significantly higher in the fruits and vegetable group with regard to the control group. Preliminary statistical analyses indicate no difference between the two treatments for all other parameters.

4 Conclusion

These results indicate that four weeks' consumption of 500 gram fruits and vegetables in comparison with 100 gram fruits and vegetables has positive effect

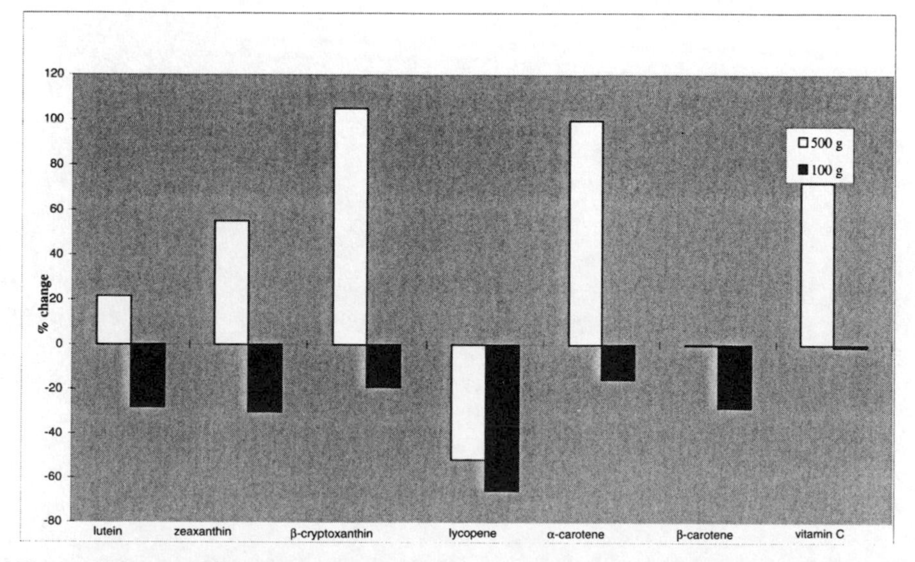

Figure 1 *Percentual change of carotenoids and vitamin C after four weeks of respectively 500 grams of fruits and vegetables and 100 grams of fruits and vegetables*

on plasma carotenoids lutein, β-cryptoxanthin, α-carotene and vitamin C. Consumption of fruits and vegetables demonstrated no effect on plasma concentrations of α-tocopherol, blood lipids, haemostatic parameters, DNA damage, immune status and antioxidant capacity as measured by biomarkers used in this study.

References

1 J.H. Weisburger, *Am. J. Clin. Nutr.*, 1991, **53**, 226S–237S.
2 K.A. Steinmetz and D. Potter, *Cancer Causes Control*, 1991, **2**, 325–357.
3 K.F. Gey, U.K. Moser, P. Jordan, H.B. Stahelin, M. Eichholzer and E. Ludin, *Am. J. Clin. Nutr.*, 1993, **57**, 787S–797S.
4 M.G.L. Hertog, E.J.M. Feskens, P.C.H. Hollman, M.B. Katan and D. Kromhout, *Lancet*, 1993; **342**, 1007–1011.

8.3

Effects of Reduction of Dietary Sucrose on Fecal Bile Acids: An Intervention Trial

G. Caderni, G. Morozzi, D. Palli, R. Fabiani, L. Lancioni, C. Luceri, G. Trallori, L. Manneschi, C. Saieva, A. Russo and P. Dolara

DEPARTMENT OF PHARMACOLOGY, UNIVERSITY OF FLORENCE, FLORENCE, DEPARTMENT OF CELLULAR AND MOLECULAR BIOLOGY, UNIVERSITY OF PERUGIA, PERUGIA AND PUBLIC HEALTH SYSTEM AND EPIDEMIOLOGY UNIT, CSPO, FLORENCE, ITALY

1 Introduction

Epidemiological studies have suggested that high levels of dietary sucrose increase the risk of colon cancer.[1-2] In rodents, we and others have also demonstrated that dietary sucrose increases colon carcinogenesis, proliferative activity in the colon mucosa and the levels of fecal bile acids.[3-5] Fecal bile acids have been reported to act as promoters in experimental colon carcinogenesis.[6] Moreover, epidemiological studies comparing high with low risk populations for colorectal cancer have indicated, although not consistently, that high levels of bile acids in the feces are a risk factor for colon carcinogenesis.[7,8]

Given these considerations we though it of interest to investigate whether a low-sucrose diet, supposed to be protective against colon cancer, might reduce fecal bile acids in a group of subjects who had at least two colonic adenomas removed and thus were considered at higher risk of developing colon cancer.

2 Results and Discussion

We recruited subjects (both sexes aged: 40–70 y) who had undergone an endoscopic polypectomy 6–24 months before the study. Subjects were identified in the Registries of the Gastroenterology Unit of the Regional Careggi Hospital and the Digestive Endoscopy Unit of Camerata Hospital of Florence, Italy. Written informed consent was obtained from each patient in accordance with the Helsinki declaration of 1975.

The effect of a low-sucrose diet was evaluated by comparing fecal bile acid concentration in subjects before and after a low-sucrose diet for one month. The recruited patients were interviewed by a trained dietician at the beginning of the study using a validated food frequency questionnaire, which has been recently developed in the framework of the EPIC-Italy study and assessed for validity and reproducibility.[9] On the same day subjects provided fecal samples kept in refrigerated bags. Fecal samples were then cooled to −80 °C and kept frozen until bile acid analysis. The dietician instructed each patient on how to implement a low-sucrose diet by eliminating soft drinks, sweets, chocolate and other sucrose-rich foods. Saccharine replaced sugar in coffee or tea. In order to keep the low-sucrose diet isocaloric (in comparison to the usual diet of each subject) the dietician advised patients to replace sucrose with an isocaloric amount of starch-rich foods. At the end of the 30-day study fecal samples were again provided together with a second dietary interview which focused on the diet followed in the last month.

The results of the dietary questionnaire indicated that subjects consumed significantly less sucrose during the dietary modification (data not shown). Fecal content of bile acids at the beginning and at the end of the study was analysed by gas-chromatography as previously described.[10] The results of this analysis indicated (Table 1) that no variations in primary or secondary bile acids occurred after the low-sucrose diet.

One possible explanation for the lack of effect after a low-sucrose diet might have been poor compliance, although the dietary questionnaire indicated that treated subjects had truly lowered their sucrose intake. Another possible explanation might be that at variance with some experimental studies, in which the effect of sucrose on fecal bile acids was determined comparing the effects of diets markedly different in their sucrose content,[3] the dietary variation in our trial was not extreme. In this respect it is interesting to note that an increase in secondary bile acids in the feces has been reported in humans consuming a high sucrose diet for two weeks.[12] In that study Kruis *et al.*[12] compared fecal bile acid concentration in humans consuming a low sucrose diet (60 g day^{-1}) and then shifting to an high sucrose diet consisting in 165 g of sucrose day^{-1}, a value much higher than the sucrose intake in our patients (36.5 g day^{-1} $\pm$ 17.2; mean $\pm$ SD). Therefore, one possible explanation for the lack of effect of the low-sucrose diet may be that, given the low baseline sucrose intake of our

Table 1 *Bile acids in the feces of 36 subjects consuming a low sucrose diet for one month. Values are means $\pm$ SE*

Fecal biliary acids ($mg\ g^{-1}$ dry weight)	*Start of the study*	*End of the study*
Primary bile acids	2.32 $\pm$ 0.63	1.93 $\pm$ 0.76
Secondary bile acids	12.27 $\pm$ 1.32	11.43 $\pm$ 1.52
Total bile acids	14.59 $\pm$ 1.47	13.34 $\pm$ 1.68

patients, dietary intervention induced only a small reduction in sucrose intake, possibly not enough to affect fecal bile acids. In conclusion, we did not find any evidence that a low-sucrose diet followed for one month may affect bile acid fecal levels.

Acknowledgements

Supported by Projects AIR2 CT940933, FAIR CT95/0653, 97 SOC 200302 05F02 and by MURST and AIRC, Italy.

3 References

1 J.B. Bristol, P.M. Emmett, K.W. Heaton and R.C.N. Williamson, *Br. Med. J.*, 1985, **291**, 1467.
2 A.J. Tuyns, M. Haelterman and R. Kaaks, *Nutr. Cancer*, 1987, **10**, 181.
3 G. Caderni, P. Dolara, T. Spagnesi, C. Luceri, F. Bianchini, V. Mastrandrea and G. Morozzi, *J. Nutr.*, 1993, **123**, 704.
4 G. Caderni, C. Luceri, M.T. Spagnesi, A. Giannini, A. Biggeri and P. Dolara, *J. Nutr.*, 1994, **124**, 517.
5 D. Stamp, X.M. Zhang, A. Medline, W.R. Bruce and M.C. Archer, *Carcinogenesis*, 1993, **14**, 777.
6 B.S. Reddy and K. Watanabe, *Cancer Res.*, 1979, **39**, 1521.
7 B.S. Reddy and E.L. Winder, *J.N.C.I.*, 1973, **50**, 1437.
8 T. Kamano, Y. Mikami, T. Kurasawa, M. Tsurumaru, M. Matsumoto, M. Kano and K. Motegi, *Dis. Colon. Rectum*, 1999, **42**, 668.
9 P. Pisani, F. Faggiano, V. Krogh, D. Palli, P. Vineis and F. Berrino, *Int. J. Epidemiol.*, 1997, **26**, S152.
10 C. Luceri, G. Caderni, L. Lancioni, S. Aiolli, P. Dolara, V. Mastrandrea, F. Scardazza and G. Morozzi, *Nutr. Cancer*, 1996, **25**, 187.
11 W. Kruis, G. Forstmaier, C. Scheurlen and F. Stellaard, 1991, *Gut*, **32**, 367.

8.4

Influence of CYP1A2 and NAT2 Phenotypes on Urinary Mutagenicity after a Hamburger Meal

G. Gabbani,[1] S. Pavanello,[1] P. Simioli,[2] A. Bordin[1] and E. Clonfero

[1] INSTITUTE OF OCCUPATIONAL HEALTH, UNIVERSITY OF PADOVA, ITALY
[2] DEPARTMENT OF CLINIC AND EXPERIMENTAL MEDICINE, SECTION OF HYGIENE AND OCCUPATIONAL MEDICINE, UNIVERSITY OF FERRARA, ITALY

High temperature cooking of animal protein-rich food has been shown to lead to the formation of heterocyclic aromatic amines (HAAs), which are reported to be mutagenic in short term tests and carcinogenic in animals.

In man, urinary mutagenicity increase after consumption of fried meat and this mutagenicity is mainly due to HAAs and/or their metabolites.[1,2] The metabolism of HAAs involves a first phase of *N*-hydroxylation catalysed by CYP1A2 followed by NAT2-dependent *O*-acetylation, which gives rise to the ultimate carcinogen.[3] A systematic exploration of caffeine metabolism indicated empirically that analyses of caffeine metabolites in urine can yield estimates of the activities of CYP1A2 and NAT2 enzymes.[4]

Our proposal was to study the influence of metabolic phenotype CYP1A2 and NAT2 on urinary mutagenicity of subjects after a meal of pan-fried beef hamburgers. The aim of the study was to identify individuals or a subgroup of higher internal mutagen exposure.

Slow and rapid NAT2 non-smoking volunteers, previously phenotyped by HPLC analysis of urinary caffeine metabolites,[5] were chosen. Participants were asked to cook at high temperature and for a long time three 100 g beefburgers and to eat two of them; the third beefburger was used to evaluated mutagen intake. Over-night pre-meal (control), 7 hours post-meal urine samples, and meat samples were collected from 13 subjects and the XAD-2/acetone extracts were assayed by a bacterial mutagenicity test on YG1024 *Salmonella typhimurium* strain (with S9 fraction).[6] CYP1A2 metabolic phenotypes (poor and

Table 1 *Mutagenicity of human volunteer urine samples following ingestion of pan-fried meat according to NAT2 and CYP1A2 phenotypes*

		Number	*Urine samples mutagenic activity net rev. mmol⁻¹ creatinine*
NAT2[a]	slow	7	10072 ± 6476
	rapid	5	10339 ± 5651
CYP1A2[b]	poor	7	9464 ± 5720
	extensive	5	11190 ± 6606

[a] NAT2 index [AFMU/(AFMU + 1U + 1X)] ≥ 0.3 = rapid.
[b] CYP1A2 index [(AFMU + 1X + 1U)/17U] ≥ 4.0 = extensive.

extensive) were determined by HPLC analysis of caffeine metabolites in urine samples collected just before the meat ingestion.[5]

All subjects but one (who was not included in this study) were highly exposed to mutagens.[6] Mean levels of mutagenic activity of pre-meal (control) and of the post-meal urine were 493 ± 330 and 10183 ± 5874 net revertants/mmol creatinine (Wilcoxon test, $z = 3.02$, $P < 0.01$), respectively. Table 1 shows urinary mutagenic levels after the meat meal according to NAT2 and CYP1A2 phenotypes. The values of urinary mutagenicity of NAT2 slow subjects ($n = 7$) showed a non-significant difference in comparison with those of NAT2 rapid ($n = 5$) ones (net rev. $mmol^{-1}$ creatinine = 10072 ± 6476 *versus* 10339 ± 5651). Also the levels of promutagens from CYP1A2 poor ($n = 7$) subjects were comparable with those of CYP1A2 extensive ($n = 5$) ones (net rev. $mmol^{-1}$ creatinine = 9464 ± 5720 *versus* 11190 ± 6606).

Our results, although not yet conclusive due to the small number of subjects studied, suggest that the greater urinary mutagenicity after a beefburger meal is not influenced by metabolic phenotype CYP1A2 and NAT2.

References

1 R. Baker, A. Arlauskas, A. Bonin and D. Angus, *Cancer Lett.*, 1982, **16**, 81.
2 H. Hayatsu, T. Hayatsu and Y. Ohara, *Jpn. J. Cancer Res. (Gann)*, 1985, **76**, 445.
3 R.J. Turesky, N.P. Lang, B.A. Butler, C.H. Teitel and F.F. Kadlubar, *Carcinogenesis*, 1991, **12**, 1839.
4 W. Kalow and B-K. Tang, *Clin. Pharmacol. Ther.*, 1991, **50**, 508.
5 F. Berthou, D. Ratanasavanh, D. Alix, D. Carlhant, C. Riche and A. Guillouzo, *Xenobiotica*, 1989, **19**, 401.
6 G. Gabbani, B. Nardini, A. Bordin, S. Pavanello, L. Janni, L. Celotti and E. Clonfero, *Mutagenesis*, 1998, **13**, 187.

8.5

The Occurrence of Cancer and of Adenomae in the Large Bowel and the Concentration of *N*-Nitrosamines in the Gastric Juice

Z. Kopanski,[1] M. Schlegel-Zawadzka,[2] A. Plaszczak,[1] A. Bruchnalska[1] and T. Wojewoda[1]

[1] MILITARY CLINICAL HOSPITAL, KRAKOW, POLAND
[2] DEPARTMENT OF FOOD CHEMISTRY AND NUTRITION, JAGIELLONIAN UNIVERSITY, KRAKOW, POLAND

1 Summary

The analysis included 52 patients with cancer or adenomae of the large intestine in whom the concentration was chromatographically determined of *N*-nitrosamines in the gastric juice. The results of the analysis were related to the location of the neoplastic changes and to their histological form.

It was confirmed that as the cancer or the adenomae develops more and more peripherally in the large intestine, so the concentration of *N*-nitrosamines in the gastric juice increases. It was also shown that a statistically significantly high concentration in the gastric juice of *N*-nitrosamines is associated with villous adenomae.

2 Introduction

Studies conducted during the last few years indicate that among the *ca.* known nitrosamines about 40 show carcinogenic properties. Those compounds are widely spread in nature. According to some authors they can also be generated *in vivo* in the animal or human alimentary canal.[1,2]

Much attention has been paid to the link between the concentration in the gastric juice of *N*-nitrosamines and the developing pathology of the gastric mucosa. We can, however, ask to what extent an increase of the concentration in the gastric juice of *N*-nitrosamines may provoke the development of pathological alterations in the farther parts of the alimentary canal, including in the large intestine. In looking for an answer to that question we analysed the changes of the concentration in the gastric juice of *N*-nitrosamines in patients with neoplasms (cancer or adenomae) in the large intestine.

3 Material and Method

The analysis included 52 patients (30 men and 22 women) aged 26 to 75 years in whom a cancer or an adenomae was confirmed of the large intestine. In histologically estimating the cancer or the adenomae of the large intestine the 1989 WHO classifications were used.[3] In determining the location of the neoplasm we differentiated: the right half of the colon (caecum, ascending, liver flexture), transverse colon, the left half of the colon (splem flexture, descending), sigmoid and rectum.

In all patients a chromatographic estimation was carried out of the concentration in the gastric juice of *N*-nitrosamines (*N*-nitroso-diethanolamine, *N*-nitrosodimethylamine, *N*-nitrosomorpholine). The measurements were made using a high pressure liquid chromatograph (KONTRON) with piston pump (model 420) and 422 spectrophotometric detector. A wave length of 245 mm was used and the concentration of *N*-nitrosamines expressed in μmol l^{-1}.

3.1 Method of Statistical Analysis

The results were compiled using SAS for IBM PC rel. 6.03. The hypothesis of distribution normality was verified with the W test. In estimating the reality of the differences the Wilcoxon test was employed. All decisions were taken on the critical level $p < 0.05$.

4 Results

Among the 52 patients examined the concentration in the gastric juice of *N*-nitrosamines remained within the limits 0.002–1.970 μmol l^{-1}, averaging. 0.221 ± 0.076 μmol l^{-1}. The changes of the concentration in the gastric juice of *N*-nitrosamines in relation to the location of the neoplastic alterations in the large intestine are presented in Table 1 and in relation to the histological form of the neoplasm in Table 2.

Table 1 *Concentration in the gastric juice of* N-*nitrosamines in relation to the location of the neoplasm (cancer or adenomae) in the large intestine*

Location of the neoplasm	*Number of cases studied n*	*Concentration in the gastric juice of* N-*nitrosamines (average ± SD) (μmol l^{-1})*
Right half of the colon	7	0.152 ± 0.038*
Transverse	6	0.162 ± 0.052*
Left half of the colon	10	0.189 ± 0.069*
Sigmoid	13	0.269 ± 0.071**
Rectum	16	0.333 ± 0.058**

* to ** differences statistically significant.

Table 2 *Concentration in the gastric juice of N-nitrosamines in relation to the histological form of the neoplasm in the large intestine*

Type of histological alterations	*Number of cases studied n*	*Concentration in the gastric juice of N-nitrosamines (average ± SD) (μmol l^{-1})*
Adenocarcinoma	21	0.239 ± 0.052*
Tubular adenomae	13	0.126 ± 0.039*
Mixed adenomae	8	0.132 ± 0.056**
Villous adenomae	10	0.387 ± 0.067

*** to ** differences statistically significant.

5 Discussion

The ethiopathogenesis of cancer of the large intestine is not at all clear but certainly genetic as well as environmental factors play a role. Some authors emphasise the fact that the pathological processes underway in all the parts of the alimentary canal, *e.g.* in the stomach, may have a great significance in the development of neoplasms in the large intestine. It is likely that infection by *Helicobacter pylori* may lead to an increase of the concentration of gastrin in the blood; that alterations of the bacterial flora of the stomach may lead to the breakdown of the bile acids into metabolites recognized as carcinogenic for the mucosa of the large intestine. As a consequence of the changes of the gastric microflora, endogenic *N*-nitroso-compounds can also form.[5]

From our own studies it appears that the concentration in the gastric juice of *N*-nitrosamines statistically change in relation to the location of the neoplasm in the large intestine. High average concentrations of *N*-nitrosamines are associated mainly with neoplasms developing in the distal part of the large intestine (rectum and sigmoid). Our observations indicate that a significantly high average concentration in the gastric juice of *N*-nitrosamines coexisting with the villous adenomae may be recognized as characteristic. Some of our observations are supported by data in the literature; others require, we believe, further clinical studies.

6 References

1 L. Fishbein, *Sci. Total Environ.*, 1979, **13**, 157.
2 D. G. Gatehouse and D. J. Tweats, *Carcinogenesis*, 1982, **3**, 597.
3 J. R. Jass and L. H. Sobin, 'Histological Typing of Intestinal Tumours', WHO, Springer Verlag, Berlin, Heidelberg, 1989.
4 J. R. Goldblum, J. J. Vicari, G. W. Falk, T. W. Rice, R. M. Peak, K. Easley and J. E. Richter, *Gastroenterology*, 1998, **114**, 633.
5 G. Nardone, A. Rocco and G. Budillon, *Ital. J. Gastroenterol. Hepatol.*, 1998, **30**, 134.

8.6

Production of Butyrate or Bioavailability as Constraints on Antineoplastic Properties in the Colon

J.A. Robertson, R.L. Botham, P. Ryden and S.G. Ring

FOOD BIOPOLYMERS, INSTITUTE OF FOOD RESEARCH, NORWICH RESEARCH PARK, COLNEY, NORWICH NR4 7UA, UK

1 Introduction

Short chain fatty acids (SCFA), and in particular butyrate, are considered to be important contributors to colonic health. Fermentation of non-starch polysaccharides (NSP) and resistant starch provides a potential diet-related mechanism to increase the amount of butyrate produced in the colon. The problem is to rationalise SCFA production with food source, effects of upper gut transit, colonic pH, mucosal requirements for butyrate and antineoplastic properties. Isotopically (^{13}C) labelled foods are being used to monitor the persistence of 'diet-derived' SCFA in the colon and in relation to the fermentation behaviour of ileal effluents reaching the colon.

Materials and Methods

Eight volunteers (age 20–55) were recruited for a whole body study and three ileostomists (age 71–76) provided ileal effluents from the test meals. Each study had ethical approval from the Norwich District Ethics Committee. Test meals, formulated to provide 50 g starch, were prepared from plants labelled in a $^{13}CO_2$ enriched atmosphere as mushy peas, mashed potato, or pasta. Ileal effluents, as test meal residues and as low NSP and starch diet residues, were fermented *in vitro*, with pH controlled at 6.5 to measure net acidogenesis.[1] Recovery of ^{13}C in faecal samples was monitored using GC-Isotope Ratio Mass Spectrometry.

3 Results and Discussion

From the whole body study concentrations of faecal SCFA from the low starch and NSP period and test meal period were similar (100–160 mmol kg^{-1} fresh weight) and similar to the expected for faecal SCFA. Faecal butyrate (20–50 mmol kg^{-1} fresh weight) was not apparently affected by the test meal. A peak in label recovery between 24 and 36 hours corresponded to the expected whole gut transit time for a test meal and indicated that not all diet-derived SCFA have been been absorbed in the colon, *i.e.* either production is excess to requirements or uptake is limited.

Polysaccharide concentration in ileal effluent may represent less than 50% of the effluent dry weight but the yield of SCFA during *in vitro* fermentation was similar to that from corresponding *in vitro* digested substrates and polysaccharide isolates,[1] *i.e*, production of SCFA from non-polysaccharide substrates can be an important source of SCFA. The *in vitro* digested substrates provided a higher proportion of butyrate ($>20\%$) compared to the ileal effluents ($<20\%$). This may reflect a lower butyrate and higher propionate production from non-polysaccharide substrates.

Acidogenesis during fermentation has the potential to lower colonic pH and hence promote absorption of butyrate.[2] Acidogenesis from ileal effluents was low (1–3 mmol g^{-1}) compared to *in vitro* digested substrates (~ 5 mmol g^{-1}), although the yield of SCFA was similar for effluents and isolated substrates (4–5 mmol g^{-1}). Thus, net acidogenesis and hence the potential to lower pH is not dependent on SCFA yield but is influenced by the fermentation of 'non-acidogenic' substrates which can buffer the fermentation system. The persistence of diet-derived polysaccharide in ileal effluents does increase the acidogenic potential but how much this affects uptake and contributes to the antineoplastic properties of butyrate is unclear.

From modelling faecal concentrations of butyrate with uptake and oxidation kinetics from published data[2–4] then faecal butyrate concentration is apparently approaching saturation kinetics, or V_{max}, for oxidation and although a decrease in pH may reduce uptake the concentration apparently remains greater than requirements for oxidation.

4 Conclusion

Stable-isotope-labelled foods provide the potential to monitor the contribution of diet to SCFA metabolism in the colon and show that diet-derived SCFA can persist in the colon. Non-polysaccharide sources can also make a significant contribution to SCFA metabolism in the colon. Fermentation of polysaccharides may be more important for acidogenesis to promote butyrate uptake than as a source of butyrate. Concentrations of faecal butyrate indicate either uptake is limiting or butyrate is produced in excess of requirements for oxidation by colonocytes, and possibly also antineoplastic properties.

Acknowledgements

Funding from The Ministry Of Agriculture Fisheries and Food and The Biotechnology and Biological Research Council is gratefully acknowledged.

5 References

1 Robertson JA, Ryden P, Botham L and Ring S (1999) *J. Env. Path., Toxicol. Oncol.*, **18**, 141–146.
2 Ritzhaupt A, Ellis A, Hosie KB and Shirazi-Beechey SP (1998) *J. Physiol.*, **507**, 819–830.
3 Mascolo N, Rajendran VM and Binder HJ (1991) *Gastroenterol.*, **101**, 331–338.
4 Jorgensen JR, Clausen MR and Mortensen PB (1997) *Gut*, **40**, 400–405.

8.7

Chemopreventive Studies of Tea on Oral Cancer

Ning Li, Zheng Sun,[1] Chi Han and Junshi Chen

INSTITUTE OF NUTRITION AND FOOD HYGIENE, CHINESE ACADEMY OF PREVENTIVE MEDICINE, AND [1]BEIJING DENTAL HOSPITAL, BEIJING 100050, CHINA

1 Introduction

Tea is one the most popular beverages consumed worldwide. Many laboratory studies have demonstrated that tea has antimutagenic and anticarcinogenic effects in various laboratory testing systems and animal models including colon, esophagus, liver, lung, mammary gland and skin,[1,2,3] but no evidence on oral cancer has been reported. However, whether tea has a preventive effect on human cancer is an unsolved issue. Although several epidemiological studies suggested a protective effect of tea consumption on certain types of human cancer, other studies have indicated an opposite effect and no clear-cut conclusions could be drawn.[4,5] The inconsistency may be attributed to some confounding factors, such as very hot tea, tobacco and alcohol.[6] It is widely agreed that to conduct intervention trials of tea on human cancer is an important approach to elucidate protective effects of tea on human cancer.

Oral cancer is the sixth commonest cancer throughout the world, particularly in some developing countries, such as India, Sri Lanka, Vietnam, Philippines and parts of Brazil, where up to 25% of all cancers are oral cancer.[7,8] In recent decades, oral cancer incidence and mortality rates have been increasing in the United States, Japan, Germany and China.[9] There is an apparent need to find chemopreventive agents for oral cancer. Oral leukoplakia is a well-established precancerous lesion of oral cancer, and people with oral leukoplakias are at high risk for oral cancer.[10] Oral leukoplakia therefore provides an ideal model for studying the effects of chemopreventive agents in the prevention of oral cancer. In previous short-term screening tests on tea ingredients, we found that the effects of any single tea ingredient (tea polyphenols, tea catechins, tea pigments, tea polysaccharide, etc.) on the initiation, promotion or progression phase of carcinogenesis were not as strong as the whole tea water extracts.[11] A mixed tea product was developed in our laboratory in which tea polyphenols, and tea pigments were added to the whole water extracts of tea in a proportion based on short-term screening tests. We examined the preventive effects of green tea, tea pigments and mixed tea (whole water extract of green tea, tea polyphenols, and tea pigments) on 7,12-dimethylbenz(a)anthracene (DMBA)-induced oral carcinogenesis in golden hamsters; possible mechanisms were investigated at the molecular levels. A randomized, placebo-controlled intervention trial was conducted on oral leukoplakia patients.

2 Materials and Methods

2.1 Animal Study

2.1.1 Animals and treatment. Seventy-four male golden Syrian hamsters were randomly divided into five groups, *i.e.* 16 hamsters in the positive control group, green tea group, tea pigments group and mixed tea group, and 10 hamsters in the negative control group. The animals in group 1 (positive control group) were treated three times per week with DMBA on the right buccal pouches by a topical application of 50 μl of 0.5% DMBA in acetone with a No. 4 sable brush. The animals in groups 2–4 (tea treated group), in addition to the DMBA application as in group 1, received 1.5% green tea, 0.1% tea pigments or 0.5% mixed tea, respectively, as the sole source of drinking water for two weeks before starting the DMBA treatment and until the end of the experiment. In group 5 (negative control group), same amount of acetone was applied topically to the right pouch. Animals in group 1 and group 5 were given tap water as drinking water. All hamsters in group 1–5 were sacrificed after 15 weeks of DMBA treatment. The right cheek pouch were excised, fixed in 10% formalin, dehydrated and embedded in paraffin. Several sections were cut at 5 μm for histopathological examination and micronucleated cells, nucleolar organizer regions (AgNOR) analysis, proliferating cell nuclear antigen (PCNA) and epidermal growth factor receptor (EGFR) analysis. The number of tumor in the oral cavity was counted and the length, width and height of each tumor were measured. The tumor volume was calculated by the formula: vol. = $4/3\pi r^3$. Mean tumor burden was determined by multiplying the number of tumors in each group by the mean tumor volume in millimeters.

2.1.2 Histopathological examination. Three deparaffinized sections were stained with hemotoxylin and eosin for routine histopathological examination, the lesion changes of oral epithelia were classified into hyperplasia, dysplasia and carcinoma (including carcinoma *in situ* and invasive carcinoma) based on the established criteria.[12]

2.1.3 Micronucleated cells analysis. As described by Wargovich *et al.*[13]

2.1.4 Morphometric analysis of AgNORs. Two deparaffinized sections were used for AgNORs analysis, using one-step silver colloid method for AgNORs staining.[14] Morphometric analysis of AgNORs was performed with an MPIAS-500 image analyzer and processing system. One hundred cells per slide were examined. The number of AgNOR dots per nucleus and the volume of single AgNOR dots (VNOR) and the volume of total AgNOR dots (TVNOR) were calculated automatically. The values of volume were expressed in μm^2.

2.1.5 Immunohistochemical analysis of PCNA and EGFR. Two deparaffinized sections were used for PCNA and EGFR staining, using the standard Avidin-Biotin Peroxidase Complex immunoperoxidase method (ABC).[15] The cells with stained nuclear were served as PCNA positive cells, while the cells with stained cytoplasm membrane were identified as the EGFR expression positive cell. Ten random fields were counted using a 40 objective. The labeling index of PCNA was expressed as the number of cells with positive staining per 100 counted cells. The results of the EGFR staining reaction were classified into grade (1–25%), grade (26–50%), grade (51–75%) and grade (76–100%), according to the number of positively stained cells per 100 counted cells.

2.2 Human Intervention Trial

2.2.1 Subjects and treatment. Sixty-four cases of oral leukoplakias were chosen for the intervention trial. They were randomly divided equally into tea-treated and placebo groups, with 32 subjects in each group, but only 29 subjects in the tea-treated group and 30 subjects in the placebo group completed the trial. Twenty non-smoking dental patients without oral leukoplakia matched for age and sex were chosen as the healthy control group. The characteristic of subjects in two groups are showed in Table 1. Subjects in the tea-treated group took 3g mixed tea and at the same time painted on the lesions topically with mixed tea in glycerin at the concentration of 10% every day. Subjects in the placebo group received the same amount of starch capsules and painted the same amount of starch containing glycerin as the tea-treated group. During the trial, subjects were asked not to take any special medicines, including vitamins

Table 1 *General characteristics of subjects*

	Tea-treated (n = 29)	*Placebo-controls (n = 30)*	*Healthy-controls (n = 20)*
Age (years)	53.7	55.4	51.5
Males : Females	18 : 11	17 : 13	10 : 10
Smokers	24	22	0
Alcohol-drinkers	6	5	0
Tea-drinkers	7	9	4
Homogeneous leukoplakia	24	24	0
Non-homogeneous leukoplakia	5	6	0
Hyperplastic leukoplakia	22	25	0
Dysplastic leukoplakia	7	5	0

and to avoid exposure to X-rays. The amount of vegetables and fruits consumed each day was recorded.

2.2.2 Clinical examination. The size and number of lesions of each subject at the baseline and at the end of trial were recorded. A complete regression was defined as the complete disappearance of the lesions. A partial regression was defined as 30% or more reduction in the size of a single lesion or in the sum of sizes of multiple lesions. Lesions with no change in size were recorded as no change. Deterioration was referred to the occurrence of new lesions.

2.2.3 Micronuclei in exfoliated oral mucosa cells. Exfoliated cells were obtained by scraping the buccal mucosa from sites with lesions by a moistened wooden tongue depressor at baseline and at the end of 3 months and 6 months of the trials. Exfoliated cells from normal mucosa of the healthy control subjects were also obtained. The slides were stained with Feulgen staining and counter-stained with Fast Green. 2000 cells were counted blindly, and the number of micronucleated cells per 1000 cells was reported.

2.2.4 Morphometric analysis of AgNORs and immunohistochemical analysis of PCNA and EGFR. Oral mucosa biopsies were conducted at the beginning and at the end of the trial from similar sites. Tissues were fixed in 10% formalin, dehydrated and embedded in paraffin. Several sections were cut at 4 μm and the deparaffinized sections were then stained for AgNOR, PCNA, and EGFR analysis using the same methods as in the animal test. The results of EGFR staining reaction were expressed as the number of positive EGFR expression cells per 100 cells counted.

2.3 Statistical Analysis

The tumor numbers, the volumes and the tumor burden were compared by the Wilcoxon-rank test. The incidence of carcinoma, the grades of EGFR expression and the change of leukoplakia lesions were compared by the chi-squared (ξ^2) test. The incidence of micronucleated cells, the number and volume of AgNOR dots and the labeling index of PCNA were compared by the Student's t-test.

3 Results

3.1 Animal Study

3.1.1 Tumor formation. The results of tumor formation and histopathological examination are listed in Tables 2 and 3. At the end of 15 weeks, all the hamsters in the positive control group (group 1) developed oral tumor and carcinoma, while the animals in the negative control group (group 5) did not develop any tumor. Animals that consumed 1.5% green tea, 0.1% tea pigments and 0.5% mixed tea (groups 2–4) had significantly less tumors and carcinoma than that of

Table 2 *Effects of tea on the histopathology of oral lesions in DMBA-treated hamsters*

Groups	No. of animals	Dysplasia Mild	Moderate	Severe	Carcinoma
1. Positive control	16	0 (0.0)[a]	0 (0.0)	0 (0.0)	16 (100)
2. Green tea, 1.5%	16	0 (0.0)	3 (18.8)	2 (12.5)	11 (68.7)[b]
3. Tea pigments, 0.1%	16	0 (0.0)	3 (18.8)	3 (18.8)	10 (62.5)[c]
4. Mixed tea, 0.5%	16	1 (6.3)	5 (31.2)	2 (12.5)	8 (50.0)[c]
5. Negative control	10	0 (0.0)	0 (0.0)	0 (0.0)	0 (0.0)

[a] Figures in parentheses are percentages.
[b] $p < 0.05$, [c] $p < 0.01$, compared with group 1 by chi-squared test.

Table 3 *Effects of tea on oral tumor formation in DMBA-treated hamsters*

Groups	No. of animals	No. of tumor-bearing animals	No. of tumor per hamster	Mean tumor volume (mm^3)	Mean tumor burden (mm^3)
1. Positive control	16	16	3.81 ± 2.13	98.2 ± 63.0	374.2 ± 239.9
2. Green tea, 1.5%	16	13	2.18 ± 1.17^a	32.3 ± 31.2^a	77.0 ± 8.1^a
3. Tea pigments, 0.1%	16	13	1.88 ± 1.30^a	22.8 ± 14.8^a	42.9 ± 27.8^a
4. Mixed tea, 0.5%	16	12	1.25 ± 1.12^a	13.5 ± 12.1^a	16.9 ± 15.1^a
5. Negative control	10	0	0	0	0

All values were expressed as mean $\pm$ SD.
[a] $p < 0.01$, compared with group 1 by Wilcoxon test.

the positive control group ($p < 0.01$), and the mean volume and the mean tumor burden was also dramatically decreased ($p < 0.01$). When compared with the positive control group, the mean tumor burden was reduced by 79%, 89% and 95% and the rate of carcinoma was reduced by 31.3%, 38.5% and 50% in the green tea, tea pigments and mixed tea groups ($p < 0.01$), respectively. Mixed tea showed a stronger inhibition of tumor and carcimoma development than green tea and tea pigments.

3.1.2 Frequency of micronucleated cells, AgNORs, labeling index of PCNA and EGFR expression. The results of micronucleated cells frequency are listed in Tables 4 and 5. After 15 weeks of DMBA treatment, the incidence of micronucleated cells, the number of AgNOR dots per nucleus, TVNOR, the labeling index of PCNA and EGFR expression in the buccal mucosa in groups 1–4 was noticeably increased over the negative control group ($p < 0.01$). Oral administration of 1.5% green tea, 0.1% tea pigments and 0.5% mixed tea significantly decreased the micronuclei formation, the number of AgNOR dots

Table 4 *Effect of tea on micronuclei formation (per 1000 cells), AgNOR and the labeling index of PCNA in oral mucosa of DMBA-treated hamsters*

Groups	*Micronuclei frequency*	*No. of AgNOR/per nucleus*	*VNOR*	*TVNOR*	*Labeling index of PCNA(%)*
1. Positive control	36.8 ± 3.9	6.46 ± 1.10	1.95 ± 0.81	8.79 ± 0.94	81.4 ± 12.4
2. Green tea, 1.5%	20.2 ± 5.8[a]	5.51 ± 0.94[a]	1.54 ± 0.75	6.94 ± 0.79[a]	64.1 ± 11.5[a]
3. Tea pigments, 0.1%	23.3 ± 5.4[a]	4.35 ± 0.75[a]	1.47 ± 0.64	6.39 ± 0.75[a]	62.1 ± 9.4[a]
4. Mixed tea, 0.5%	17.6 ± 3.8[a]	4.29 ± 0.9[a]	1.39 ± 0.55	5.40 ± 0.84[a]	55.2 ± 12.7[a]
5. Negative control	1.2 ± 0.92	3.10 ± 0.36[a]	0.94 ± 0.31	1.97 ± 0.42[a]	8.4 ± 3.4[a]

All values are mean ± SD.
[a] $p < 0.01$, compared with group 1 by Student's t-test.

Table 5 *Effect of tea on EGFR expression in oral mucosa of DMBA-treated hamsters*

Groups	*No. of animals*	*1–25%*	*26–50%*	*51–75%*	*76–100%*
1. Positive control	16	0 (0.0)[a]	0 (0.0)	2 (20.0)	8 (80.0)
2. Green tea, 1.5%	16	1 (10.0)	3 (30.0)	4 (40.0)	2 (20.0)[b]
3. Tea pigments, 0.1%	16	2 (20.0)	3 (30.0)	4 (40.0)	1 (20.0)[c]
4. Mixed tea, 0.5%	16	1 (10.0)	2 (20.0)	5 (50.0)	2 (20.0)[b]
5. Negative control	10	0 (0.0)	0 (0.0)	0 (0.0)	0 (0.0)

[a] Figures in parentheses are percentages.
[b] $p < 0.05$, [c] $p < 0.01$, compared with group 1 by chi-squared test.

per nucleus, TVNOR, the labeling index of PCNA and EGFR expression ($p < 0.01$). These results showed that tea preparations had a protective effect on DNA damage and cell proliferation induced by DMBA. The values of VNOR in the three tea treated groups were less than those in the positive control group, but the differences were not statistically significant ($p > 0.05$). There was no EGFR expression of oral epithelial cells in hamsters of the negative control group.

3.2 Human Intervention Trial

3.2.1 Clinical manifestations. After 6 months of tea intervention, partial regression of the lesions was observed in 11 of the 29 (37.9%) cases, no change in 17 (58.6%) cases and deterioration in 1 (3.4%) case. While in the placebo group, partial regression of the lesions was found in 3 of the 30 (10.0%) cases, no change in 25 (83.3%) cases and deterioration in 2 (6.7%) cases. The partial regression rate was significantly higher in the tea-treated group than in the placebo group ($p < 0.05$) (Table 6).

Table 6 *Changes of clinical manifestations in leukoplakia patients after 6 months of tea treatment*

	Tea-treated (n = 29)	*Placebo controls (n = 30)*
Partial regression[a]	11 (37.9)	3 (10.0)
No change	17 (58.6)	25 (83.3)
Deterioration	1 (3.4)	2 (6.7)

Figures in parentheses are number of subjects.
[a] The rate of partial regression was significantly different between the two groups by chi squared test, $p < 0.05$.

Table 7 *The number of micronucleated exfoliated buccal cells in leukoplakia patients per 1000 cells at baseline, 3 month and 6 six month of trial*

	Tea-treated (n = 29)	*Placebo controls (n = 30)*	*Healthy controls (n = 20)*
Baseline	10.50 ± 5.29^{a}	10.10 ± 4.07^{a}	1.40 ± 0.61
After 3 month	$6.68 \pm 3.21^{b,c}$	10.35 ± 4.07	
After 6 month	$5.39 \pm 3.05^{b,c}$	11.30 ± 4.29	

All values are mean ± S.D. Figures in parentheses are number of subjects.
[a] Significantly different from healthy control by Possion test, $p < 0.01$.
[b] Significantly different from baseline by Possion test, $p < 0.01$.
[c] Significantly different from placebo group by Possion test, $p < 0.01$.

3.2.2 Frequency of micronuclei formation. The data in Table 7 show that frequency of micronucleated exfoliated buccal cells in lesions sites were higher than those of the healthy subjects ($p < 0.01$). After 3 months and 6 months of tea treatment, micronuclei formation in exfoliated buccal cells in lesions sites decreased significantly ($p < 0.01$), while no significant changes were found in the placebo group.

3.2.3 Morphometric analysis of AgNORs, labeling index of PCNA and EGFR expression. After 6 months of tea treatment, the number of AgNOR dots per nucleus, TVNOR and the proliferating index of PCNA decreased significantly ($p < 0.01$) in the tea treated group, while no significant changes were found in the placebo group (Table 7). No changes in VNOR were observed in both the tea-treated and placebo groups. The percentage of EGFR positive cells was reduced after 6 months of tea treatment; however, it was not statistically significant ($p > 0.05$) (Table 8).

4 Discussion

The results of the present animal study indicated that oral administration of 1.5% green tea, 0.1% tea pigments and 0.5% mixed tea as drinking fluid

Table 8 *The number and the volume of AgNOR dots per nucleus in leukoplakia lesions of oral mucosa before and after trial*

	Tea-treated (n = 22)			*Placebo controls (n = 21)*		
	No. of AgNOR	*VNOR*	*TVNOR*	*No. of AgNOR*	*VNOR*	*TVNOR*
Baseline	6.34 ± 2.19	1.87 ± 0.71	16.3 ± 2.8	6.24 ± 2.01	1.61 ± 0.45	13.94 ± 4.41
6 months	4.44 ± 3.80[a,b]	2.00 ± 1.72	11.78 ± 2.71[a]	6.10 ± 2.71	1.91 ± 0.91	14.37 ± 5.10

All values are mean ± S.D. Figures in parentheses indicate number of subjects.
[a] Significantly different from basline by t-test, $p < 0.01$.
[b] Significantly different from placebo controls by t-test, $p < 0.05$.

Table 9 *Proliferating Index of PCNA and Percentage of EGFR positive cells in leukoplakia lesions of oral mucosa before and after trial*

	Tea-treated (n = 22)		*Placebo controls (n = 21)*	
	PCNA	*EGFR*	*PCNA*	*EGFR*
Baseline	36.2 ± 22.9	36.4 ± 25.8	37.3 ± 22.8	35.8 ± 26.5
After 6 month	24.3 ± 16.5[a,b]	32.2 ± 20.4[a]	39.0 ± 23.4	36.7 ± 26.5

All values are mean ± S.D. Figures in parentheses are number of subjects.
[a] Significantly different from baseline by t-test, $p < 0.05$.
[b] Significantly different from placebo controls by t-test, $p < 0.05$.

significantly inhibited the formation of DMBA-induced oral tumor and decreased tumor numbers and size. In the human intervention studies, after 6 months of trial, 37.9% of the tea-treated group showed a reduction in the size of leukoplakia lesions, while only 10.0% of subjects in the placebo group showed a reduction. These results indicate that tea treatment had significant inhibitory effect on oral tumorigenesis induced by DMBA in hamsters and improved the clinical manifestations of the oral precancerous lesions. Thus, animal and human data show consistent results.

DNA damage is considered as a crucial mechanism in cancer development.[16] Micronuclei formation reflects the extent of ongoing DNA damage and has been shown to correlate with cancer risk at several sites, such as oral, esophagus, lung and bladder, and, therefore, was often used as a biomarker of early-carcinogenesis.[17–19] Micronuclei of oral buccal mucosa cells are formed in the basal cells and migrate to the epithelial surface and are detected in exfoliated cells and was thought to reflect the increased oral cancer risk. The frequency of micronucleated exfoliated buccal cells in oral leukoplakias were higher than in cells from normal mucosa sites, especially in smokers.[20] Several previously reported intervention trials in high risk population for oral cancer supported the use of micronuclei frequency in exfoliated buccal cells as an intermediate biomarker.[21,22] Abnormal cellular proliferation is another important mechanism in carcinogenesis. Indicators of proliferation may be used as intermediate biomarker for chemoprevention research. It has been suggested that the number and size of AgNOR dots in a nucleus and PCNA could reflect the status of cell activation and therefore reflect cell proliferation.[23–25] EGFR is a product of the erb oncogene and the overexpression of EGFR is involved in the pathogenesis of certain epithelial neoplasms.[26,27]

In the animal study, oral administration of 1.5% green tea, 0.1% tea pigments or 0.5% mixed tea significantly reduced micronuclei formation, the number and volume of AgNOR dots per nucleus, the labeling index of PCNA and the level of EGFR expression. In our human trial, after 6 months of mixed tea intervention, the micronuclei formation in exfoliated oral cells, the numbers of AgNORs/per nucleus and TVNOR, as well as the proliferating index of PCNA in the tea-treated group were significantly decreased, while there were no significant changes in the placebo group. These results show that tea has a significant chemopreventive effect on DNA damage and has a significant inhibitory effect on oral mucosa cell proliferation. In this study, the rate of EGFR expression in the tea-treated group decreased after 6 months of treatment and was also lower than that of the placebo group, but the differences were not statistically significant. This may be due to the large individual variation of EGFR expression rates and relatively small sample size. Again, animal and human data are consistent.

The mechanisms of the inhibitory effects of tea preparations on oral cancer have not been fully elucidated. Our results showed that protecting DNA damage and suppression of cell proliferation might be important mechanisms. In addition, the antioxidative and free radical-scavenging activities of tea preparations may possibly play a more important role in protecting against oral cancer.

In conclusion, our present research indicated that green tea, tea pigments and mixed tea have significant inhibitory effect on oral tumorigenesis induced by DMBA in hamsters; mixed tea had stronger inhibitory effect than any single tea ingredients. At the same time, the result from various intermediate markers in this clinical trial showed that mixed tea is able to improve the precancerous changes in oral leukoplakia patients by protecting DNA damage and inhibiting proliferation of oral mucosa cells, and thus suggests that the mixed tea might have certain preventive effects in human oral cancer. This is in line with our animal laboratory studies. The results from this study have provided some encouraging and direct evidence on the preventive effects of tea on human cancer.

Acknowledgements

This study was supported by a research grant from the Chinese National Science Foundation, No. 39330180. The authors are grateful to the doctors and nurses of the Department of Oral Medicine, Beijing Dental Hospital for collaboration in clinical observation and for their help in preparation of blood and tissue samples; The authors would also like to thank Dr. Xiaoyong Liu of

Beijing Dental Hospital for her help in morphlogic analysis of AgNOR. The authors also thank Prof. Qikun Cheng of the Institute of Tea Science and Research, Chinese Academy of Agricultural Science for providing the green tea, tea pigments and mixed tea samples.

5 References

1 Mukhtar H, Wang ZH, Katiyar SK and Agareal R, Tea components: Antimutagenic and anticarcinogenic effects. *Prev Med* 1992, **21**: 351–360.

2 Yang CS and Wang ZY, Tea and cancer. *J Natl Cancer Inst*, 1993, **85**: 1038–1049.

3 Wang ZY, Cheng JS and Zhou ZC, Antimutagenic activity of green tea polyphenols. *Muta Res*, 1989, **223**: 273–285.

4 Katiyar SK and Mukhtar H, Tea in chemoprevention of cancer: Epidemiological and experimental studies (review). *Int J Oncol*, 1996, **8**: 221–238.

5 Kohlmeier L, Weterings KGC, Steck S and Kok FJ, Tea and Prevention: an evaluation of the epidemiological literature. *Nutr Res*, 1997, **27**: 1–13.

6 Victora CG, Munoz N and Hong JGE, Hot beverages and esophageal cancer in southern Brazil: a case-control study, *Int J Cancer*, 1987, **39**: 710–716.

7 Parkin SM, Laara E and Muir CS, Estimates of the worldwide frequency of sixteen major cancers. *Int J Cancer*, 1988, **41**: 184–197.

8 Magrath I and Litvakk J, Cancer in developing countries: Opportunity and challenge. *J Natl Cancer Inst*, 1993, **85**: 862–874.

9 Tanaka T, Chemoprevention of oral carcinogenesis. *Oral Oncol, Eur J Cancer, 1995*, **31B**: 3–15.

10 Vecchia CL, Tavani A, Franceschi S, Levi F, Corrao G and Negri E, Epidemiology and prevention of oral cancer, *Oral Oncology*, 1997, **33**: 302–312.

11 Liu LJ, Han C and Chen JS, Short term screening of anticarcinogeneic ingredients of tea by cell biology assays. *J Hyg Res*, 1998, **27**: 53–56 (in Chinese).

12 Banoczy J, Csiba A and Hungary B, Occurrence of epithelial dysplasia in oral leukoplakia. *Oral Surg*, 1976, **42**: 766.

13 Stich HF and Rosin MP, Quantitating the synergistic effect of smoke and alcohol consumption with the micronucleus test on human buccal mucosa cells. *Int J Cancer*, 1983, **31**: 305–308.

14 Ploton D, Menager M and Jeannesson P, Improvement in the staining and in the visualization of the argyrophilic proteins of the nucleolar organizer region at the optical level. *Histochemical Journal*, 1986, **18**: 5–14.

15 Hsu SM, Raine L and Fanger H, Use of avidin-biotin-peroixdase complex (ABC) in immunoperoxidase techniques: a comparision between ABA qbd unlabelled PAP procedures. *J Histochem Cytochem*, 1981, **29**: 577–580.

16 Weinstein IB, The origins of human cancer: molecular mechanisms and their implications for cancer prevention and treatment. *Cancer Res*, 1988, **48**: 4135–4143.

17 Poppel GV, Kok FJ and Hermus RJJ, Beta-carotene supplementation in smokers reduces the frequency of micronuclei in sputum. *Br J Cancer*, 1992, **66**: 1164–1168.

18 Reali DF, Dimarino S, Bahramandpour A and Carducci R, Micronuclei in exfoliated urothelial cells and urine mutagenicity in smokers. *Mutation Res*, 1987, **192**: 145–149.

19 Garewal HS, Ramsey L, Kaugars G and Boyle J, Clinical experience with the micronucleus assay. *J Cell Biochem, 1993, suppl.* **17F**: 206–212.

20 Desai SS, Ghaisas SD, Jakhi SD and Bhide SV, Cytogenetic damage in exfoliated

oral mucosal cells and circulating lymphocytes of patients suffering from precancerous oral lesions. *Cancer Letters*, 1996, **109**: 9–14.

21 Stich HF, Stich W, Rosin MP and Vallejera MO, Use of the micronucleus test to monitor the effect of vitamin A, beta-carotene and canthaxanthin on the buccal mucosa of betel nut/tobacco chewers. *Int J Cancer*, 1984, **34**: 745–750.

22 Benner SE, Lippman SM and Wargovich MJ, Micronuclei, A biomarker for chemoprevention trials, results of a randomized study in oral premalignancy. *Int J Cancer* 1994, **59**: 457.

23 Fakan S and Hernadez-Verdun D, The nucleus and the nucleola organizer regions. *Biol cell*, 1986, **56**: 186–206.

24 Hall PA, Levinson DA and Woods AL, Proliferating cell nuclear antigen (PCNA) inmmunolocalization in paraffin sections: an index of cell proliferation with evidence of deregulated expression in some neoplasm. *J Pathol*, 1990, **162**: 285–294.

25 Muzio LL, Mignogna MD, Staibano S and Vico GD, Morphometric study of nucleolar organizer regions(AgNOR) in HPV-associated precancerous lesions and microunvasive carcinoma of the oral cavity. *Oral Oncology*, 1997, **33**: 247–259.

26 Storkel S, Reichert T and Reiffen KA, EGFR and PCNA expression in oral squamous cell carcinomas-a valuable tool in estimating the patient's prognosis. *Oral Oncol Eur J Cancer*, 1993, **29B**: 273–277.

27 Shin DM, Ro JY and Hong WK, Dysregulation of Epidermal Growth Factor Receptor expression in premalignant lesions during head and neck tumorigenesis. *Can Res*, 1994, **54**: 3153–3159.

8.8

Risk Perception and Healthy Eating – Implications for the Development of Effective Communication Strategies

Nigel Lambert and Lynn J. Frewer

CONSUMER SCIENCE SECTION, INSTITUTE OF FOOD RESEARCH, NORWICH RESEARCH PARK, COLNEY, NORWICH NR4 7UA, UK

1 Introduction

This paper describes common barriers encountered in interventions aimed at promoting healthy eating. The application of two psychological models for facilitating effective risk communication about poor diet is summarised.

1.1 Diet as a Major Risk Factor in Cancer Development

In the US and Northern Europe it is estimated that one-third of all cancers are attributable to poor diet.[1] Dietary recommendations from national and international health organisations have consistently emphasised plant-based diets with high levels of fruits, vegetables and cereal products as optimal for health.[1]

In 1997 cancer accounted for 25% of all UK fatalities. In England and Wales this translated to 135 420 deaths.[2] Although cancer is primarily a disease of old age (46% of fatalities were people aged 75 or over), 52% of fatalities occurred in the 35–74 age group, with 23% in the 35–64 age group. It is the number of *premature* deaths from cancer that the UK Government are committed to reduce. In a recent White Paper[3] the target was to reduce the death rate from cancer amongst people aged under 65 years by at least 20% by 2010 from a baseline at 1996. A significant drop in the incidence of cancer would have enormous financial and social benefits for the nation.

1.2 Promoting Healthy Eating and Barriers to Dietary Change

Through public health campaigns, governments in the UK and elsewhere have attempted to encourage healthy eating, but barriers to dietary change have resulted in little improvement in peoples' dietary choices. Throughout the 1990s there has been a large number of studies, especially in the US, exploring the issues involved in promoting dietary change.[4] Research activities can generally be split into 4 areas: behavioural research within clinical trials, self-help and minimal contact intervention strategies, school nutrition programmes and development of innovative methodologies.

The main difficulties encountered when attempting to change patterns of dietary choice may be summarised in the following points:

(a) There is no simple 'good-bad' message. There are no good or bad foods *per se*, optimum diet being a quantitative concept. The message of '5 portions of fruit and vegetables a day' has been widely used but this is very ambiguous: which fruits and vegetables? Cooked or uncooked? Processed or fresh? What is a portion?
(b) Knowledge, beliefs and attitudes were found to vary markedly with age, sex, ethnicity, education and income indicating the need to carefully target nutritional interventions to various population sub-groups.[5]
(c) The perceived cost of eating a healthful diet is a commonly reported barrier.[5–7]
(d) People indicate that time and inconvenience are factors mitigating against dietary change.[7,8]
(e) People report confusion about conflicting dietary advice from 'experts'.[5]
(f) People have no perceived need to change. This was reported for over 60% of EU consumers[7] and similar 'low priority' perceptions were reported in the US.[8]
(g) Sensory barriers (such as fear of giving up foods people enjoyed or not liking the taste of fruits and vegetables) represent an important barrier to dietary change.[7]
(h) There is little social pressure to change dietary choices.[6]

2 Methods

It is important to understand the psychology which drives food choices made by individuals in order to design more effective intervention strategies. Various psychological models have been used to understand problems associated with developing effective risk-benefit information aimed at improving health through nutritional improvement. The merits of two of the most common will be briefly discussed.

2.1 Stages of Change Model

This provides a theoretical framework for approaches to accelerating the rate of behavioural change in a population.[9] Developed initially for studying the cessation of smoking, it has been used in many other settings including dietary change.[10] The model proposes that change is a dynamic variable with 5 stages:

(1) Precontemplative – where an individual is not intending to make a behavioural changes in the foreseeable future.
(2) Contemplative – where an individual is considering a behaviour change but not yet making a firm commitment to change.
(3) Preparation – an individual has made a commitment to changing behaviour in the next 30 days but has not yet changed behaviour.
(4) Action – the individual has successfully changed behaviour.
(5) Maintenance – the behavioural change is sustained over 6 months.

At each stage the person has a different set of attitudes and beliefs, and consequently each stage requires a different type of intervention. To understand a study sample in this manner offers enormous advantages; if for example most of one's subjects are in the precontemplative or maintenance stages of change, there is little point using action-orientated interventions. Using an inappropriate approach at different stages of change may explain the failure of many dietary interventions.[10]

2.2 Elaboration Likelihood Model

This is a theory of persuasive communication used widely in many settings.[11] The utility of the model has been demonstrated in the field of risk communication about different food hazards.[12–14] The theory proposes that there are two routes to persuasion: a central route using careful and thoughtful assessment of arguments (where long-term attitude change is likely to result from consideration of risk information) and a peripheral route (unlikely to result in long-term attitudinal or behavioural change) where people use simple inferences about the merits of the argument contained in the information without recourse to complex cognitive processing. Factors such as information source, personal relevance and persuasive power of the risk communication have been explained with this theory. These have clear links to the promotion of dietary change. Conflicting dietary advice from experts may act as a cue which results in peripheral processing of information. Lack of persuasive content, perceptions of low personal relevance and other source factors such as trust in information providers may also influence whether or not central processing routes are used. Personal relevance seems a common factor in most dietary interventions, as that most people feel they already eat healthily. In the risk communication literature this effect is termed 'optimistic bias' or 'unreal optimism', where people tend to rate their own risks as being less than for a comparable member of society. It is important to develop ways to target information to ensure it reaches vulnerable groups, as well as testing information content and other contextural effects within this model.

3 Conclusion

Reducing the incidence of UK cancer by promoting healthy eating is a priority within health education. The use of psychological theory to develop effective risk communication may overcome some of the barriers to dietary change identified in the health communication literature.

4 References

1 M. Nestle, *Nutr. Rev.*, 1998, **56**, 127.
2 Office of National Statistics, Department of Health. 1997, London.
3 Department of Health, 'Our Healthier Nation', White Paper, 1999, London.
4 K. Glanz, *Prev. Medicine*, 1997, **26**, S43.
5 L. Harnack, G. Block, A. Subar and S. Lane, *J. Nutr. Ed.*, 1998, **30**, 131.
6 D.N. Cox, A.S. Anderson, M.E.J. Lean and D.J. Mela, *Pub. Health Nutr.*, 1998, **1**, 61.
7 Pan-EU survey on consumer attitudes to food, nutrition and health, Report No. 3, Institute of European Food Studies London, U.K.
8 G.I. Balch, K. Loughrey, L. Weinberg, D. Lurie and E. Eisner, *J. Nutr. Ed.*, 1997, **29**, 178.
9 J.O. Prochaska, C.C. DiClemente and J.C. Norcross, *Am. Psychol.*, 1992, **47**, 1102.
10 C. NiMhurchu, B.M. Margetts and V.M. Speller, *Nutr. Rev.*, 1997, **55**, 10.
11 R.E. Petty and J.T. Cacioppo, 'Communication and Persuasion', Springer-Verlag, 1986.
12 L.J. Frewer, C. Howard and R. Shepherd, *Risk Analysis*, 1997, **17**, 269.
13 L.J. Frewer, C. Howard, D. Hedderley and R. Shepherd, *Risk Analysis*, 1998, **18**, 95.
14 L.J. Frewer, C. Howard, D. Hedderley and R. Shepherd, *Pub. Und. Sci.*, 1999, **8**, 35.

Subject Index